*Also in the Variorum Collected Studies Series:*

**TREVOR H. LEVERE**
Chemists and Chemistry in Nature and Society, 1770–1878

**W.H. BROCK**
Chemistry and the Chemical Industry in the 19th Century
The Henrys of Manchester and Other Studies

**ANDREW WEAR**
Health and Healing in Early Modern England
Studies in Social and Intellectual History

**SILVIO A. BEDINI**
Patrons, Artisans and Instruments of Science, 1600–1750

**Y. TZVI LANGERMANN**
The Jews and the Sciences in the Middle Ages

**EDWARD S. KENNEDY**
Astonomy and Astrology in the Medieval Islamic World

**DAVID M. KNIGHT**
Science in the Romantic Era

**DAVID J. JEREMY**
Artisans, Entrepreneurs and Machines
Essays on the Early Anglo-American Textile Industries, 1770–1840

**IAN INKSTER**
Scientific Culture and Urnbanisation in Industrialising Britain

**ROY MACLEOD**
The 'Creed of Science' in Victorian England

**JACK MORRELL**
Science, Culture and Politics in Britain, 1750–1870

**WILFRED VERNON FARRAR (Ed. R.L. Hills & W.H. Brock)**
Chemistry and the Chemical Industry in the 19th Century
The Henrys of Manchester and Other Studies

**WILLIAM H. BROCK**
Science for All
Studies in the History of Victorian Science and Education

**CRAIG G. FRASER**
Calculus and Analytical Mechanics in the Age of Enlightenment

VARIORUM COLLECTED STUDIES SERIES

# Sciences of the Earth

Professor David Oldroyd
(Photo by Stephen Reinhardt)

David Oldroyd

# Sciences of the Earth

## Studies in the History of Mineralogy and Geology

Ashgate
VARIORUM



Aldershot · Brookfield USA · Singapore · Sydney

**Published in the Variorum Collected Studies Series by**

Ashgate Publishing Limited
Gower House, Croft Road,
Aldershot, Hampshire GU11 3HR
Great Britain

Ashgate Publishing Company
Old Post Road,
Brookfield, Vermont 05036–9704
USA

ISBN 0–86078–770–2

**British Library Cataloguing-in-Publication Data**
Oldroyd, David.
Sciences of the Earth: Studies in the History of Mineralogy and Geology.
(Variorum Collected Studies Series: CS628)
1. Earth sciences–History. 2. Mineralogy–History. 3. Geology–History.
I. Title.
550. 9

**US Library of Congress Cataloging-in-Publication Data**
Oldroyd, D.R. (David Roger).
Sciences of the Earth: Studies in the History of Mineralogy and Geology/
David Oldroyd.
p. cm. – (Variorum Collected Studies Series: 628). Includes index.
1. Mineralogy–History. 2. Geology–History.
I. Title. II. Series: Collected Studies: CS628.
QE359. A1043 1998 98–27217
549' 09–dc21 CIP

Printed on acid free paper in Great Britain by
WBC Book Manufacturing, Bridgend, Mid Glamorgan.

VARIORUM COLLECTED STUDIES SERIES CS628

# CONTENTS

This volume contains xiv + 340 pages

# PREFACE

At Cambridge, in the 1950s, I took a rather odd degree: natural sciences and chamber music – just as some people did medicine and rugger, or classics and chess. As a result the formal part of my degree was not very good, and, as a research career was then blocked, there were at the time apparently but three career options, at least in the short run: the armed forces for two years; mining; or science teaching. The third alternative seemed the most attractive; and so I became a chemistry teacher in London, even though geology was the science I had enjoyed most as a student.

Early in my new job, I attended a teachers' conference in Oxford called 'Science as a General Study in the Sixth Form', and I was introduced to the idea, then commonly held, that history and philosophy of science could serve as a bridge that might somehow link the 'two cultures'; and also to the idea that such bridging was a 'good thing'. As a musician and a science graduate, I felt that I already had a foot in both of C.P. Snow's cultures, and I well remember that a splendid conference lecture on Galileo by Michael Hoskin from Cambridge suddenly revealed how interesting the history of science might be. I also learnt that there was an M.Sc. course available at University College, London, for which one could attend lectures in the evenings – so I embarked on the programme. Rather optimistically, or naïvely, I thought it might one day help me gain preferment in my profession, or even lead on to something in academia.

Then one day, walking down the Strand, I happened past New Zealand House and noticed that there were teaching jobs on offer all over New Zealand, with fares paid, accommodation provided, furniture moved. My wife and I had always wanted to travel, and so off we went with our two small children. We spent seven years in New Zealand, half at a government school in the North Island (which was mostly an awful experience) and half in a prestigious private school in Christchurch (which was delightful, except for the quaint metaphysics to which the staff subscribed). I sat the written papers for the M.Sc. in Napier and then had to find a thesis topic. University College could do nothing to provide supervision, and so I worked alone, writing a very unsophisticated thesis on 'Geology in New Zealand prior to 1900'. To my relief and surprise, the examiners had no complaints. I was spared a viva. Looking back at the thesis now, I discover that it was excessively Whiggish. This was perhaps hardly surprising. The course that I had done in London included no mention of historiographic issues, and I had never heard of Herbert Butterfield or his critique of Whig history.

Be that as it may, now armed with my M.Sc., my ambition swelled and I thought I might indeed contemplate a university position. It seems incredible today, but when I applied for a lectureship in the School of History and Philosophy of Science at The University of New South Wales I was the only applicant with a higher degree in the discipline; and so I got the job. It was as easy as that! But then I had to find a Ph.D. thesis topic. I wrote to Victor Eyles, who had been one of my thesis examiners, and he put me in touch with a remarkable scholar from Sydney University, the petrologist, the late Tom Vallance. Tom knew *all* there was to know about the history of geology, it seemed, and had a library of thousands of rare and wonderful books. He became my model of what I hoped I might one day become – but of course I never have and never will.

After a little casting around, I lighted on a topic: 'The influence of the Chemical Revolution on mineralogy'. It seemed to offer a useful linkage between what I knew about geology and chemistry, and the histories of these subjects, as I had studied them in London. The Professor of History and Philosophy of Science at The University of New South Wales approved and suggested that one could usefully look at the topic through the lens of the work of T.S. Kuhn. I had never heard of this personage (news travelled slowly to New Zealand), and his name had never been mentioned at University College. (Besides being innocent of historiographical issues, the historical and philosophical subjects in Part I of the M.Sc. degree were entirely divorced from one another, and the philosophy of science consisted, amazingly, of truth tables, Carnap, and such, *only*.) I recall looking for Kuhn under the head 'Coon' in the library – such was my ignorance.

So, beginning from a very low base, I started on my thesis, with the kindly guidance of Dr Robert Gascoigne, an organic chemist who had turned to history and philosophy of science in mid-career. I also received much advice from Vallance and loans of many of his splendid books. But then I began to fall into the trap of many doctoral students. I became a victim of what might be called the 'Leibniz Syndrome' (a term I commend to the literature). Leibniz, it is said, was employed to make an historical enquiry into the competing claims of two powerful families, in a territorial dispute. He traced the history of the region back and back and ended up writing a history of the Earth from its first formation – his famous *Protogaea*. I found myself in a somwhat similar fix. Before one could deal with mineralogy and the 'Chemical Revolution', one had to understand 'phlogistic mineralogy'; then what came before, namely 'mechanical-philosophy mineralogy'; and before that Renaissance ideas about minerals, which I found were imbued with Neo-Platonic and Stoic notions. So I had to read about neo-Platonism and Stoicism – and thus Plato too; not to mention medieval ideas about minerals, which meant I had to 'get up' Aristotle a bit. Eventually I cut off the regress at Paracelsus, and my thesis was entitled 'From Paracelsus to Haüy: The Development of Mineralogy in Relation to Chemistry' (1974). What had initially been 'the topic' became but one chapter of the thesis. Some of the various 'stations of the cross' can be seen in the papers that I published in the early 1970s, and which I am now pleased to see appear in the present volume in orderly form. They give, I think, quite a useful overview of the development of ideas about minerals and earths from the Renaissance to the end of the eighteenth century.

At about the time my thesis was being finished, a very able and industrious young scholar, Randall Albury, arrived from the States as a new appointee in my department. His recently completed Ph.D. (from Johns Hopkins) studied the work of Condillac, Lavoisier, Bichat, Vicq d'Azyr, and Haüy from the perspective of Michel Foucault's *Order of Things*. Randall's work thus introduced me to Foucault, and perhaps to our mutual surprise we discovered that many of the 'facts' that I had painstakingly assembled in my reading for my thesis fitted into a Foucauldian framework remarkably well. Hence we published our joint paper of 1977. (Readers will notice the difference of style in this paper from others in this volume, resulting from Randall's collaboration.)

In the mean time, I had been teaching history of the philosophy of science for a while and had even learnt a bit about it in the process. I became interested in the procedures of 'analysis' and 'synthesis', an interest that is revealed in the paper on Stahl in the present collection (and was subsequently developed more fully in a book, *The Arch of Knowledge*, 1986). I found that it was a central methodological issue in the work of Robert Hooke, and that it was best illustrated in his geological writings. To my surprise, no one had examined this linkage before to any degree, and so my Hooke paper of 1972 came into being. (Subsequently, I wrote quite a lot on Hooke, even his music theory; so my 'two culture' inclination perhaps came in useful again.)

The paper on Clarke emerged from discussions with Vallance, who, as ever, lent me the relevant books. One of my philosophical colleagues at The University of New South Wales thought it hilarious that anyone should write a paper about blow-pipes. But, as we now know, the study of scientific instruments and practice, be they ever so simple, has become a most 'respectable' field. Moreover, there is now a Cambridge Ph.D. on Clarke, which (the author told me) used my old paper as a starting-point.

I attended the Lyell conference in London in 1975, well known to historians of geology, and presented a paper on the historicism and geology around the turn of the nineteenth century and on to the work of Lyell. To my chagrin, it was not accepted for publication in *The British Journal for the History of Science* (which in 1976 published selected papers from the conference). So during my next sabbatical, I pursued this topic with some energy and the outcome was the paper 'Historicism and the Rise of Historical Geology'.

I am not sure now that all the ideas in this long paper are quite correct. Its arguments perhaps need too many qualifications. For example, a recent book by Rhoda Rappaport (one of my Ph.D. examiners), *When Geologists were Historians: 1665–1750* (1997), would make me want to rethink the earlier period, for which she argues for a closer connection between work in history and studies of the earth than I acknowledged. However, Martin Rudwick has several publications for the early nineteenth century, which work around the same theme as that which I was developing, showing, for example, in greater detail how geologists like Lyell were influenced by particular historians such as Niebuhr. Rudwick also deploys the distinction between 'genetic' and 'historical' accounts of the earth (also proposed by Jacques Roger), and sees the emergence of geo-historical thinking in the work of Cuvier as the 'point' that marked the emergence of geological science, as opposed to 'theories of the earth', mineralogy, 'geognosy', and so on. (Rudwick, incidentally, prefers the term 'epigenetic' to 'genetic', and I think this is an improvement.) My 'historicism' paper, republished here, remains my most frequently cited journal article.

In the early 1980s, news about the 'strong programme' in the sociology of knowledge was reaching Australia, hastened by the arrival of another new colleague, David Miller from Edinburgh. He introduced me to Mary Douglas's grid/group theory, and I discovered how David Bloor and others had sought to deploy it – in Bloor's case specifically in relation to the sociology of science. I had a day-long conversation with him in Edinburgh, and got to know much more about the 'strengths of the strong programme' and about grid/group theory. I had been asked to review Nicolaas Rupke's book on William Buckland, and I thought it would be interesting to frame the essay in terms of a discussion of Rudwick's use of Douglas's work to analyse the social field of early nineteenth-century geologists. The result appears in the present volume in a 'joint essay review' (1979). Subsequently, I looked at grid/group theory more closely and traced the history of its development (*History of Science*, 1986). I came to the conclusion that it had severe problems and was, I contended, of little use to historians of science, being much too 'fuzzy'. Rudwick later told me that he had given up Douglas's theory (or method of sociological analysis), but I believe he did so quite independently of my criticisms of the doctrine.

As readers may have gathered, I have been something of an autodidact in history of geology; and such a person must always rely on the nineteenth-century pioneering history by Sir Archibald Geikie. So I was well-read in his historical work, if nothing else. Reading Geikie's history made me interested in his geological work. Also, by the late 1970s I had become caught up with historiographical questions, for I found myself teaching a course on historiography of science, where Whig history was the first topic for consideration; so I *had* read Butterfield by then. Thus it was that geological and historiographical interests came together in the paper on Geikie in the present volume.

I should mention, perhaps, that since 1980, I have taken a special interest in Geikie, and have done a good deal of archival work on him, including a selective study of some his 10,000 official letters, discovered only fairly recently at Haslemere. He is a prominent figure in several of my more recent papers, and a book (1990). Closer study of Geikie suggests that he was a rather less attractive man than his superb prose had earlier led me to think. But I revere him as a (Whiggish) historian of geology, master synthesiser, and distinguished geologist. He was a paradigmatic 'two-culture' man, with great gifts as artist, writer, lecturer, bureaucrat, and classicist, along with his science.

Finally (or initially) in the present collection there is the little paper on a Latin poem, of uncertain authorship, about Mount Aetna. A colleague in the medical school, Peter Paisley, an avid reader of Loeb Classics and much besides, drew it to my attention, and thought that I should be interested in it. I was, and so we wrote our *Centaurus* paper. We were amazed by: (i) the vituperation we discovered betwixt and between classicists in their publications; and (ii) the fact that while they knew everything about the text of *Aetna* and its history, they seemed quite uninterested in its 'scientific' content. This defect, with much hubris, we sought to remedy.

A little anecdote about *Aetna* is worth recounting. There was in my class at Luton Grammar School a brilliant boy named Frank Goodyear. After doing O-level Latin, he taught himself Greek in eighteen months and obtained a major scholarship in Classics at Trinity, Cambridge. A few years later when he was beginning his Ph.D. (he was by then a year ahead of me) we had tea and crumpets in my rooms at Emmanuel, and he told me he was going to write his thesis on an anonymous Latin poem of just a few hundred lines. I could scarce believe that *that* could be a Ph.D. topic; but it was, and Frank soon obtained his doctorate and secured a chair at Bedford College, London. I never saw him again and I have heard that he died some years ago. I was always dazzled by Frank's brilliance, but I think it would be fair to say that his prodigiously learned volume on *Aetna* took scant notice of the 'scientific' aspects of the poem. (However, I have not heard that he or any other classicist ever noticed our effort. Perhaps its hubris was just too much for them.)

So those are the papers. I should like to thank my colleagues, Dr Peter Paisley and Professor Randall Albury, for allowing me to include the items that they co-authored in the present volume, as well as the many other colleagues, referees and editors who assisted in the preparation of the papers in this book.

In the years following the papers represented in the present volume, I have been increasingly drawn to nineteenth-century geology, and have developed a minor cottage industry in following in the footsteps of major geologists, yielding, among other things, a book on a major controversy in the history of Scottish geology, a long series of papers on controversies in Welsh geology, and a paper on a Chinese topic; but, I admit, there is precious little Australian material. (Single-handed historical fieldwork in the Australian bush and desert is not so appealing as it is in Britain, where one is never far from a cosy pub or a generous bed and breakfast establishment.) I am currently working on a book on Lakeland geology.

All this is very self-indulgent (as my friends tell me); but history and philosophy of science can be an indulgent, as well as exacting, field. You can combine an interest in any of the sciences with inquiries into almost anything you like: philosophy, art, music, law, painting, geography, sociology, literature, politics, architecture, philology, . . . For me *geology* is the best of all the fields: for it *requires* me to go to wonderful places, and meet delightful people. It took me some years to find the field to suit my tastes and inclinations exactly, but at least I claim to have done *that*.

DAVID OLDROYD

*Sydney*
*March, 1998*

## ACKNOWLEDGEMENTS

Grateful acknowledgement is made to the following persons, journals, institutions, and publishers for their kind permission to reproduce the studies included in this volume: Munksgaard International Publishers Ltd., Copenhagen (for studies I and VII); Black Bear Press, Cambridge (II, III and V); Taylor and Francis Ltd., London (IV, IX, XIII and XIV); The University of Chicago Press, Chicago (VI); John W. Moore, Editor, *Journal of Chemical Education* (VIII); Janet Brown, on behalf of the Council of the British Society for the History of Science, London (X and XI); Science History Publication Ltd., Chalfont St Giles, Bucks. (XII).

# PUBLISHER'S NOTE

The articles in this volume, as in all others in the Variorum Collected Studies Series, have not been given a new, continuous pagination. In order to avoid confusion, and to facilitate their use where these same studies have been referred to elsewhere, the original pagination has been maintained wherever possible.

Each article has been given a Roman number in order of appearance, as listed in the Contents. This number is repeated on each page and is quoted in the index entries.

# I

# Science in the Silver Age: *Aetna*, a Classical Theory of Volcanic Activity.

P. B. PAISLEY & D. R. OLDROYD

*Introduction*

Among the Classical works published in the well-known Loeb edition is a poem of 645 lines, entitled simply *Aetna.*[1] This work, the authorship of which cannot be determined with complete certainty (though we offer our own views on this question below), has survived from antiquity in the form of fourteen currently known manuscripts, the earliest dating from the tenth century, and was first issued in printed form in Paris in 1507.[2] Besides the Loeb edition, two other scholarly editions have been published fairly recently.[3]

Over the years, *Aetna* has been the object of considerable scholarly interest,[4] for in points of textual interpretation it offers ample scope for academic debate, there being many discrepancies in the surviving manuscripts. The problem of authorship has also provided a challenge to a number of writers. But despite the considerable degree of exegetic skill that has been displayed by the various editors and commentators, their interest has been focussed largely on textual problems, and the interesting questions thrown up by the 'scientific' content of the poem have received much less attention. Moreover, the particular niche occupied by this poem in the history of geological thought has been almost entirely overlooked. Indeed, so far as we are aware, apart from a very brief mention in Zittel's *History of geology and palaeontology,* [5] the poem has been quite ignored by historians of science.

For this reason, therefore, we think it instructive to examine this poem, not from the point of view of the philologist or the Classical scholar, but with the intention of displaying its particular hypotheses with respect to Earth processes, its character as an example of Roman 'science', its relationship to the other 'geological' opinions of the ancient world, and its place in the general history of ideas concerning the Earth. We shall not be overly concerned with textual problems. These fall within the domain of the Classical scholar, upon whose labours we gratefully rest.

## *The 'Geological' Doctrines of* Aetna

We begin our discussion with a straightforward exposition of the 'geological' doctrines of *Aetna.* It is, of course, a considerable anachronism to speak of the *geological* theories of the ancient world, for the science of geology cannot be said to have been constituted until the end of the eighteenth century.[6] Nevertheless, the poem *Aetna* takes as its central concern a topic that is of high significance for modern geology, and it is not therefore wholly inappropriate to enquire into the nature of the 'geological' theories that its anonymous author espoused. Besides, we have no other term that may readily be used in its stead.

The poem begins with a claim that its author intends to limit himself to *naturalistic* explanations of phenomena:

> Let none be deceived by the fictions poets tell – that Aetna is the home of a god, that the fire gushing from her swollen jaws is Vulcan's fire, and that the echo in that cavernous prison comes from his restless work.[7]

Explanations couched in terms of the actions of the gods are to be shunned; only natural phenomena are to be invoked. Despite this intention, the author finds himself unable to eschew mythological *allusions* entirely. For example, we read:

> Once Aetna burst open its caverns and glowed white-hot: as though its deep-pent furnaces were shattered, a vast waye of fire gushed forth afar upborne by the heat of the lava-stone, just as when the ether lightens under the fury of Jupiter and plagues the bright sky with murky gloom.[8]

But where, as in this passage, the author embellishes his account with mythical metaphors or similes, the reader is kept aware that myth and fact are not to be confused; though they are sometimes conflated. We might say that the author seeks to proceed beyond what Comte was to call the 'theological' stage of science.

The Earth, our author informs us, is not solid throughout.[9] It is filled with hollows and passages, somewhat similar to those found within the body of a living organism. The passages may have been formed at the very creation of the Earth, when lumps of rock, falling together, could have spaces left between them, similar to those obtained when a heap of stones is loosely thrown together. Alternatively, air, water, or hot vapours (or a combination of these), working through the Earth, may themselves have channelled out the passages. But whatever their cause, the existence of such passages cannot be gainsaid. Springs and underground rivers, caves and grottoes, fully attest their presence.

Next, we are informed that fire is more vigorous in its action in a confined space than in the open; and winds are every bit as vehement, if not more so.[10] Here, it may be presumed, the poet is drawing upon the commonplace observation that a heaped-up fire, or one within an enclosed space, usually burns fiercely; whereas if the fuel is pulled apart the flames tend to die down.

We are now introduced to *Aetna*'s major theoretical premise – that earthquakes and volcanoes arise from the action of winds within the Earth. Aetna itself, seemingly, bears direct witness to this hypothesis, for it contains 'vast openings which terrify and plunge in an abyss'.[11] 'Facts and your eyes instruct you: facts unaided compel belief'.[12] Here the author enters into a lyrical description of the volcano in action.

As stated, the cause of earthquakes is deemed to be *wind*. But wind is also held responsible for the *fire* associated with volcanic activity. Line 213 reads: 'Spiritus inflatis nomen, languentibus aer', which has been translated by Ellis as: 'The winds when inflated are called *spirit*, when in subsidence *air*';[13] and by Wight Duff as: 'Winds when swollen are called »spirit«, but »air« when sunk to rest',[14] both translations taking the noun 'winds' as grammatical subject. However, a looser translation that might render the poet's meaning more clearly for us would be: 'Winds when blowing actively are called »spirit«, and when inactive [not blowing] »air«'. Allowing such a translation, it appears that the author may have been seeking to employ terms that would allow a clear distinction to be drawn between air in motion (or perhaps in 'tension', as the Stoics woult have it) and air when still or relaxed.

Whatever the most appropriate translation, it is clear that the author has in mind a kind of bellows-action of the air. It is a commonplace that a blast of air from a pair of bellows will cause a fire to increase, while still air does not have a like effect. So 'spirit', says our author, commands; fire obeys.[15] Or, as we might say more prosaically: 'spirit' is the cause and fire the effect. Moving air – blasts of air – cause fire.

This being settled, the poet turns to consider the motive cause of the activity of the wind (or 'spirit'). Several possibilities are canvassed:[16] i) deep subterranean caverns storing the winds; ii) the high peak of Aetna absorbing winds through its pores from without; iii) rain driving air into these pores; iv) rock falls within the Earth causing motion of the internal air; v) the movement of waters (or watery vapours) within the Earth causing the motion. (In relation to the third suggestion, analogies are drawn with the action of mechanical contrivances such as the hydraulic organ.)[17] But whatever the initial cause of the motion might be (the author gives no firm opinion), the winds are supposedly driven on and on until finally they break out to the surface through the orifice of Aetna with an action similar to that of a fire-hose.[18]

The winds, we are told,[19] do not enter the Earth through the same orifice as that by which they leave – witness the cloud that forever hangs over the summit of Aetna. Also, the winds lose their heat-generating power once they have passed into the open and become still. However, it seems that it is really neither here nor there whether the winds are to be attributed to 'internal' or 'external' causes. A forest fire, it is argued, can be fanned just as effectively by a south or a north wind. We should *not* suppose that the winds just die down naturally within the Earth, and are regenerated simply by time alone. Rather, a quiet spell of the volcano should be attributed to temporary blockages of the passages within the Earth; and once a blockage is shifted the activity occurs once again with renewed vigour.

In all this, then, we find elaboration of the first hypothesis upon which the author's theory of volcanic action is based. The winds – by their blasts – generate heat and fire.

The second major theoretical assumption is that there is a particular fuel which serves as proper nutriment for Aetna's flames.[20] Liquid sulphur, alum, and oily bitumen are suggested as likely candidates, all of which are indeed found in association with volcanic activity. There also seems to be a reference to some 'nameless' fusible substance that may likewise act as volcanic fuel – but the text may be corrupt here.[21]

More importantly, lava-stone (*molaris lapis*) is said to be a source of the volcanic fire. It is fusible, and strikes a spark when struck with iron. There is some indication that the author thinks it can be converted to obsidian – Ellis giving as a translation, a 'coal-like mass of close-packed strength';[22] but the text is again obscure at this point. The author seems to be saying that the lava-stone can be melted again and again, yielding each time the same fire-containing product, though eventually it yields ashes when exhausted. It would appear that the author expects a combustible substance to produce an ash, and he has difficulty in reconciling this expectation with the observation of what in fact happens when lava-stone is melted and then cooled. (In any case, he is mistaken in thinking that lava may be regarded as a volcano's fuel or aliment.)

The poem now offers a vivid description of the volcano Aetna in eruption,[23] and includes an account of the formation of pumice and the flow of molten lava in waves down the mountain-side. One rendering of an uncertain passage refers to a fusible stone which the Silicians call '*rhyta*'.[24] This could be a reference to rhyolitic lava flows, but the evidence is by no means certain.

Completing the 'scientific' portion of the poem, the author summarizes his theory as follows:

> [Aetna's] fire is not of the limited heat within our own experience, but more akin to that of heaven or the kind of flame with which Jupiter himself is armed. With these mighty forces is allied the gigantic volcanic spirit forced out of straitened jaws, as when mechanics hasten to pit their strength against masses of natural iron, they stir the fires and, expelling the wind from panting bellows, rouse the current in close array ['rouse the wind with a close-continued series of puffs'[25]]. Such is the manner of its working: so goes

> far-famed Aetna's burning. The Earth draws in forces through her perforations; volcanic spirit compresses these into narrow space, and the path of conflagration lies through the mightiest rocks ['the fire works itself a passage through rocks of the largest size'[26]].[27]

Or, in summary, the theory is that winds in the Earth are blown as if by a giant bellows, finding their way to the outside world by volcanic orifices such as Aetna. Fire is generated, and substances such as brimstone and bitumen are enflamed in the process. Lava-stone likewise is an aliment for the volcanic fires, as is shown by the fact that it strikes fire when struck by iron. We might say that the theory is partly physical-*cum*-mechanical and partly chemical; but such an interpretation would involve considerable anachronism, for there was no clear-cut distinction in antiquity between physical and chemical changes.

*Date and Authorship*

Current general critical agreement exists to place the composition of *Aetna* at about the middle of the first century A.D., on stylistic and philological grounds. Certainly the poem cannot have been composed after 79 A.D., the year of the disastrous eruption of Vesuvius that cost Pliny his life and destroyed Pompeii and Herculaneum by a deluge of volcanic ash. For despite the particular concern of the poem with Mount Aetna, it is virtually inconceivable that a work dealing with general theories of volcanic action could have ignored such an event. Further than this, any refinement of opinion on the poem's date rests to a large degree on one's views as to its authorship, which is a less straightforward matter.

A verdict on the authorship of *Aetna* by one who demands nothing less than certainty must remain '*incerti auctoris*'. Nevertheless, we believe that a strong case can be made for Seneca's correspondent, the Procurator of Sicily Lucilius Junior, particularly if the 'scientific' and philosophical content of the poem is given due consideration. His possible authorship was first proposed by J. C. Wernsdorf as far back as 1785,[28] and subsequently he has come to be regarded as the most likely candidate.

Broadly speaking, three approaches are possible in support of Lucilius Junior, namely:

i) those supporting him on philological grounds;
ii) those discrediting alternative suggestions, on similar grounds;
iii) those supporting him on the 'scientific' and philosophical evidence of *Aetna*, and the Seneca correspondence.

We should like to emphasize the third approach, while making some comment on the first two.

A number of other possible authors besides Lucilius have been suggested in the critical literature that has accumulated round *Aetna* over some centuries, and Vergil, Cornelius Severus, Manilius, and Seneca himself have all received some support. And the most recent thorough review of the question by F. R. D. Goodyear dismisses ascription to Lucilius as being 'unsupported by anything which could be called evidence'.[29] However, Goodyear also dismisses all the other suggestions in turn; and, as was pointed out by Waszink,[30] all the arguments that can be brought to bear against Lucilius can also be brought to bear, in similar fashion but with greater force, against his suggested rivals. So it seems that we have a residual argument in favour of Lucilius, albeit of a somewhat negative character.

The positive evidence in favour of Lucilius Junior emerges from a reading of Seneca's *Epistulae morales* and his *Naturales quaestiones*, both of which were actually addressed to Lucilius. In the case of the *Epistulae* it is tantalizing that we lack Lucilius's side of the correspondence. Nevertheless, we can use the extant letters to reconstruct something of the wide-ranging interest in literary, philosophical, and 'scientific' matters of Seneca's correspondent, and in addition we can ascertain some of his particular activities in these areas.

Most importantly, it appears that Lucilius *did* embark on a work about Aetna. The subject is first mentioned by Seneca in *Epistula LI*, thus:

> You have over there Etna; that lofty and most celebrated mountain of Sicily (although I cannot make out why Messala – or was it Valgius? for I have been reading in both, – has called it »unique«, inasmuch as many regions belch forth fire, not merely the lofty ones where the phenomenon is most frequent, – presumably because fire rises to the greatest possible height, – but low-lying places also.)[33]

And in a later letter, *Epistula LXXIX*, the Aetna subject is raised again. The crucial point here, as Ellis pointed out,[34] is that Seneca's letter can only be interpreted as referring to a poem on which Lucilius is *already* engaged. Also, we should note that the subject of Aetna is raised in *Epistula LI* not as an isolated phenomenon, but as a matter for inclusion in some more general discussion of the generation of terrestrial heat. Consequently, it is natural that Seneca should reinforce his *Epistula LXXIX* with more explicit exhortations to prosecute the project implicitly suggested in *Epistula LI*:

> Nay, what am I to offer you *not* merely to describe Aetna in your poem, and *not* to touch lightly upon a topic that is a matter of ritual for all poets? Ovid could not be prevented from using this theme simply because Vergil had already fully covered it; nor could either of these writers frighten off Cornelius Severus.[35]

That is to say, Seneca urges his friend not to treat Aetna merely as an episode in some larger design, as many had done. And also he says:

> Besides, he who writes last has the best of the bargain; he finds already at hand words which, when marshalled in a different way, show a new face . . . Now if Aetna does not make your mouth water I am mistaken in you.[36]

It is clear, therefore, that Lucilius was engaged in (or may have finished) a work on Aetna at the time of writing of Seneca's 79th letter. The question, however, is whether it was our *Aetna*. One cannot answer this with confidence simply on philological or stylistic grounds, but, as Ellis points out,[37] there are certain strongly suggestive points in favour of such an identity.

In summary, then, we have the information that Lucilius Junior worked on an Aetna composition, and this *after* the mountain had been discussed with Seneca as part of a general 'scientific' field of volcanic activity. Also we know Seneca's point of view that such a poem would best succeed if it restricted itself to Aetna as its only theme. Since this is exactly what we have in the extant *Aetna*, the burden of proof lies very much with those who wish to discount Lucilius's authorship, the more so in view of the many stylistic similarities between *Aetna* and the Senecan corpus addressed to Lucilius.

We should also emphasize more than have other commentators the 'scientific' and philosophical content of *Aetna* when seen in the light of the Lucilian works of Seneca. For we believe that this, above all, strongly supports the claims of Lucilius Junior as probable author.

First, it is beyond argument that Lucilius and Seneca shared a common interest in 'scientific' matters. As Procurator of Sicily, Lucilius had ample opportunity to observe Mount Aetna; and the information concerning it that Seneca asks for in *Epistula LXXIX* is quite detailed, and essentially of a 'scientific' character. Moreover, since Lucilius was himself born near the other famous volcano, Vesuvius, he is naturally a promising candidate for authorship. We also know from the *Epistulae* as a whole that he was inclined towards Epicurian opinions. For example, in *Epistula CVI* we learn that Lucilius had written to Seneca in some detail concerning the material nature of the soul. There is a persistent factor running through the whole correspondence, namely Seneca's efforts to wean his friend from the atheistic materialism of the Epicureans and towards the 'natural theology' of Stoicism (which, like that of Paley,[38] did not preclude the working out of a mechanistic answer to Nature's problems, so long as a parallel final cause was also to be operative).

Now in *Epistula XLVI*, we are told that Lucilius composed a philosophical work (presumably in prose) which Seneca deems worthy of comparison, from a stylistic point of view, with the writing of Epicurus himself. Then, in *Epistula LI*, the subject of Aetna is raised in a way that at least implies the writing of a general treatise on volcanic activity. But is there any reason to suspect that this would suggest the writing of a poem? There is – since that same *Epistula LI* involves reference to the pleasing style of Lucretius. For an author who had already been compared to Epicurus, the undertaking of a poem on some 'scientific' topic would very likely suggest itself. In addition, we know that Lucilius wrote a verse treatise, in hexameters, on fountains, since Seneca quotes from it in his *Quaestiones*.[39] We also know[40] that *Quaestiones* was largely complete before the bulk of the *Epistulae* were written, and that Book III is an early part of it. So the poem on fountains was presumably complete by the time *Epistula LXXIX* reveals the writing of the *Aetna* poem. It must therefore be a separate work, predating the proposed *Aetna*. The point is that neither prose nor poetry was new to Lucilius when Seneca wrote *Epistula LXXIX*, nor could Lucilius fail to see the possibility of a 'Lucretian' work.

Turning now to the internal evidence supplied by *Aetna*, we find certain inconsistencies, which, however, are understandable if Lucilius is assumed to be the author. Like Lucretius's *De rerum natura, Aetna* avows an intention of maintaining purely naturalistic explanations of phenomena. But unlike it, *Aetna* studiously avoids any overt attack on religious belief, and even includes a number of gestures towards popular mythology, without any sarcasm or censoriousness.[41] In a work that owes so much to Epicurean influence, and which as Ellis points[42] out contains abundant verbal parallels with *De rerum natura*, this is at first sight surprising. But it is not at all unexpected if Lucilius is the author, for Seneca had already written to him (*Epistula XXIV*): 'You are always impressive, but never more pungent than when you are putting the truth in appropriate words'[43] – quoting as example a line from yet another poem of Lucilius.[44] Nothing would be more natural than that Lucilius should temper his Epicurean 'science' to please his Stoic friend.

There are, in addition, specific passages in *Aetna* that give support to Lucilius's authorship. Some of these are more than mere coincidences in the choice of vocabulary. For example, we have seen that, as used in *Aetna*, the term *'spiritus'* is one to be used with some precision;[45] and its similar use by Seneca[46] is striking.

Finally, there are times when textual criticism cannot settle a point, but where consideration of 'scientific' content can shed some light. For example, the Latinists cannot agree on the use – or not – of '*Charybdis*' in Line 107. But those who would reject it have done so on inconsequential grounds, namely that *as a metaphor* it is grotesque.[47] Grammatically, however, the passage in question can be read as a separate metaphor, with '*Charybdis*' agreeing (as half a twin subject) with '*vacuata*'.[48] In the poem we have the development of a simple theory of the formation of a porous Earth, and the comparison with a whirlpool ('*Charybdis*') to express the idea of parts of the Earth standing up so as to form a series of channels is not necessarily grotesque. Besides, anyone (like Lucilius) who lived in Sicily would not fail to be impressed by the geographical juxtaposition of Aetna and the famous whirlpool of the Straits of Messina, and might readily wonder whether there was some causal connection between the two. So the appearance of '*Charybdis*' as a metaphor in *Aetna* should not be surprising. And, to seal the matter, we find that *Epistulae XIV* and *LXXIX* request detailed information on Sicily's *Charybdis* – at a time when, according to the forgoing argument, Lucilius was engaged in writing his work on Aetna. We recall also Lucilius's work on fountains.

The foregoing arguments notwithstanding, one can still make a case for other authors such as Vergil or Cornelius Severus, on the basis of the passage in Seneca's *Epistula LXXIX*, quoted above.[49] Indeed, the earliest extant manuscript of *Aetna* ascribes authorship to Vergil,[50] and, as we have seen, early printed versions of Vergil's works, which contain the poem, suggest either Vergil or Cornelius Severus as author.[51] But against Vergil it must be said that the detailed ancient account of his work that one finds in Macrobius's *Saturnalia* makes no mention of *Aetna*.[52] And in his *Epistula LXXIX* Seneca refers (in relation to Cornelius Severus) to a part of a poem, not a whole work – which counts against Severus.

Arguments for Seneca as author of *Aetna* largely rest on his known interest in Earth phenomena, and the similarity between some of his known theoretical ideas and those to be found in *Aetna*. But this is inconclusive, and points to Lucilius as much as Seneca, given their known friendship and correspondence. A suggestion that the writer was Manilius, the first-century author of *Astronomicon*,[53] seems to have received negligible support.

Thus, *if* Lucilius is accepted as author of *Aetna*, and *if* we accept the prevailing view that Seneca's *Epistulae* to Lucilius were begun in A.D. 63,[54] we can place the date of composition in the range 63–79 A.D. Nevertheless, the question of authorship does remain open for further debate on the part of those who may be interested in such matters.

## *Precedents and Parallels for the Theoretical Concepts of* Aetna

Although *Aetna* deals with the problems of volcanic phenomena in much more detail than does any other work that has survived from antiquity, it does not offer the reader anything strikingly original so far as 'geological' theory is concerned. Rather, the author may be said to have been attempting a general synthesis of the theoretical opinions of his day, providing thereby a thoroughly eclectic perspective. But in no way does this detract from its interest as an example of Roman

| Underground passages or pores within the Earth | Air/wind the cause of Earthquakes | Active wind the cause of fire | Air may be driven into the Earth by the Sea | Falling material may be the cause of movement of Air within the Earth, or of Earthquakes | Theory of volcanic 'fuel' |
|---|---|---|---|---|---|
| Democritus[55] | Anaximander[67] | Aristotle[81] | Aristotle[89] | Anaximenes[93] | Ovid[97] |
| Metrodorus[56] | Anaxagoras[68] | Epicurus[82] | Callisthenes[90] | Epicurus[94] | Strabo[98] |
| Aristotle[57] | Archelaus[69] | Lucretius[83] | Lucretius[91] | Lucretius[95] | Seneca[99] |
| Epicurus[58] | Democritus[70] | Ovid[84] | Seneca[92] | Seneca[96] | Pliny[100] |
| Lucretius[59] | Metrodorus[71] | Strabo[85] | *Aetna* | *Aetna* | Apollonius[101] |
| Ovid[60] | Aristotle[72] | Seneca[86] | | | *Aetna* |
| Diodorus[61] | Callisthenes[73] | pseudo-Aristotle[87] | | | |
| Strabo[62] | Epicurus[74] | Apollonius[88] | | | |
| Seneca[63] | Lucretius[75] | *Aetna* | | | |
| Pliny[64] | Posidonius[76] | | | | |
| pseudo-Aristotle[65] | Strabo[77] | | | | |
| Apollonius[66] | Seneca[78] | | | | |
| *Aetna* | Pliny[79] | | | | |
| | pseudo-Aristotle[80] | | | | |
| | *Aetna* | | | | |

'geological' knowledge. We have no other work dealing specifically with a particular theoretical problem relating to the Earth – no other 'case study' of a particular 'geological' phenomenon.

Given the eclectic character of *Aetna*, it is an interesting exercise to attempt to identify the likely sources of the major theoretical concepts that it contains. Thereby, one may form a general conspectus of some of the leading ideas concerning the Earth that are to be found among the ancient writers. For convenience, these may be placed in tabular form as on page 12.

Certain other theories concerning earthquakes and volcanoes, that are to be found in the ancient writers, may also be listed:

| Mythical explanations of volcanoes (chiefly associated with the legend of 'Typhoeus') | Analogies drawn with thunder and lightning | Employment of the theory of exhalations | Earthquakes ascribed to water | Earthquakes ascribed to fire |
|---|---|---|---|---|
| Hesiod[102]<br>Pindar[103]<br>Vergil[104]<br>Ovid[105]<br>Manilius[106]<br>Lucan[107] | Anaxagoras[108]<br>Epicurus[109]<br>Seneca[110]<br>Pliny[111] | Aristotle[112]<br>Diodorus[113]<br>Seneca[114]<br>Pliny[115] | Thales[116]<br>Democritus[117]<br>Epicurus[118]<br>Lucretius[119]<br>Seneca[120] | Anaxagoras[121] |

But these ideas were not employed in *Aetna*.

It will be seen from these tables that most of the authors surveyed sought to provide naturalistic explanations, invoking simple mechanical or 'chemical' processes, or offering analogies with aspects of everyday experience, rather than turning to myth. Where myth was invoked, it was generally the story of Typhoeus (or Typhon, also called Enceladus), a hundred-headed monster said to be born of the

Earth and Zeus, who had rebelled against the gods, but had been vanquished by his father and lay imprisoned under the Earth. When he stirred, the mountain supposedly began its eruptions.

It is, of course, impossible to determine the precise sources for all the ideas that were employed in *Aetna*, but assuming that Lucilius was the author it seems likely that Seneca himself was a major stimulus, for his *Quaestiones* provide a fairly detailed survey of earlier opinions on the causes of earthquakes (and accordingly has been used extensively in the compilation of the details above). In addition, an important passage in Lucretius's *De rerum natura*[122] was almost certainly utilized; and one can be confident also that Aristotle's *Meteorologica* was consulted. The adoption of the theory of volcanic 'fuel' suggests that ideas may have been drawn from Ovid, Strabo, or Pliny, though Seneca may also have been the source for this. Whether *Aetna* influenced pseudo-Aristotle's *De mundo*, or *vice versa*, cannot now be determined, though there are certainly points in common, most obviously a story of two brothers saving their aged parents from an eruption of Aetna.[123] The two works are thought to have been written at about the same time, and there are several theoretical similarities, as the first table shows.

## *The Historical Influence of* Aetna *and its Theoretical Doctrines*

A number of Classical writings can be identified from the period following 79 A.D. which give some attention to volcanic phenomena, but very little is added to the theoretical ideas that we find in *Aetna*, and several of the authors revert to mythological explanations. The only theoretical addition that we have found occurs in the following passage by Philostratus the Elder (b. *ca*. 190 A.D.), in support of the hypothesis of the existence of subterranean passages and cavities, with flames breaking out through external ducts:

> If one wishes to speculate about such matteres, the island [Sicily] provides natural bitumen and sulphur; and when these are mixed by the sea, the island is fanned into flame by many winds, drawing from the sea that which sets the fuel aflame.[124]

The last writer of antiquity that we have noticed who referred explicitly to volcanic phenomena was Nonnos of Panopolis, in his *Dionysiaca*, a work that sought to refurbish the ancient mythologies of Greece and Rome. Naturally, therefore, we find mention once again of the legend of Typhon (Tophoeus, Enceladus);[125] and also there appears the tradition of earthquakes arising from blasts of air within the Earth.

After Nonnos, there seems to have been a long silence on the question of earthquakes and volcanic action, though doubtless speculation on the subject continued. When the question was revived in the Middle Ages, much the same theories were entertained as those of antiquity. The 14th-century writer, Conrad von Megenberg, for example, ascribed earthquakes to the movement of vapours within caverns of the Earth,[126] as had so many of his ancient predecessors.

It is remarkable that although *Aetna* has been a continuing source of interest, speculation and academic controversy, and the vehicle for some dazzling displays of erudition on the part of generations of Classicists, it has had virtually no influence on the development of *geological* thought in the Renaissance and after, so far as we have been able to determine. The only specific reference to the poem that we have been able to identify in a 'scientific' work is that which appears in Athanasius Kircher, who quotes two passages (Lines 94–97 and 418–422) from it in his *Mundus subterraneus*, curiously ascribing the former to Cornelius Severus and the latter to Vergil.[127]

As is well known, Kircher believed in a deep internal fire within the body of the Earth, as well as subterranean water channels.[128] An eclectic, he sought to synthesize earlier opinions respecting volcanic action by reference to the four Aristotelian causes, proposing fire itself as the formal cause; sulphur, salt [*sic*], nitre, bitumen, and inflammables activated by the sea as the material cause; and winds as the efficient cause, with volcanic vents acting like the nozzles of bellows.[129]

By contrast, when in the 16th century Agricola made a detailed search through the ancient literature for references to the Earth he seems to have failed to examine the *Aetna* poem. Nevertheless, his theory of volcanic action was based on ideas similar to those which had been put forward many times in the ancient world. His theory was that volcanic heat was generated by burning bitumen and sulphur, with volcanoes acting as vents for the subterranean fires. He subscribed also to the Aristotelian theory of exhalations, and supposed that mountains might be formed by winds.[130]

Thus as a specific source for later theories of volcanic action *Aetna* seems to have dropped out of sight, so far as scientific writers were concerned, in the modern period, the reference of Kircher being so deeply buried in *Mundus subterraneus* as to receive no further attention. Nevertheless, it is worth noting that a number of 18th-century writers adopted theories of volcanic action that had a good deal in common with those of the *Aetna* poem of antiquity. For example, Buffon wrote of the combustion of sulphur and bitumen which were supposedly the 'food' of subterranean fire.[131] Spallanzani, while rejecting the idea that air passing through subterranean passages could, *per se*, be the cause of volcanic activity, suggested that the sea might percolate through passages in the rocks to produce violent effects when it came into the hot material below.[132] Similarly, John Michell ascribed earthquakes to deep-seated subterranean fires, and to the sudden vaporization of water coming into contact with such fires in underground caverns.[133] In this, he was offering a forerunner to the theory of 'craters of elevation', as stated by Breislak and von Buch.[134]

This paper does not seek to offer a general history of theories of volcanic action. Such an enquiry has yet to be undertaken. But for any scholars that may wish to embark upon such an enquiry we would urge that they give careful consideration to the doctrines that were espoused in antiquity. For the historical roots of some of the modern ideas run very deep in this area. And for a general conspectus of Classical theories of volcanic action the *Aetna* poem is of prime importance.

## REFERENCES

Many of the following references are drawn from the volumes of the Loeb Classical Library (London & Cambridge (Mass.)). Apart from the first reference,where we have drawn on this standard Loeb edition we shall merely give the indication *L. C. L.* and the appropriate volume number and date.

1. Anon., *Aetna*, in: *Minor Latin poets: with introductions and English translations by J. Wight Duff . . . and Arnold M. Duff . . .* (London & Cambridge (Mass.), 1961), pp. 349–419.
2. 'P. Vergilii Maronis aethna: quae tamen a quibusdam Cornelio Severo tribuitur', in: *Opera Vergiliana docte et familiariter exposita: Docte quidem Bucolica & Georgica a Servio Donato & Mancinello: cum adnotationibus Beroaldinus: Aeneis vero ab iisdem praeter Mancinellum: praeterea quaedam ab Dominico Calderino: Familiariter vero omnia tam opera quae opuscula ab Iodoco Badio Ascensio* ([Paris, ] 1570), pp. 160–171. A possibly earlier version is: 'P. V. M. Aetna quae a quibusdam Cornelio tribuitur', in: *P. Virgilii Maronis Bucolica . . .* [J. & B. Honate, Milan, 1480 (?)], sigs. cc4–dd4.
3. [*Vergil*] *Aetna: Herausgegeben und übersetzt von Will Richter* (Berlin, 1963); *Incerti auctoris Aetna: edited with an introduction and commentary by F. R. D. Goodyear* (Cambridge, 1965).
4. The literature is surveyed in: F. R. D. Goodyear, *op.cit.* (note 3), pp. 10–22.
5. K. von Zittel, *History of geology and palaeontology to the end of the century . . . translated by Maria M. Ogilvie-Gordon* (London, 1901), p. 10.
6. For a general discussion of the establishment of geology as a science, see: R. Porter, *The making of geology: Earth science in Britain 1660–1815* (Cambridge, 1977).
7. *Aetna*, Lines 29–32: *L.C.L.* (1961), pp. 360–361.
8. *Aetna*, Lines 606–610: *L.C.L.* (1961), pp. 416–417.
9. *Aetna*, Line 96: *L.C.L.* (1961), pp. 366–367.
10. *Aetna*, Lines 146–147: *L.C.L.* (1961), pp. 370–373.
11. *Aetna*, Line 181: *L.C.L.* (1961), pp. 374–375.
12. *Aetna*, Line 192: L.C.L. (1961), pp. 376–377.
13. R. Ellis, *Aetna: a critical recension of the text, based on a new examination of the MSS. with prolegomena, translation, textual and exegetic commentary, excursus, and complete index of words* (Oxford, 1901), p. 26.
14. *Aetna, L.C.L.* (1961), p. 379.
15. *Aetna*, Lines 217–218: *L.C.L.* (1961, pp. 378–379.
16. *Aetna*, Lines 283–321: *L.C.L.* (1961), pp. 384–389.
17. For descriptions of the workings of these ancient instruments, see, for example: W. Chappell, *The history of music. (Art and science).* Vol. I. *From the earliest records to the fall of the Roman Empire. With explanations of ancient systems of music, musical instruments, and of the true physiological basis for the science of music, whether ancient or modern* (London, 1874), pp. 325–370.
18. *Aetna*, Line 328: *L.C.L.* (1961), pp. 388–389.
19. *Aetna*, Lines 330–332: *L.C.L.* (1961), pp. 390–391.
20. *Aetna*, Lines 386–387: *L.C.L.* (1961), pp. 394–395.
21. See: F. R. D. Goodyear, *op.cit.* (note 3), p. 175.
22. R. Ellis, *op.cit* (note 13), p. 50.
23. *Aetna,* Lines 467–507: *L.C.L.* (1961), pp. 402–407.
24. *Aetna*, Lines 532–534: *L.C.L.* (1961), 408–409.
25. R. Ellis, *op.cit.* (note 13), p. 66.
26. *Ibid.*
27. *Aetna*, Lines 559–568: *L.C.L.* (1961), pp. 410–411.

28. J. C. Wernsdorf, 'Lucilii Iunioris Aetna', in: *Poetae Latini minores* (Altenburg, 1785), Vol. 4, Part I.
29. F. R. D. Goodyear, *op.cit.* (note 3), p. 57.
30. J. H. Waszink, 'De Aetne carminae auctore', *Mnemosyne,* 2 (ser. 4) (1949), pp. 224–241.
31. Seneca, *Epistulae morales, L.C.L.* (3 vols., 1961–1962).
32. Seneca, *Naturales quaestiones, L.C.L.* (2 vols., 1971–1972).
33. Seneca, *Naturales quaestiones*, L.C.L. (1971–1972), vol. i, p. 337.
34. R. Ellis, *op.cit.* (note 13), p. xxxvii.
35. Seneca, *Naturales quaestiones, L.C.L.* (1971–1972), vol. ii, p. 203.
36. *ibid*.., p. 205.
37. R. Ellis, *op.cit.* (note 13), pp. xxix–xlviii.
38. W. Paley, *Natural theology: or, evidences of the existence and attributes of the deity, collected from the appearances of nature* (London, 1802; facsimile edn., Farnborough 1970).
39. Seneca, *Naturales quaestiones, L.C.L.* (1971–1972), vol. i, p. 213.
40. See: *Physical science in the time of Nero: being a translation of the Quaestiones Naturales of Seneca by John Clarke . . . with notes on the treatise by Sir Archibald Geikie* (London, 1910). Clarke's Introduction discusses in detail the dating of the *Epistulae* and the *Quaestiones* (both of which Seneca addresses to Lucilius). The sum of his views is that the bulk of the *Quaestiones* was complete before the *Epistulae* were written, a view in accordance with those of Gummere (*op.cit.*, note 31) and Corcoran (*op.cit.*, note 32). Book III of the *Quaestiones* was formerly thought to be the opening section (a view not now generally held) and is still thought to be a very early section of the work. Accordingly, a poem quoted in Book III of the *Quaestiones* can be assumed to pre-date a poem referred to in *Epistula LXXIX* – and from which Seneca cannot yet quote, because clearly he has not yet read it.
41. *E.g., Aetna* Lines 1–23: *L.C.L.* (1961), p. 358–361.
42. R. Ellis, *op.cit.* (note 13), p. xxvii.
43. Seneca, *Naturales quaestiones, L.C.L.* (1971–1972), vol. i, p. 179.
44. 'Not single is the death which comes; the death which takes us off is but the last of all.' This fragment of Lucilius offers a very Senecan Stoic thought.
45. See above, pp. 3–4.
46. Seneca. *Naturales quaestiones, L.C.L.* (1971–1972), vol. ii, pp. 100 & 172.
47. F. R. D. Goodyear, *op.cit* (note 3), p. 127.
48. *Aetna,* Line 107: *L.C.L.* (1961), p. 368.
49. See p. 8.
50. Cambridge University Library, MS. Kk.v.34. (This is assigned to the 10th C.)
51. See note 2.
52. Macrobius (tr. and ed. P. V. Davies), *The Saturnalia* (New York & London, 1969).
53. See: R. Ellis, *op.cit.* (note 13), p. xxiv.
54. Seneca, *Epistulae morales, L.C.L.* (3 vols., 1961–1962), vol. i, p. ix.
55. Quoted in Seneca, *Naturales quaestiones, L.C.L.* (1971–1972), vol. ii, p. 185.
56. *Ibid*., p. 183.
57. Aristotle, *Meteorologica, L.C.L.* (1952), p. 209.
58. *Epicurus: the extant remains with short critical apparatus translation and notes by Cyril Bailey* (Hildesheim & New York, 1975), p. 73.

59. *Lucretius: on the nature of the universe translated and with an introduction by Ronald Latham* (Harmondsworth, 1951), p. 233.
60. Ovid, *Metamorphoses, L.C.L.* (1960), vol. ii, p. 389.
61. Diodorus of Sicily, *L.C.L.* (1961), vol. iii, p. 115.
62. Strabo, *Geography, L.C.L.* (1959–1961), vol. ii, p. 457.
63. Seneca, *Naturales quaestiones, L.C.L.* (1971–1972), vol. ii, p. 193.
64. Pliny, *Natural history, L.C.L.* (1938–1962), vol. i, p. 325.
65. Pseudo-Aristotle, *De mundo, L.C.L.* (1965), pp. 373 & 375.
66. *Philostratus: the life of Apollonius of Tyana, L.C.L.* (1912), vol. i, p. 501.
67. Quoted by Ammianus Marcellinus, *L.C.L.* (1935), vol. i, p. 347.
68. Quoted by Aristotle, *Meteorologica, L.C.L.* (1952), p. 199.
69. Quoted by Seneca, *Naturales quaestiones, L.C.L.* (1971–1972), vol. ii, p. 163.
70. Quoted by Seneca, *Ibid.*, p. 183.
71. Quoted by Seneca, *Ibid.*, p. 183.
72. Aristotle, *Meteorologica, L.C.L.* (1952), p. 205.
73. Quoted by Seneca, *Naturales quaestiones, L.C.L.* (1971–1972), vol. ii, p. 195.
74. Epicurus, *op.cit* (note 58), p. 71.
75. Lucretius, *op.cit.* (note 59), p. 234.
76. Quoted by Diogenes Laertius, *Lives of eminent philosophers, L.C.L.* (1958–1959), vol. ii, p. 154.
77. Strabo, *Geography, L.C.L.* (1959–1961), vol. iii, p. 25.
78. Seneca, *Naturales quaestiones, L.C.L.* (1971–1972), vol. ii, p. 189.
79. Pliny, *Natural history, L.C.L.* (1938–1962), vol. i, p. 323.
80. Pseudo-Aristotle, *De mundo, L.C.L.* (1965), p. 375.
81. Aristotle, *Meteorologica, L.C.L.* (1952), p. 237.
82. Quoted by Seneca, *Naturales quaestiones, L.C.L.* (1971–1972), vol. ii, p. 187.
83. Lucretius, *op.cit* (note 59), p. 238.
84. Ovid, *Metamorphoses, L.C.L.* (1960), vol. ii, p. 389.
85. Strabo, *Geography, L.C.L.* (1959–1961), vol. iii, p. 95.
86. Seneca, *Naturales quaestiones, L.C.L.* (1971–1972), vol. ii, p. 101.
87. Pseudo-Aristotle, *De mundo, L.C.L.* (1965), p. 371 (in reference to lightning).
88. Philostratus, *The life of Apollonius of Tyana, L.C.L.* (1912), vol. i, p. 501.
89. Aristotle, *Meterologica, L.C.L.* (1952), p. 207.
90. Quoted by Seneca, *Naturales quaestiones, L.C.L.* (1971–1972), vol. ii, p. 195.
91. Lucretius, *op.cit.* (note 59), p. 238.
92. Seneca, *Naturales quaestiones, L.C.L.* (1971–1972), vol. ii, p. 195.
93. Quoted by Seneca, *Naturales quaestiones, L.C.L.* (1971–1972), vol. ii, p. 159.
94. Epicurus, *op.cit.* (note 58), p. 71.
95. Lucretius, *op.cit.* (note 59), p. 234.
96. Seneca, *Naturales quaestiones, L.C.L.* (1971–1972), vol. ii, p. 193.
97. Ovid, *Metamorphoses, L.C.L.* (1960), vol. ii, p. 389.
98. Strabo, *Geography, L.C.L.* (1959–1961), vol. iii, p. 91.
99. Seneca, *Naturales quaestiones, L.C.L.* (1971–1972), vol. ii, p. 101.
100. Pliny, *Natural history, L.C.L.* (1938–1962), vol. i, p. 363.

101. *Philostratus: the life of Apollonius of Tyana, L.C.L.* (1912), vol. i, p. 501.
102. *Hesiod [Theogony]: translated by Richard Lattimore* (Ann Arbor, 1959), pp. 174–175.
103. Pindar, *The odes, L.C.L.* (1961), p. 157.
104. *Vergil's Aeniad translated by John Dryden: with introduction and notes* (New York, 1909), p. 150.
105. Ovid, *Metamorphoses, L.C.L.* (1960–1964), vol. i, p. 263.
106. *M. Manilius: Astronomicon ... recensit et enarravit A. E. Housman* (London, 1903–1920), vol. ii, p. 99.
107. Lucan, *Pharsalia, L.C.L.* (1969), pp. 245–247.
108. Quoted by Seneca, *Naturales quaestiones, L.C.L.* (1971–1972), vol. ii, p. 157.
109. Quoted by Seneca, *Ibid.*, p. 187.
110. *Ibid.*, p. 189. (Seneca states his agreement with Epicurus on several points of theory.)
111. Pliny, *Natural history, L.C.L.* (1938–1962), vol. i, p. 325.
112. Aristotle, *Meteorologica, L.C.L.* (1952), p. 231.
113. Diodorus of Sicily, *L.C.L.* (1961), vol. iii, p. 115.
114. Seneca, *Naturales quaestiones, L.C.L.* (1971–1972), vol. ii, p. 207.
115. Pliny, *Natural history, L.C.L.* (1938–1962), vol. i, p. 339.
116. Quoted by Seneca, *Naturales quaestiones, L.C.L.* (1971–1972), vol. ii, p. 147.
117. Quoted by Seneca, *Ibid.*, p. 185.
118. Quoted by Seneca, *Ibid.*, p. 187.
119. Lucretius, *op.cit.* (note 59), p. 234.
120. Seneca, *Naturales quaestiones, L.C.L.* (1971–1972), vol. ii, p. 153.
121. Quoted by Seneca, *Ibid.*, p. 157.
122. Lucretius, *op.cit.* (note 59), pp. 233–238.
123. *Aetna,* Lines 625–646: *L.C.L.* (1961), pp. 416–419; pseudo-Aristotle, *De mundo, L.C.L.* (1965), p. 403.
124. Philostratus the Elder, *Imagines, L.C.L.* (1960), p. 199.
125. Nonnos, *Dionysiaca, L.C.L.* (1955–1962), vol. i, p. 91.
126. Conrad von Megenberg, quoted in : F. D. Adams, *The birth and development of the geological sciences* (2nd edn., New York, 1954), p. 404.
127. A. Kircher, *Mundus subterraneus, in XII libros digestus;* . . . (3rd ed., Amsterdam, 1678), Book IV, pp. 185 and 201 (1st edn., 1665).
128. See: K. F. Mather & S. F. Mason, *A source-book of geology* (New York & London, 1939), pp. 17–19.
129. A. Kircher, *op.cit.* (note 127), p. 204.
130. G. Agricola, *De ortu et causis subterraneorum Lib. V.* (Basle, 1546) pp. 37–38.
131. G. L. L. de Buffon, *Histoire naturelle, générale et particulière* . . . (44 vols., Paris, 1749–1804), vol. i, p. 502.
132. L. Spallanzani, *Travels in the two Sicilies, and some parts of the Appenines* (London, 1798), vol. iii, p. 354.
133. J. Michell, 'Conjectures concerning the cause, and observations upon the phaenomena of earthquakes; . . .', *Philosophical transactions*, 51 (1761), pp. 566–634.
134. L. von Buch, *Geognostische Beobachtungen auf Reisen durch Deutschland und Italien* . . . (Berlin, 1802–1809), vol. ii, p. 208. This theory of 'craters of elevation' was, in fact, much older than the ideas of Michell. It is similar to suggestions made in antiquity by Seneca : *Naturales quaestiones, L.C.L.* (1971–1922), vol. ii, p. 189.

# II

# SOME NEO-PLATONIC AND STOIC INFLUENCES ON MINERALOGY IN THE SIXTEENTH AND SEVENTEENTH CENTURIES

On making a cursory examination of the writings of the sixteenth and seventeenth centuries on matters relating to the mineral kingdom, one soon comes across terms such as "sperm", "seed", "idea" or "word" which suggest that the writers concerned were making considerable use of Neo-Platonic and Stoic notions, and other ancient beliefs, in their attempts to give accounts of the structure and composition of the earth. It is the purpose of the present paper to examine such writings in some detail, to display their dependence on notions derived from the ancient world-views, and to show how these beliefs provided a characteristic pattern of explanation in the period prior to the general acceptance of the "mechanical philosophy". The writings with which we shall be concerned here have not often been examined from the point of view of the historian of the earth sciences, though most of them are of recognized importance in fields such as the history of chemistry.[1] It is true that F. D. Adams, in his well known *Birth and Development of the Geological Sciences*,[2] and elsewhere,[3] gave considerable attention to the mineral theories of the period which we wish to consider, but he did not pay much attention to the philosophical background of the writers concerned. Rather, he took the view that "philosophical speculation" could be held responsible for slow progress towards the establishment of satisfactory mineral theories:

> If . . . we ask why the ancients entertained such quaint and curious views concerning the origin of the metals and their ores, we find that their views were quaint because they were based almost entirely on philosophical speculation and not on observation.[4]

[1] A number of the writers considered in the present paper are discussed in A. G. Debus, *The English Paracelsians*, London, 1965; and R. P. Multhauf, *The Origins of Chemistry*, London, 1966. These works provide a suitable *entrée* to the primary literature considered in this paper.

[2] F. D. Adams, *The Birth and Development of the Geological Sciences*, Baltimore, 1938 (republished New York, 1954).

[3] F. D. Adams, "The Origin and Nature of Ore Deposits: An Historical Study", *Bulletin of the Geological Society of America*, **45**, 374–424, 1934.

[4] Adams (2), 328.

In this paper, then, an attempt will be made to place the early mineralogical writings in their proper philosophical context, and, as stated above, to examine the general pattern of explanation which they provided for problems raised by observations of mineral substances. Some attempt will also be made to evaluate the adequacy of the explanations proposed.

The Neo-Platonic and Stoic beliefs which provided the basis of so many of the explanations of the mineral kingdom in the sixteenth and seventeenth centuries will be sketched in very briefly here, though it is obvious that I cannot do them justice within the confines of the present paper. Fortunately, however, there is no shortage of secondary literature on these topics,[5] though the relevant fragments of those Stoic writings which bear on the thesis of this paper are not all available in English translation.[6]

Plato's doctrine of "Ideas" is too well known to require exposition here. We may remind ourselves, however, that in the *Timaeus*, the Cosmos is depicted as a "Living Creature endowed with soul and reason owing to the providence of God".[7] Moreover, an underlying substance is postulated, which can, by having various "figures" stamped upon it, according to certain "Ideas" in the "intelligible" realm, give rise to the actual objects which we see around us in our everyday experience.

> And the figures that enter and depart are copies of those that are always existent, being stamped from them in a fashion marvellous and hard to describe. . . .[8]

That which receives the "stamp" is likened to the "Mother"; that which is the "Source" of the "Ideas" imprinted on the recipient "Mother" is likened to the "Father"; and the resultant "engendered between these two . . . [is] the

[5] On Neo-Platonism, see, for example, T. Whittaker, *The Neo-Platonists. A Study in the History of Neo-Platonism*, 4th ed., Hildesheim, 1961; the introduction to MacKenna's translation of *The Enneads* of Plotinus (Plotinus, *The Enneads Translated by Stephen MacKenna Third Edition, Revised by B. S. Page* . . ., London, 1962); J. K. Feibleman, *Religious Platonism* . . ., London, 1959. On Stoicism, see, for example: E. V. Arnold, *Roman Stoicism* . . ., Cambridge, 1911 (re-issued London, 1958); R. D. Hicks, *Stoic and Epicurean*, New York, 1962; S. Sambursky, *Physics of the Stoics*, London, 1959. A work containing an important review of the "world-picture" relevant to this paper is W. J. O. Pagel, *Das medizinische Weldbild des Paracelsus. Seine Zusammenhänge mit Neuplatonismus und Gnosis* . . ., Wiesbaden, 1962.

[6] For translations of some of the relevant texts, see J. L. Saunders, *Greek and Roman Philosophy after Aristotle*, New York and London, 1966. For original sources, see J. Von Arnim, *Stoicorum Veterum Fragmenta*, 4 vols, 1st ed., 1903–24, reprinted in facsimile, Stuttgart, 1964.

[7] Plato, *Timaeus*, 30B. R. G. Bury (ed.), *Plato with an English translation Timaeus, Critias, Cleitophon, Menexenus, Epistles*, London and Cambridge (Mass.), 1966, 55.

[8] *Ibid.*, 50C (117).

Offspring".[9] By the imprint of "Ideas" on the material substrate, the four traditional elements, or "roots of things", arise.

Plato, of course, was not the originator of animistic descriptions of the universe, and his own account is but one version of the venerable doctrine of macrocosm and microcosm.[10] The metaphor of the earth as a "womb" is one of very long standing,[11] but there can be little doubt that Plato, through the powerful influence of the *Timaeus*, provided a source of inspiration for much subsequent thinking about the mineral kingdom; and evidence to support this contention will be mustered below.

The Stoics, following their founder, Zeno,[12] were also avowed supporters of hylozoism, the universe being viewed as a living whole, with each material specimen being in some sense "animated". (Conversely, souls were held to be corporeal.) The active principle, or *Logos* of the universe, supposedly contained within itself active forms of all the things which were to come into being. These active principles were likened to "seeds", and hence were known by the term *Logoi spermatikoi*, sometimes translated as "seminal intelligences" or "seminal reasons". It was believed that the universe passed through a series of identical cycles, beginning with a fiery vapour (equivalent to God), which was transformed into air, and then into water. From "seeds" within the water there then developed the four traditional elements, earth, water, air and fire. In a later phase of the cycle, however, water was reconverted into air, and air into fire, yielding a final conflagration.[13] The similarity to certain Biblical passages[14] (not to mention Burnet's *Sacred Theory of the Earth*[15]) needs no emphasis though, unlike the Stoic system, the Biblical scheme was not cyclical.

As might be expected, the details of the theory varied from one writer to another. Eusebius, for example,[16] following Zeno, thought of the primary fire

[9] *Ibid.*, 50D (119).

[10] For a detailed account of the history of the evolution of this concept, see G. P. Conger, *Theories of Macrocosms and Microcosms in the history of Philosophy*, New York, 1922 (republished New York, 1967).

[11] For a discussion of the ancestry of this notion, see A. A. Barb, "Diva Matrix", *Journal of the Warburg and Courtauld Institutes*, **16**, 193–238, 1953.

[12] *i.e.*, Zeno the Stoic (*ca.* 340–265 B.C.), not Zeno of Elea (*ca.* 490–30 B.C.), well known for his "paradoxes".

[13] See Arnold (5), 195–6.

[14] *e.g.*, *The Second Epistle General of Peter*, Chapter 3.

[15] T. Burnet, *The Theory of the Earth: Containing an Account of the Original of the Earth, and of all the General Changes which it hath already undergone, or is to undergo Till the Consummation of all Things* . . ., 2nd ed., London, 1690–1 (republished with an introduction by Basil Willey, London and Fontwell, 1965).

[16] Saunders (6), 92 (*cf.* J. Von Arnim, *op. cit.*, vol. I, 27).

as a kind of seed. Diogenes Laertius, also citing Zeno's ideas, stated[17] that the "seminal reason" of the world was within water, and fashioned the primordial substances into the four elements. It is doubtful, however, that such ideas influenced the sixteenth and seventeenth century writers on the mineral kingdom directly. The later Roman Stoics, such as Epictetus and Marcus Aurelius, do not emphasize the notion of "seminal intelligences", and the principal Stoic work on geological phenomena, Seneca's *Quaestiones Naturales*,[18] does not mention this doctrine, though it makes much of the presumed cyclic character of the universe, with catastrophic floods supposedly preceding a conflagration at the end of each cycle. Moreover, the doctrine of "seminal reasons" is also to be found in some passages of the *Enneads* of Plotinus,[19] though Armstrong has argued[20] that such passages are not in keeping with the overall philosophical position of the Neo-Platonist. It is probable, therefore, that it was through the writings of Plotinus that the scholars of the Renaissance assimilated elements of the Stoic philosophy, along with the better known doctrines of Plato.

The leading spirit in the revival of interest in Platonism in Renaissance Europe is usually taken to be Marsilio Ficino (1433–99),[21] head of the so-called Platonic Academy of Florence, established under the patronage of Cosimo de' Medici. Ficino, believing that the Christian faith and Platonism mutually supported each other, undertook the strenuous task of making the whole of the writings of Plato and Plotinus available in Latin translation,[22] and of developing his own particular philosophical stance, a blend of Christian and Platonic theism, in his principal work the *Theologica Platonica* of 1482.[23] Ficino is also important in that he played a part in the transmission of the

[17] Diogenes Laertius, *Lives of the Eminent Philosophers*, VII, 135–6. (Translated R. D. Hicks, 2 vols, London and Cambridge (Mass.), 1958, vol. II, 240–1.)

[18] *cf.* J. Clarke, *Physical Science in the Time of Nero Being a Translation of the Quaestiones Naturales of Seneca*, London, 1910.

[19] *e.g.* Plotinus, *The Enneads*, IV, 3, 10 (MacKenna (5), 269).

[20] A. H. Armstrong, *The Architecture of the Intelligible Universe in the Philosophy of Plotinus An Analytic and Historical Study*, Cambridge, 1940, Amsterdam, 1967, 63.

[21] For an account of the philosophy of Ficino, see, for example: J. L. Burroughs in E. Cassirer, P. O. Kristeller and J. H. Randall (Jr.) (eds.), *The Renaissance Philosophy of Man*, Chicago, 1948, 185–92; or P O. Kristeller, *The Philosophy of Marsilio Ficino . . . Translated into English by Virginia Conant*, New York, 1943, and Gloucester (Mass.), 1964. A fascinating account of Ficino's mystical rituals, in which he attempted to commune with the "world spirit", is given in: D. P. Walker, *Spiritual and Demonic Magic from Ficino to Campanella*, Studies of the Warburg Institute, vol. XXII, London, 1958.

[22] M. Ficino, *Prohemium Marsilii ficini florẽtini in libros Platonis ad Laurentium medicem Virum Magnanimum. Marsilii Argumentum in Hipparchum*, Florence, 1484–5. M. Ficino, *Prohemium M. Ficini Florentini in Plotinum, &c. . . .*, Florence, 1492.

[23] M. Ficino, *Platonica Theologica . . .*, Florence, 1482.

Hermetic tradition by preparing Latin translations of certain Hermetic tracts,[24] thus providing another pathway for the passage of ideas from antiquity.

A leading feature of the system developed by Ficino was a particular version of the theory of the "Great Chain of Being", with a hierarchy extending from God, through the angels, to man, to other living creatures, and finally to inanimate objects. For Ficino, all attributes of being arose from God, and it was the task of man's soul to strive to possess the infinite truth and goodness of God. We may also remark here that in the view of the Renaissance Neo-Platonists (following Plotinus) the Platonic "Ideas" were construed as ideas in the mind of God, and this, in some sense at least, removed the ambiguity which had always faced the readers of the *Dialogues*.

Thus, by the end of the fifteenth century, the ancient philosophies of Platonism and Neo-Platonism were freely available to European scholars, and consequently certain Stoic notions, contained within the corpus of writings of Plotinus, were also widely known, independently of any direct influence which Stoic philosophy might have had. There was, besides, the continuing influence of the alchemists, who had also assimilated the doctrine of "seminal reasons" into their metallurgical theories, by a route which is now difficult to trace in detail. The doctrine of "seminal reasons" is also to be found in Augustinism, and Stoic ideas were transmitted to European scholars in the writings of Galen. Thus a multiplicity of ancient beliefs could be drawn upon in the sixteenth century by writers on the mineral kingdom, and it is scarcely possible to tease apart the many strands in the philosophical thread which linked them with the ancient world. Our thesis is merely that the ancient beliefs provided the basis for mineralogical explanations in the period prior to the general acceptance of the mechanical philosophy, with perhaps Neo-Platonic and Stoic notions being of particular significance. An account of some of the ways in which these early ideas were utilized now follows.

It is perhaps convenient to begin with Paracelsus (1493–1541).[25] As is well

[24] M. Ficino, *Mercurii Trismegisti liber de potestate et sapientia. Dei e Graeco in Latinum traductus a M. Ficino Florentino ad Cosmum Medicem Patriae Patrem. Pimander incipit*, Treviso, 1471. For a resumé of the Hermetic tradition and its influence in the sixteenth and seventeenth centuries, see F. Sherwood Taylor, *The Alchemists Founders of Modern Chemistry*, Melbourne, London and Toronto, 1951, chapter 14.

[25] The literature on Paracelsus is very considerable, but items which are of particular relevance to the present paper are: W. Pagel, *Paracelsus An Introduction to Philosophical Medicine in the Era of the Renaissance*, Basle and New York, 1958; W. Pagel, "Paracelsus and the neo-Platonic and Gnostic Tradition", *Ambix*, **8**, 125–66, 1960; W. Pagel, "The Prime Matter of Paracelsus", *Ambix*, **9**, 117–35, 1961. Paracelsus's mineralogical ideas have been discussed in: J. Schroeter, "Die Stellung des Paracelsus in der Mineralogie des 16. Jahrhunderts", *Schweizerische mineralogische und petrographische Mitteilungen*, **21**, 313–31, 1941.

known, he drew on an extraordinary diversity of sources; he epitomized eclecticism. In his profound study of Paracelsus and his work, Pagel has noted the influence of Jewish mysticism and cabbalism, Neo-Platonism, pantheism, gnosticism, the doctrine of macrocosm and microcosm, naturalism and empiricism, astrology, humoralism, alchemy and magic, the doctrines of sympathies and correspondences, and Christian trinitarianism, as well as the influences of individual scholars of the Renaissance.[26] But among this remarkable range of influences, Pagel emphasizes the particular importance of Neo-Platonism. He declares that:

> The Neo-Platonic trend in the concept of Nature supplies the overriding doctrinal basis for the work of Paracelsus. . . . He [Paracelsus] . . . filled the conceptual basis of Neo-Platonism with a wealth of naturalistic observations and allegorical interpretations of natural phenomena—and it is this which gives his work its original flavour.[27]

Moreover, Pagel has emphasized[28] that one can only thread one's way through the maze of Paracelsus's writings by constantly bearing in mind that he viewed matter as a kind of corporification of spirit—a characteristically Neo-Platonic position. For Paracelsus, matter was but spirit made manifest to us by some process of corporification. Spirit held the first place in the Paracelsian scheme. In the work entitled *The Philosophy Addressed to the Athenians*,[29] Paracelsus postulates an initial *Mysterium Magnum*, from which, by a process of "separation", the four traditional elements arose. In *The Philosophy of Theophrastus Concerning the Generation of the Elements*,[30] a similar view is

[26] Pagel (25, 1958), chapter 5.

[27] *Ibid.*, 226.

[28] *Ibid.*, 82–9.

[29] In the following notes, reference will be made first to A. E. Waite's English translation of part of the Paracelsian corpus of writings, and then to the Geneva collected edition of Paracelsus's works, *viz*: A. E. Waite, *The Hermetic and Alchemical Writings of Aureolus Philippus Theophrastus Bombast, of Hohenheim, called Paracelsus the Great. Now for the first time faithfully translated into English*, 2 vols, London, 1894; Paracelsus, *Aur. Phil. Theoph. Paracelsi Bombast ab Hohenheim, Medici et Philosophi Celeberrimi, Chemicorúmque Principis Opera Omnia Medico-Chemico-Chirurgica, Tribus Voluminibus Comprehensa. Editio Novissima et Emendatissima, ad Germanica & Latina exemplaria accuratissimè collata: Variis tractatibus & opusculis summa hinc inde diligentia conquisitis, ut in Voluminis Primi Praefatione indicatur, locupletata: Indicibusq̃; exactissimis instructa*, Geneva, 1658. *The Philosophy Addressed to the Athenians* is found on pages 249–81 of the second volume of Waite's translation, and on pages 239–52 of the second volume of the *Opera Omnia*. Some authorities do not regard *The Philosophy Addressed to the Athenians* as having been written by Paracelsus himself, though it contains a proper exposition of his thought. (See Pagel (25, 1958), 89.)

[30] Waite (29), vol. I, 201–34; Paracelsus, *Opera Omnia*, vol. II, 254–87. (The contents of these differ fairly substantially.)

expressed, though in somewhat different terms. A divine, immaterial spirit, the *Iliaster*,[31] was, at the time of creation:

> . . . originally distributed into four parts—the air, which is a heaven embracing all things; fire, which is a firmament producing day and night, cold and heat; earth, which affords fruits of all kinds and a solid foundation for our feet; and water, from whence are given forth all minerals and half the means of nutriment of things.[32]

Each region (selected to be in keeping with the traditional element theory) would give rise to characteristic substances, according to the appropriate matrix provided by the element of that region. We are reminded of Plato's underlying substance, likened to a "Mother" and "Nurse of becoming". In Paracelsus, however, there are now four matrices, which can themselves give imprints to matter in a characteristic way, like a seal on wax. The *Mysterium Magnum* is also likened to a matrix. Material substances, of course, are for Paracelsus constituted of the celebrated *tria prima*.[33]

Paracelsus also makes considerable use of the notion of "seeds", both in his mineralogical writings and elsewhere, and this suggests a continuing influence of the Stoic tradition, though it does not necessarily indicate that the theory was drawn directly from Stoic sources. The "seminal" notions also form part of the Hermetic tradition.

The Paracelsian mineralogical ideas are given their clearest exposition in two works whose contents largely overlap, and which may consequently be treated together: *The Economy of Minerals or the Genealogy of Minerals*[34] and *A Book about Minerals*.[35] Here the "seminal" notions are well developed. It is suggested that mineral formation is in some way analogous to the development of fruits on plants, which have themselves developed from seeds. Clearly, plants develop from seeds within the element earth into the element air, where fruits are born. Earth, then, serves as a matrix for the seed of the plant, providing it with appropriate nourishment. The branches of this plant extend upwards into the neighbouring element, air. Similarly, thinks Paracelsus, the matrix element, water, nourishes the seeds of minerals and metals, which grow into mature specimens within the earth. The matrix of minerals—the

[31] Also called *Illeias, Eliaster, Iliadum*, etc.

[32] Waite (29), vol. I, 203; Paracelsus, *Opera Omnia* vol. II 255.

[33] For a general discussion of the Paracelsian matter theory see R. Hooykaas, "Die Elementenlehre des Paracelsus", *Janus*, **39**, 175–88, 1935 and *Ibid.*, **41**, 1–28, 1937.

[34] Waite (29), vol. I, 89–113. This work is not included in the Geneva folio edition of 1658.

[35] Waite (29), Vol. I, 237–56; Paracelsus, *Opera Omnia*, vol. II, 341–50.

element water—forms a "tree" within the body of the earth,[36] which deposits its fruits in due season, later to be harvested by man. And, just as a chestnut must be extracted from its skin, so too must metals be extracted from their ores. (Man also has a worthless coat, which should likewise be stripped off.) By contrast, the fruit of a cherry is exposed and naked. Gold, too, is similarly exposed and requires no extraction.

Paracelsus believed that there was an aqueous interior to the earth, anastomosing through the whole, and a source of mineral fruits:

> For, as the element of water lies in the middle of the globe, so, the branches run out from the root in its circuit on all sides towards the plains and towards the light. From this root very many branches are born. One branch is the Rhine, another the Danube, another the Nile, etc.[37]

All these branches terminate in the seas, which form a kind of "canopy" over the whole "tree". This, like a normal arboraceous plant, would in time wither and die, having deposited its fruits within the earth. Ultimately, minerals would reach their "death" in the terminus of the waters—the oceans. Doubtless, this was intended to give some account of the mineral veins within the earth, the decay (or weathering, as the moderns say) of rocks and minerals, of the springs and mineral solutions issuing from the earth, and of the salinity of the oceans, but the debt of the chosen mode of explanation to earlier thinking requires no emphasis.

In other passages, Paracelsus makes it clear that his Salt, Sulphur and Mercury are put together within the matrix water, to form the various minerals and metals. The water, he says, is like a "bag filled with seeds of all kinds to be sown".[38] We are again reminded of the *logoi spermatikoi* of the Stoics. As a further complication, however, Paracelsus later states that there are different kinds of Salt, Sulphur and Mercury, which give rise to the wide variety of mineral substances. Just as a painter can use one colour to depict numerous figures, so there may be many different Salts, Sulphurs and Mercuries, though all of the same essence. The artificer for all these processes is that mysterious entity, God's agent, the *Archeus*, perhaps most easily thought of as Nature personified. The Paracelsian *Archeus* appears to be derived from the Gnostic *Archon*,[39] and may also have affinities with the Demiurge of the *Timaeus*.

[36] Waite (29), vol. I, 92–3 and 241. The former of these references is not represented in the *Opera Omnia*. The latter is found in vol. II, 343. Adams (3), 385–8, drew attention to the prevalence, in the sixteenth and seventeenth (and even the eighteenth) centuries, of the idea of a mineral "tree" within the body of the earth.

[37] Waite (29), vol. I, 232; Paracelsus, *Opera Omnia*, vol. I, 277.

[38] Waite (29), vol. I, 244; Paracelsus, *Opera Omnia*, vol. II, 344.

[39] *cf.* Pagel (25, 1960), 126.

Certainly they seem to have the same function—that of a principle of organization. The *Archeus* was to be ". . . a dispenser, who will arrange and set in order what ought to be joined together, so that what should be done may find accomplishment".[40]

The Paracelsian corpus devotes considerable attention to the extraction of medicaments from mineral sources, but these matters are well known, and will not be pursued here.

In the writings of the celebrated French potter, Bernard Palissy (1510?–90),[41] one finds a theory of metals and minerals which is in many ways similar to that of Paracelsus, though there are also significant differences. That there should be such similarities in the work of a man who had little knowledge of classical culture, and probably little direct contact with the Paracelsian writings, is indicative that the notion of the growth of metals from "seeds" in an aqueous "matrix" was common currency in the sixteenth century, and that it was doubtless derived as much from the folk-lore of alchemy as direct influences arising from the intellectual revival of the Renaissance. This should warn us against attempting to ascribe too much to direct Neo-Platonic influences, though it in no way conflicts with the broader view that the sixteenth-century writers on the mineral kingdom drew on a variety of ideas derived from antiquity in their attempts to give adequate accounts of mineral substances. But let us examine Palissy's theory of metals, expressed in his *Discours Admirables* of 1580.[42]

Like Paracelsus, Palissy believes that metals are formed in some manner within water. Within this water is a salt, which has the power of "attracting to itself the things it likes, to congeal them and reduce them to metals".[43] Also in the water there may be a kind of oil, which when congealed in the process of metal formation, gives rise to sulphur, one of the components of metals.[44] The role of mercury in the scheme is by no means clear. It is on

[40] Waite (29), vol. I, 247; Paracelsus, *Opera Omnia* vol. II 346.

[41] On Palissy, see, for example: W. Kirsop "The Legend of Bernard Palissy', *Ambix*, **9**, 136–54, 1961; H. R. Thompson, "The Geographical and Geological Observations of Bernard Palissy the Potter", *Annals of Science*, **10**, 149–65, 1954.

[42] The title of the first French edition was *Discours Admirables, de la Nature des Eaux et Fonteines, tant naturelles qu'artificielles, des metaux, des sels et salines, des pierres, des terres, du feu et des esmaux, avec plusieurs autres excellens secrets des choses naturelles, Plus un traité de la marne fort utile et necessaire pour ceux qui se mellent de l'agriculture. Le tout dressé par dialogues, esquels sont introduits la Theorique et la Practique. Par M. Bernard Palissy . . .*, Paris, 1580. I have, however, largely used the readily available English translation by Aurèle la Rocque: B. Palissy *The Admirable Discourses of Bernard Palissy*, Urbana, 1957.

[43] *Ibid.* (1957), 97.

[44] *Ibid.*

one occasion referred to as "the beginning of a metal".[45] Elsewhere,[46] it appears to be another material, concealed within the aqueous matrix, which is drawn together with the sulphur by the salty "seed" to produce metals. A general similarity to the Paracelsian scheme may be discerned (although, as indicated above, Palissy probably was not directly acquainted with Paracelsus's writings). We have a version of the Paracelsian chemical trichotomy, but it is interwoven with the older Sulphur/Mercury theory of the Jābirian school, utilized by the alchemists. We also have the doctrine of water functioning as an aqueous matrix for the development of metallic seeds. Palissy's theory bears a considerable resemblance to that of Becher, to be described below (page 147), and it is clear that they both drew largely on alchemical traditions. This does not conflict with the view that Neo-Platonic and Stoic ideas played a role of particular importance in the sixteenth- and seventeenth-century mineral theories, though it does indicate that an examination of these influences alone does not provide the full picture.

In the exposition of the Paracelsian mineralogical scheme given above, we have, perhaps, erred by making the account too prosaic, so that the mystical vision of the Swiss philosopher has become somewhat distorted as we attempt to make it intelligible to ourselves. Yet, as will be demonstrated below, a somewhat similar process occurred among Paracelsus's successors. As more literal and matter-of-fact versions of his theories were proposed, the mysticism was stripped away, leaving what today may appear as a somewhat incongruous mineralogical doctrine. But this emancipation from mysticism did not occur immediately. Mystical influences are still clearly evident in the writings of Sendivogius and Van Helmont, for example, and it is to the work of these two men that we now turn our attention.

The Polish (or Moravian) alchemist, Michael Sendivogius (1556–1636 or 1646), provides a useful example of the direct influence of Paracelsus on ideas concerning the mineral kingdom, with employment of the "organic" metaphor. According to Lévi,[47] Sendivogius transcribed some of Paracelsus's manuscripts, so it is not surprising that fairly clear evidence for the influence of the Swiss philosopher on Sendivogius may be found. His principal work[48] ran through

[45] *Ibid.*, 98.

[46] *Ibid.*, 101.

[47] Waite (29), vol. I, 288 (citing chapter 19 of E. Lévi, *Dogme et Rituel de la Haute Magie* . . ., 2nd ed., Paris, 1861—not seen).

[48] Sendivogius's theories were given their first expression in an anonymous volume (with his name hidden in the anagram *Divi Leschi Genus Amo*) published in 1604, with the title *Tractatus Duodecim de Lapide Philosophorum*. It is believed that the work was published in Prague and Frankfurt-on-Main in the same year. (See W. Hubicki, "Michael Sendivogius's Theory, its Origin and Significance in the History of Chemistry", *Proceedings of the Tenth International Congress of the History of Science, Ithaca, New York, 1962*, Paris,

many editions and was translated into several languages, including English. It had considerable influence on seventeenth-century ideas concerning the mineral kingdom. John Webster, in particular, mentions Sendivogius several times in his *Metallographia*. (See below, p. 150.)

It was mentioned above that Paracelsus thought of minerals as having been born in some manner within the womb of the element water. Sendivogius adopts a somewhat similar doctrine, except that the terrestrial globe itself is visualized as the womb, within which metals or minerals might grow from some seed or sperm. (There is some inconsistency in the use of the terms "seed" and "sperm" as the explanatory principle.) It is suggested that there is an empty space at the centre of the globe, into which the four elements "send forth their qualities", just as man "sends forth his seed into the entrance of a wombe of a woman". In a similar way, each element "sendeth forth its own thinnesse and subtlety".

Like Paracelsus, Sendivogius postulates an organizing principle, or *Archeus*, controlling the processes occurring within the earth. The *Archeus*, "the servant of nature . . . mixeth . . . [the] spermes, and sends them forth".[49] By a kind of distillation process, it propels the "vertues" up from the central cavity into the superficies of the earth.[50] That which ascends is likened to a moist vapour, and according to the different situations into which it penetrates, it gives rise to metals or other mineral substances. We might think this rather strange—that the nature of the substance produced would depend on the location in which it was formed, rather than the matter from which it was constituted. But for Sendivogius it was the perfectly natural mode of explanation. Does not water, if spilled on a table, become salty if it runs over salt, or red if it runs over a red pigment? The place changes the water, rather than the water changing the place.

Bringing his theory into line with earlier alchemical thinking on the nature of metals, Sendivogius likens his ascending vapour to mercury; it is the first matter of metals. It comes as no surprise, then, to find sulphur entering into the account:

> For when that vapour is sublimed from the center of the earth, it passeth through places that are hot, and pure, where the fatnesse of

---

1964, 829–33.) The work was republished in Paris in 1608 by Beguin, under the title *Cosmopolitani Novum Lumen Chymicum*. Hubicki states that the work ran through over thirty editions, in Latin, German, French and English. I have used the English edition of 1650: Micheel [*sic*] Sendivogius, *A New Light of Alchymie: Taken out of the fountain of Nature, and Manuall Experience. To which is added a Treatise of Sulphur: . . .*, London, 1650.

[49] Sendivogius (48, 1650), 8.

[50] *Ibid.*, 11.

> the Sulphur sticks to the walls; I say that vapour which the Philosophers have called the Mercury of Philosophers applyes it self to, and is joined to that fatnesse, which then it sublimes with it self.[51]

If, then, the vapour is fortunate enough to pass through hot pores, cleansed by previously ascending vapours, then the "fatnesse" may produce gold. Less well favoured situations yield the inferior metals.

An interesting proposal to account for the formation of stones and rocks is also offered:

> When the four Elements distill their vapour into the center of the earth; and the Archeus of nature sends forth the same, and sublimes it; this whilst it passeth through places, and the pores of the earth, takes along with it self all the impurity of the earth into the very superficies, which afterward the air congeals . . . and so are made rocks, and stony mountains, according to great and little pores.[52]

Gems, on the other hand, are produced if the vapour distils into "a place of pure salt water", rather than a region rich in the "fatnesse of sulphur", though the colour of gems could supposedly be accounted for by contamination with sulphur.

Like Paracelsus,[53] Sendivogius entertains two levels of matter, which he designates "first" and "second". He believes that Nature receives matter from the elements, and converts this into seed, which then develops, by a kind of putrefactive process, in a suitable matrix, into other substances—metals, gems, and the like. The seed, we are mysteriously informed, is "nothing else but the air congealed in some body: or is a moist vapour".[54] The seed corresponds to the "second matter". But there is considerable ambiguity in the account, for elsewhere the matrix, in which the mineral seed will supposedly develop, is said to be air. Perhaps it is the air within the pores of the earth which causes the congelation to produce a mineral substance. But this is hard to reconcile with the statement that the seed is "nothing else but aire, which by fatnesse is congealed in water". And it is bewildering, to say the least, to read that:

> [Water] is the sperm of the world, in which the seed of all things is kept: it is the keeper of the seed of every thing. Yet wee must know that the Seed is one thing, and the Sperme another: the Earth is the receptacle of the Sperme, but Water is the receptacle of the Seed.[55]

[51] *Ibid.*, 12.
[52] *Ibid.*, 14.
[53] *cf.* Pagel (25, 1961).
[54] Sendivogius (48, 1650), 17.
[55] *Ibid.*, 85.

It should be noted that Sendivogius adopts the concept, favoured by Paracelsus, that there are waters passing through the body of the earth like an "axell tree".[56] These preserve the earth from burning, and convey the seeds through the pores of the earth, where they may subsequently develop.

It is, as has been suggested above, difficult to reconcile all the strands in the scheme which we have been considering, and it is not our task to force the alchemist's ideas into the strait-jacket of the twentieth-century mind. It is clear, however, that Sendivogius is attempting to give an account of the mineral kingdom in terms of the matter theories traditional in his time, and with an "organic" world-view permeating his whole line of argument.

Somewhat similar are the views of Van Helmont (1579–1644),[57] also an exponent of the "organic" world–view. Like Sendivogius, Van Helmont is clearly influenced by Paracelsus. Water is to act as a matrix or womb for the formation of minerals:

> Mettals, small stones, Rocky-Stones, Sulphurs, Salts, and so the whole rank of Minerals, do find their Seeds in the Matrix or Womb of the Waters, which contain the Reasons, Gifts, Knowledges, Progresses, Appointments, Offices and Durations of the same.[58]

The total theory is, however, considerably more complex than is suggested by this passage, taken in isolation. It is clear that the notion of water as matrix has been adopted, and Van Helmont continues: "Seeds . . . [issue] out of the dark Womb of the Water (which the Voice of the Word hath here deposited as durable unto the end) . . ."[59]

What interpretation should be placed on Van Helmont's seeds? The seminal notions of antiquity have by this time been elaborated into a fairly complex theory. A seed contains "information" (as we might say) determining how an object will develop; and the process is supposedly analogous to fermentation.

[56] *Ibid.*, 87.

[57] For expositions of Van Helmont's philosophy, see: W. Pagel, "The Religious and Philosophical Aspects of van Helmont's Science and Medicine", *Bulletin of the History of Medicine*, Supplement No. 2, Baltimore, 1944; and J. Mepham, "Van Helmont", in R. Harré (ed.), *Early Seventeenth Century Scientists*, Oxford, 1965, 129–57. In the account of Van Helmont's mineral theory that follows, I have used J. B. Van Helmont, *Oriatrike Or, Physick Refined. The Common Errors Therein Refuted, And the whole Art Reformed & Rectified: Being a New Rise and Progress of Phylosophy and Medicine, for the Destruction of Diseases and Prolongation of Life. Written by that most Learned, Famous, Profound and Acute Phylosopher, and Chymical Physitian, John Baptista Van Helmont. . . . And now faithfully rendred into English . . . By J[ohn]. C[handler]. of M[erton]. H[all]. Oxon.*, London, 1662. This work is the English translation of Van Helmont's *Ortus Medicinae* . . . of 1648 (Amsterdam).

[58] Van Helmont (57, 1662), 693.

[59] *Ibid.*

In some ways, the seed has an "idea" impressed upon it which determines its subsequent development. In the case of the human seed, "regular Idea's . . . are implanted on the seed by the lust of the Generater . . .".[60] In a more humdrum way, perhaps, but in an analogous manner nevertheless, minerals develop in their aqueous matrices according to the "ideas" there impressed. The "ideas" originate in the mind of God, but God does not himself enter the matrix and there "shed" His ideas. Rather, the *Archeus* acts as God's agent in the proceeding. It is difficult to construe Van Helmont's meaning precisely, but my interpretation of his intention is as follows.

Only two of the four traditional causes are accepted: the material and the efficient, the latter serving as an "exciter or stirrer up",[61] while water supposedly is the material cause of all mineral substances. The efficient cause is believed to be a "Ferment or leaven or seminal beginning". These interact (in some way not clearly specified) to produce the seed, which will, in the fulness of time, develop to produce a metal or mineral. We are told that the ferment is:

> . . . a formall created Being, which is neither a substance nor an accident, but a neutrall thing framed from the beginning of the World, in the places of its own Monarchie. . . . Ferments are gifts, and Roots stablished by the Creator the Lord, for the finishing of Ages, sufficient, and durable, by continual increase, which of water, can stir up and make Seeds proper to themselves . . . the *Ferment* holds the Nature of a true Principle. . . . [Nevertheless,] . . . the efficient cause is considered as an immediate active Principle in the thing, which is the Seed, and as it were, the moving Principle to generation, or the constitutive beginning of the thing: but the *Ferment*, is often before the Seed, and doth generate this from it self.[62]

Thus the ferment apparently has priority over the seeds, and, in some way, bears the impress of the "idea" of the Creator. The actual relationship between the ferment, seed, and *Archeus* is, however, obscure. We are informed that "the chief Workman [or *Archeus*] consists of the conjoyning of the vitall air, as of the matter, with the seminal likeness".[63] It is "the first or chief Instrument of life and feeling". Still more obscurely, the reader is informed that the:

> *Archeus*, the Workman and Governour of generation, doth cloath himself presently with a bodily cloathing: For in things soulified, he walketh thorow all the Dens and retiring places of his Seed, and begins to transform the matter, according to the perfect act of his own Image.[64]

[60] *Ibid.*, 548.
[61] *Ibid.*, 29.
[62] *Ibid.*, 31.
[63] *Ibid.*, 35.
[64] *Ibid.*

As in the writings of Paracelsus and Sendivogius that have been discussed above, the *Archeus* again appears to be the personification of some kind of organizing principle. It has a "dispositive knowledge of things to be done",[65] that is, it can so organize matter that the seeds in which it is found will develop into their appointed end-products (whether animal, vegetable or mineral). It may have been an analogue of the *Demiurge* of the *Timaeus*, often interpreted as the personification of reason.[66] Even without making this identification, however, Platonic (or Neo-Platonic) influences are apparent. "Ideas" are in some way made manifest as the corporeal objects of our experience.

Our exposition of the Helmontian mineral theory requires further development, however. The water of the aqueous matrix, having obtained "a locall, or implanted Seed", is converted by a kind of putrefactive process into a "Minerall juyce (*Bur*) according to the particular kindes, chosen by the direction of the Seedes".[67] The process could be further visualized as the impregnation of a "Sulphur" in the water, by the seed, thereby converting it into a "Mineral water", from which solid mineral matter would later be precipitated. Van Helmont was familiar with certain solutions, known to miners, which deposited mineral matter when exposed to air. He stated that the "greenish paleness . . . waxeth yellow, or grows white, or becomes more fully green", from which it may be presumed that he was referring to the oxidation of some solutions of sulphides and perhaps also of ferrous salts. For Van Helmont, however, an "organic" explanation was natural, for:

> When as the seed comes forth to light cloathed with a Liquor, and Gas hath begun to defile the Sulphur of the water, there is a middle life of the seed: But the last life is when it now waxeth hard: But the last life of the metallick seed, is the first life of the Metalls. . . .[68]

Van Helmont, it will be recalled, thought of a gas as water acted upon by a ferment. Thus this passage can be reconciled with his other statements, in that the overall process of mineral formation is here viewed as a kind of fermenation in an aqueous medium, with resulting growths of mineral matter.

It is difficult to bring all these various strands into some consistent and intelligible theory. In particular, the role of water in the Helmontian scheme is somewhat ambiguous. Is it just a *place* where minerals grow (as perhaps suggested by Paracelsus—or so Van Helmont interprets him[69]), or is it a truly

[65] *Ibid.*, 113.

[66] *cf.* R. G. Bury, *Plato with an English translation Timaeus Critias Cleitophon Menexenus Epistles*, London and Cambridge (Mass.), 1966, 9.

[67] Van Helmont (57, 1662), 49.

[68] *Ibid.*, 156.

[69] *Ibid.*, 104.

fundamental stuff of which all bodies are ultimately formed? Certainly, on occasions, water appears to have the Paracelsian character of a matrix[70] but generally it is to be construed as the ultimate constituent of bodies. However, by accepting this dual rôle for water in the Helmontian system, one has not penetrated to its most distinctive feature, which I judge to be the manner in which he attempts to account for complex inorganic phenomena by viewing them as the products of living processes. The notion of ferments and seeds developing to produce mineral substances provided the leading principle of the whole mineralogical scheme. By this approach, the crystallization of minerals in orderly array might be accounted for.

Here it may be noted that the view that all the requisite "information" for the growth of a crystal is in some way contained within its "seed" may be construed as an example of so-called "essentialist" thinking on mineralogy.[71] The "essence" of the mineral is somehow contained within its seed. Given this "essence", the eventual properties of the mineral which will later grow from it are pre-determined. This may seem a far cry from the Aristotelian belief that the properties of things may be deduced from the correct verbal definitions of their essences, and certainly Van Helmont was no Aristotelian, as the opening pages of his *Oriatrike* clearly demonstrate. Nevertheless, I suggest that the seminal theory of the Helmontians provides an example of the influence of "essentialist" thinking on the earth sciences (though not of the *logical* essentialism of the Aristotelians). The seeds contained the "essences" of minerals. If one is to be guided by what we know of the influence of "essentialism" on chemistry and biology,[72] one may well anticipate that significant changes would be made in mineralogy when this "animist-cum-essentialist" view was superseded. We shall consider below whether this anticipation is warranted.

A good example of the continuing influence of Helmontian views on ideas about the mineral kingdom is provided by Thomas Sherley (1638–1678), "Physitian in Ordinary to His MAJESTY" (Charles II), in his *Philosophical Essay: Declaring the Probable Causes, where Stones are Produced in the Greater*

[70] *e.g., Ibid.*, 693.

[71] For Sir Karl Popper's use of the term "essentialist", see his *Open Society and its Enemies*, 4th ed., 2 vols, London, 1962, vol. II, chapter 11.

[72] I have discussed the influence of essentialism on the chemical work of Stahl in a recent paper: "An Examination of G. E. Stahl's *Philosophical Principles of Universal Chemistry*", *Ambix*, **20**, 36–52, 1973. For a discussion of the influence of essentialism on biology, see D. L. Hull, "The Metaphysics of Evolution", *British Journal for the History of Science*, **3**, 303–37, 1967.

*World* . . .[73]. Sherley interested himself in the question of the growth of "stones" within the human body, to which end he undertook a general enquiry into the problem of the formation of stones in the world at large. (It is noteworthy that he believed that an investigation of the stones of the "macrocosm" was relevant to an understanding of stones within the "microcosm".) His explanations are avowedly Helmontian in character, but are far more explicit and intelligible than those of his predecessor. It may be suggested, however, that the increased clarity of the writings reveals a greater apparent incongruity in the explanations which are proposed. But we shall allow Sherley to speak for himself, that the reader may make his own judgment upon this point. With admirable clarity, Sherley states his view:

> That stones, and all other sublunary bodies, are made of water, condensed by the power of seeds, which with the assistance of their fermentive Odours, perform these Transmutations upon Matter.
>
> That is, that the matter of all Bodies is originally meer water; which by the power of seeds is coagulated, condensed, and brought into various forms, and that these seeds of things do work upon the particles of water, and alter both their texture and, figure; as also, that this action ceaseth not, till the seed hath formed it self a Body, exactly corresponding with the proper Idea, or Picture contained in it. And that the true seeds of all things are invisible Beings (though not incorporial;) this I affirm, and shall endeavour to prove.[74]

Here, Sherley's meaning is perfectly clear, and so too is his definition of the all important term "seed":

> . . . by Seed I understand a fine, subtile Substance, (imperceptible by our gross Organs of Sensation;) in which God hath impressed a Character of that thing he will have it produce from the Matter it is to work upon: which it doth perform by putting the parts of Matter into such a peculiar motion as is requisite to produce the intended Effect.[75]

Much of the mystery is removed from the Helmontian ferment. It is:

> . . . an Expansive, Elastick, or Springy power of the Seed of any thing; by which motion of its self it also moveth the smallest particles of that Matter in which it is immersed: by which motion also (which is of

[73] T. Sherley, *A Philosophical Essay: Declaring The Probable Causes, whence Stones are produced in the Greater World. From which occasion is taken to search into the Origin of all Bodies, discovering them to proceed from Water, and Seeds. Being a Prodromus to a Medicinal Tract Concerning the Causes, and Cure of the Stone in the Kidneys, and Bladders, of Men*, London, 1672.

[74] *Ibid.*, 24.

[75] *Ibid.*, in preface "To the Reader" (unpaged).

> divers kinds, according to the variety of Seeds) the particles of Matter acquire new shapes, sizes, and postures amongst themselves; and so a new texture of the whole is produced, agreeable to the peculiar Nature of the Seed, and correspondent to its Idea. . . .[76]

It should be noted, however, that in order to make the Helmontian scheme intelligible to himself, Sherley incorporates certain aspects of the mechanical philosophy in his account. The seeds have certain "ideas" impressed upon them by God. They are corporeal, though exactly what they are made of is perhaps uncertain. They are immersed in the matrix water (or, as we might say, in an aqueous medium), and they have the power to move the corpuscles of water in which they are immersed, thereby giving rise to certain configurations of these particles, and consequently producing metals, minerals, stones, *etc.* This process, whereby the particles of water are re-orientated and suitably grouped to produce solid minerals, is held to be analogous to some kind of fermentation. It will be noted, however, that whereas Van Helmont considered the ferment to be "logically" prior to the seeds, for Sherley, the ferment is merely the manifestation of the activity of the seeds, and is, in a sense, conceptually superfluous to the total scheme.

The seeds, then, would fement the water to a "Mineral-juice" or "*Bur*",[77] from which, in course of time, solid mineral substances would coagulate. Sherley accepts without question the view that stones and metals grow in the earth, yet in doing this he was citing what he took to be authoritative witnesses, and doubtless, also, was thinking of his own personal observations.[78]

It is not difficult to discern Platonic influences in Sherley's mineral theory, for he devotes a section to showing that his suggestions find a warrant in Plato's writings.[79] He interprets Plato as follows (with an evident Neo-Platonic slant):

[76] *Ibid.*, 34.

[77] The terms *Bur* and *Gur* seem to be used interchangeably in the literature. *Gur* is derived from the German word *gären*—to ferment. *Guhr* (a variant of *Gur*) is defined by the *O.E.D.* as: "A loose earthy deposit from water found in the cavities of rocks". For a discussion of the role of *Bur* or *Gur* in early mineral theories, see F. D. Adams (2), 283–4.

[78] It is incorrect to dismiss records of the "growth" of stones in the ground as being entirely fanciful. I well remember as a small boy, living on the "clay with flints" of Hertfordshire, trying to remove all the pebbles from a given area of our garden. All attempts proved futile—the pebbles were constantly replenished. More recently, I removed every pebble from the bed of the stream of my garden in Sydney, in order to construct a stone pathway. Within six months, there were as many pebbles in the stream as ever! Today, the manner in which stones tend to work their way towards the surface of the ground is attributed primarily to "frost heaving". See A. Holmes, *Principles of Physical Geology*, New and revised edition, London and Edinburgh, 1965, 408–9.

[79] Sherley (73), 117ff. It is noteworthy that Sherley claims to be able to discern similarities in the views of Plato and Moses. The following passage is of interest in that it

> By his Intelligible World, *Plato* means the Divine Decrees; which are inherent in the Mind and wisdom of God: and these Original *Idea's*, . . ., do produce a Secondary sort of *Idea's* (that is, the *Seeds* of things;) and these he makes to be the more immediate Delineation, or Image of the whole work; sometimes calling them . . . *an Exemplar*, sometimes . . . *an Image*: . . . And this last *Idea* and Seed, contains the Picture of the thing to be made; and depends upon the Primary, or Original *Idea*, and Exemplar, which is seated in God himself.[80]

Sherley's theory of the "seeds" of minerals should not, I think, be dismissed as some "quaint belief". The regularity of crystalline structure which characterized mineral substances was surely a fact which demanded explanation, and Sherley's hypothetical "seeds", which acted as "templates" for the growth of minerals, offered an explanation which was reasonable enough in its day. In modern embryological theory somewhat analogous explanations are required in order to account for the development of living organisms. For the natural philosopher of the seventeenth century, the orderly growth of mineral crystals was a problem of almost equal complexity,[81] and a purely mechanical explanation might well have been deemed inadequate.

The theory of ferments had further attractions. It could account for the problematical hot springs, of considerable interest to the iatrochemists for their supposed medicinal virtues, and investigated in some detail in the early years of the seventeenth century by Edward Jorden (1569–1632).[82]

Like Paracelsus, Jorden adopts the view that the four elements merely served as matrices—loci for the development and nourishment of various substances. He does not state clearly the nature of the matter of which sub-

---

indicates that Sherley believed that the idea of "Seminal Principles" adopted in his mineral theory had Stoic ancestry. He writes (119): "For, according to *Platoes* Philosophy (as well as that of *Moses*) God is the Executive cause, and productive Efficient of all things. . . . And as to the manner how all things were made, he [Plato] says . . . *Every thing was essentialised by certain* Prolifick, or efformative words, which the *Stoicks* call λόγον σπερματικόν, a Spermatick, or Seedy word: Which agrees exactly with *Moses* his *Fiat*, and with that of St. Paul; . . . Gods *Fiat* was the Creator of all the Seminal and Prolifick Principles of all things; and those created Seeds were . . . *the Efficients*; and . . . *water*, was the Matter of which they were all made." It may be noted that this appears to accord with the thesis that the "seminal" and "idealistic" doctrines which we have been discussing represent examples of "essentialist" thinking in mineralogy.

80 *Ibid.*, 120–1.

81 For discussions of the early history of crystallography, see H. Metzger, *La Genèse de la Science des Cristaux*, Paris, 1918; J. G. Burke, *Origins of the Science of Crystals*, Berkeley and Los Angeles, 1966.

82 For Jorden's important contributions to early analytical chemistry, see A. G. Debus, "Solution Analysis Prior to Robert Boyle", *Chymia*, **8**, 41–61, 1962.

stances are formed, but invokes a "Seminarie Spirit" of minerals, which meeting with "convenient matter", acts upon this matter in a kind of fermentatory process, which concomitantly "procures an actual heat". This

> seminary spirit of minerals hath its proper wombs where it resides, and is like a Prince or Emperour, whose prescripts both the Elements and matter must obey; and it is never idle, but alwayes in action, producing and maintaining natural substances, untill they have fulfilled their destiny.[83]

It is this heat, generated by the fermentatory process as minerals grow in the earth, which is responsible, so Jorden believes, for the heat discernible in certain springs.

A common variant of the doctrine of the growth of metals and minerals from some kind of seeds or sperms was the opinion that *both* were involved, in a kind of copulatory process. Such was the view expressed in the writings of J. J. Becher (1635–82),[84] in which one finds the traditional mercury and sulphur of the alchemists supposedly taking part in the act as female and male entities respectively. Earth provides the matrix for the growth of the produce of the generative act. Becher seems to have obtained some oily substance by distilling or compressing seeds, and he designates this by the term "sperm",[85] stating that it may be thought of as philosophical Sulphur. The hypothetical Sulphur of metals is likewise supposedly oily and viscid; it is the male actor in the drama of metal generation. The philosophical Mercury is the female counterpart. The sulphureous and mercurial vapours mingle with each other within the body of the earth; the male entity imparts form to its female counterpart, and the conception of a metal occurs accordingly. In Becher, it is clear that we see the very strong influence of the alchemical tradition, and that he is apparently more influenced by descendants of the Aristotelian theory of exhalations of the *Meteorologica*[86] than by Platonism or Neo-Platonism. Nevertheless, Becher's work provides further evidence for the continuing influence

[83] E. Jorden, *A Discourse of Natural Bathes, and Mineral Waters. Wherein, The Original of Fountains in general is declared. The nature and difference of Minerals, with Examples of particular Bathes. The Generation of Minerals in the Earth, from whence both the Actual Heat of Bathes, and their Virtues proceed. By what means Mineral Waters are to be discover'd. And lastly, of the Nature, and uses of Bathes, but especially of our Bathes at Bathe, in Somersetshire*, 3rd ed., London, 1669, 104.

[84] The following account of Becher's ideas is based on his *Institutiones Chimicae Prodromae, id est, . . . Oedipus Chimicus Obscuriorum Terminorum & Principiorum Chimicorum, Mysteria aperiens & resolvens*, Amsterdam, 1664.

[85] *Ibid.*, 137.

[86] *cf.* Aristotle, *Meteorologica*, translated H. D. P. Lee, London and Cambridge (Mass.), 1952, 29, 287 and 289.

of an "organic" world-view on seventeenth-century mineral theories, although, as is well known, the intellectual situation in Germany at that period was scarcely comparable to that in England and France, where the impact of the "mechanical philosophy" was felt considerably earlier.

One of the clearest examples of the influence of Neo-Platonism on mineral theory is to be found in the writings of the Paracelsian, Nicolas Le Févre (*ca.* 1615–69), also called Le Febure.[87] Like any good spagyrist, Le Févre was primarily concerned with the extraction of certain "vertues" from animals, vegetables, or minerals, to be used for medicinal purposes. The details of such procedures need not, however, concern us here, where we confine ourselves to an elucidation of his mineral theory, which has much in common with that of his master, Paracelsus. Revealing trinitarian influences, Le Févre maintains that there is a:

> spiritual substance . . . [that] hath a threefold denomination; . . . in respect of its natural heat and fire, it is called Sulphur; in respect of its moysture, which is the food and aliment of this fire, Mercury; and finally, in respect of the radical drought, which is, as it were, the knot and cement of the fire and moysture, it is called Salt. . . .[88]

Moreover:

> . . . as this Spirit is universal, so can he not be specificated but by the means of particular Ferments, which do print in it the Character and Idea of mixt bodies, to be made such or such determinate substances, according to the diversity of Matrixes.[89]

Putting this another way, we may say that by some kind of fermentatory process an Idea is imprinted on the Universal Spirit by the matrix in which the Spirit is present; the Spirit consequently becomes corporified, yielding a "grosse body". The matrices (of which, following Paracelsus, there are four broad categories) "specifie and corporifie" the Spirit. Le Févre postulates a continual flux of this Spirit between the earth and the stars, and this gives him a general cyclic theory of mineral formation and decay. The Spirit

[87] On Le Févre, see: J. R. Partington, *A History of Chemistry*, 4 vols, III, London, 1962, 17–24.

[88] Nicasius le Febure, *A Compendious Body of Chymistry: Teaching the whole Practice thereof by the most exact Preparation of Animals, Vegetables and Minerals, preserving their essential Vertues*, 2 vols, London, 1664, I, 15.

[89] *Ibid.*, I, 16.

emanates from the stars,[90] passes through the air, and falls upon the water and the earth, where it becomes corporified in either the aqueous or the terrestrial matrix. (Perhaps some initial corporification also occurs in the air.) But the process is reversible: ". . . [there are] as two Ladders, whereby the heavenly influences descend down, and re-ascend from the lower parts".[91]

Thus the Ideas stamped on the Spirit may be lost, so that the bodies lose their corporeity. The chemist may himself bring this about when he heats a substance on a furnace—turning it into a vaporous state, as we would say. Le Févre considers the heating of an arsenical mineral. Arsenical vapours are given off, and their tang may be detected by the nose as they are evolved from the solid material. Evidently a complete spiritualization has yet to be achieved, for some of the arsenical "Idea" still adheres, even in the vaporous condition.[92] The scheme also provides a general rationalization for attempts to transmute one substance into another. By heating, a substance may be spiritualized, and a new "Idea" may subsequently be impressed upon the Spirit. Given this theoretical framework, a transmutation of one substance into another would seem feasible.

It should be noted that, for Le Févre, water should be considered neither as a constituent of mixts, nor as a single matrix, but it supposedly contains within itself a vast number of matrices, from which animals, plants (*e.g.*, duckweed!), minerals, shells, pearls, salt, etc., can develop. Thus there is here some significant difference in approach from the other views which have been discussed above. The aqueous matrices of Le Févre have somewhat the same explanatory function as the seeds of Sherley. It may also be noted that Le Févre believes that water may serve to convey the Spirit into the interior of the earth.

Le Févre considers the knotty problem of whether or not minerals are animate, and decides, reasonably enough, given his premises, that "Minerals do live, as long as they are joyned to their Matrix and Root, since they grow there sensibly; but when they are separated from it, they are justly called *Inanimate Mixts*".[93] It is worthy of remark that an almost identical opinion had been expressed long before by Plotinus.[94]

[90] Paracelsus, likewise, believed that the development of minerals and metals occurred under the influence of the stars. See, for example, Waite (29), vol. I, 181; Paracelsus, *Opera Omnia*, vol. II, 111. For a discussion of the influence of astrology on Paracelsus's views (including his ideas on minerals) see F. R. Jevons, "Paracelsus's Two-Way Astrology I. What Paracelsus Meant by 'Stars' II. Man's Relation to the Stars", *The British Journal for the History of Science*, **2**, 139–47, 148–55, 1964.

[91] Le Févre (88), I, 37.

[92] *Ibid.*, 17.

[93] *Ibid.*, 62.

[94] Plotinus, *Enneads*, VI. 7, 11 (MacKenna (5), 569).

Thus far, we have given a general account of Le Févre's theory. Further elaboration is to be found in his account of the generation of metals. Here it is suggested that it is *light*[95] "which by its ejaculation and irradiation did imprint and stamp the Idea and Character of its vertue in water as in a general and convenient Matrix, to enable her afterwards to supply both matter and Form, Body and Spirit, Salt, Sulphur and Mercury, [and] all physical generations . . .".[96] Being thus irradiated, the water is furnished with a fermentative and generative principle, and is in a condition in which metals can be formed. It must first be "conveighed and driven into all the general and particular Matrixes, which do determine and modifie it according to the Idea and Character of their natural predestination in their Creation . . .".[97]

On reaching an appropriate matrix, the spirit of Salt in the water "receives the vitriolick stamp, character and Idea", and the internal Sulphur works by ripening, digesting, concocting, and coagulating the water to produce a Vitriol. Then, guided by that mysterious entity, the *Archeus*, the Vitriol is altered into a "sulphur of a Metallick nature" (*i.e.*, a marcasite). After further coagulation, hardening and fixation, the marcasites are eventually converted into metals. Le Févre also refers to the "Mercury or true spirit of metallick bodies [which] contains in it self, their sperme, their ingress and Tincture, and by consequence all the vertue which it did possess, and all its power and activity since the first moment of its coagulation, to the last of its fixation".[98] It is not clear, however, how the role of the Mercury is related to that of the Salt and the Sulphur in the process of metal or mineral formation. Indeed, that there are inconsistencies in the account with which we are presented cannot be denied. Nevertheless, the general tenor of Le Févre's theory is clear enough: "Ideas" are impressed on the Universal Spirit, which consequently becomes corporified in some mysterious way. The process is, however, reversible, and when a substance is vaporized, the "Ideas" are "thrown off", the corporeal cover is removed, and the Spirit is regenerated. Thus a general cyclical theory of matter is proposed, forming the basis of an interesting attempt to found a mineralogical theory on Neo-Platonic principles.

As our final example, we consider the work of John Webster (1610–82), author of one of the most influential seventeenth-century English texts on the mineral kingdom.[99] Webster displays the characteristic eclecticism of his period,

[95] The significance of the role of light in this scheme should be noted, bearing in mind the basic importance of light for the Neo-Platonists of the Renaissance. (See, for example, K. Goldammer, "Lichtsymbolik in philosophischer Weltenschauung, Mystik und Theosophie vom 15. bis zum 17. Jahrhundert", *Studium Generale*, **13**, 670–82, 1960.)

[96] Le Févre (88), II, 129.

[97] *Ibid.*, 130.

[98] *Ibid.*, 133.

and it is clear that he is trying to blend effectively the theories of his predecessors, notably Sendivogius and Van Helmont. Consider his account of the generation of metals in the earth:

> . . . the Water being sharp and salt, and falling down in the subterraneous caverns and passages of the Earth, doth meet with the drie, sulphureous, and warm steams that rise from the lower parts of the Earth, do joyn together, and so become unctuous and fat, which setling in close holes, and cavities of Rocks where the air cannot enter, as in a close womb, is in length of time thickned into a soft substance, which they call *Gur*;[100] and after by the warmth of the place, or womb, and its own internal fire, sulphur, or heat, is concocted into a metallick body, pure or impure, according as the steams were, when they joyned together, and the place in which they are generated.[101]

Vestiges of the old Sulphur/Mercury theory are to be discerned here, as also Van Helmont's *Gur*, the doctrine of "wombs" for the growth of metals, and the Sendivogian treatment of the rising "steams" which will in time yield metals in appropriate cavities within the earth. Here, also, Webster is following the opinion of Jorden.

Elsewhere,[102] it is made clear that seeds supposedly rise with the "steams", to lodge in suitable crevices within the body of the earth and there grow, via *Gur*, into mature metals. Further rising steams might ripen the baser metals into nobler varieties. But the "idealistic" doctrine has not been forgotten: ". . . the intrinsecal efficient cause [of the seeds] is contained in them, which we call the *Archaeus*, or seminal *Idea* that doth form and organize the body, according to the species from whence it was derived, if the matrix be proper, and of due kind for it".[103] Here, it may be noted , the *Idea* and the *Archeus* have become one and the same, and in some way contained within the seed. But they still have the same fundamental explanatory role—that of accounting for the apparent organization of mineral specimens.

But we have not yet done with Webster's eclecticism. The Faber[104] or *Archeus* may be identified with a "particle of vital air or *aether*"—a notion reminiscent of the doctrine of Sendivogius. And, as in Paracelsus, the cause of mineral formation is to be found in the stars:

[99] J. Webster, *Metallographia: or, an History of Metals. Wherein is declared the signs of Ores and Minerals both before and after digging, the causes and manner of their generations, their kinds, sorts, and differences; with the description of sundry new Metals, or Semi-Metals, and many other things pertaining to Mineral Knowledge* . . ., London, 1671

[100] See note 77.

[101] Webster (99), 83–4.

[102] *Ibid.*, 71.

[103] *Ibid.*, 82.

[104] *Faber* (Latin) = workman, artificer, smith, etc.

> But to come to the true efficient cause of the generation of Metals . . . the substance seems to be this. That the solar particles, celestial spirit, or internal and incombustible Sulphur (which is the true fire of Nature) hid in the viscous matter of Mercury, and excited a[n]d stirred up by the motion of the celestial bodies, central sun, or subterraneous fire or heat . . . doth generate, perfect, and ripen Metals; . . . And all do consent that Sulphur is the efficient cause, or father, and Mercury the passive or mother of all Metals.[105]

And so we get back to the stars as a kind of first cause. It cannot be claimed that any clear and coherent theory emerges from Webster's book. In this respect is is certainly inferior to the earlier volume of Sherley. In his endeavour to pay regard to all the doctrines of his predecessors, Webster becomes disconcertingly obscure. His reliance upon the traditional "idealistic" and "organic" modes of explanation is clearly manifest, however.

With the increasing acceptance of the mechanical philosophy in the seventeenth century, this "idealistic", "animistic" or "organic" basis of mineralogy gradually declined. It is noteworthy, however, that even Lemery, who possibly provides the most obvious example of an enthusiastic application of Cartesian mechanism in chemistry, still made obeisance to the notion of a "Universal Spirit", such as that of Le Févre. Under the head, *Des Principes de la Chymie*, Lemery (1645–1715) wrote: "Le premier Principe qu'on peut admettre pour la composition des Mixtes est un Esprit universel, qui estant répandu par tout, produit diverses choses selon les diverses Matrices ou Pores de la terre dans lesquels il se trouve embarassé [sic] . . .".[106] However, having thus doffed his cap to the former paradigm, Lemery proceeded to argue that it was no longer to be considered a suitable explanatory principle, for he continued: "Mais comme ce Principe est un peu Metaphysique, et qu'il ne tombe point sous les sens, il est bon d'en establir de sensibles, c'est pourquoy ie rapporteray ceux dont on se sert communément".[107] And he goes on to argue that, for practical purposes, the ultimate constituents are water, spirit (*i.e.* mercury), oil (*i.e.* sulphur), salt and earth. These supposedly provide an acceptable basis for an account of the formation of stones and crystals, without invoking any "Universal Spirit". Thus stones could be accounted for by proposing that they were formed by the action of an acid liquor on earth, with subsequent

[105] Webster (99), 83.

[106] N. Lemery, *Cours de Chymie, Contenant La Maniere De Faire les Operations qui sont en usage dans la Medecine, par une Methode facile. Avec des Raisonnements sur chaque Operation pour l'Instruction de ceux qui veulent s'appliquer à cette Science*, 2nd ed., Paris, 1677, 3.

[107] *Ibid.*

hardening due to subterranean heat, or petrification due to cold.[108] An exact mixing of the components, and their *union estroite*, would result in gem formation.

In England, Boyle, though an ardent exponent of the mechanical philosophy, was, in his early work at least, prepared to consider a seminal origin for minerals quite seriously. In the *Sceptical Chymist*, he writes cautiously: ". . . it seem[s] . . . Likely that the Mineral Earths or those Metalline steams . . . do contain in them some seminal Rudiment, or something Equivalent thereunto . . .",[109] and he reports many cases of the apparent growth of metals in mines, previously believed to be worked out. But in his later *Origin and Virtues of Gems*[110] he moves closer to the purely mechanical position, though even here he still uses the word "womb", while making it quite clear that he does not intend it to mean anything more than a mould.

Thus, by the latter half of the seventeenth century, modes of explanation alternative to the old "idealistic" concepts were being proposed, and were gradually displacing the earlier explanatory schemes. This general change in world–view had its influence upon ideas about the nature of the mineral kingdom just as it did on ideas about the animal and plant worlds. In Steno's *Prodromus*,[111] usually taken to be the herald of the new age for geological sciences, one finds no attempt to explain mineralogical phenomena in terms of seeds, ferments or spiritual essences. The accretion of crystalline matter provides the basis of the proposed explanations of crystal formation and an organic origin of mineral crystals is explicitly denied.[112] On the other hand, it may well be that Steno's interest in the places where solids crystallize, and, indeed, his

[108] *Ibid.*, 279. It may be noted that certain passages in the *Meteorologica* provide a warrant for the view that stones may be formed either by the heat of a "dry exhalation", or by cold, which drives out heat and moisture. See: Aristotle (86), 287 and 363.

[109] R. Boyle, *The Sceptical Chymist: or Chymico-Physical Doubts & Paradoxes, Touching the Spagyrist's Principles Commonly call'd Hypostatical, As they are wont to be propos'd and Defended by the Generality of Alchymists . . .*, London, 1661, 364.

[110] R. Boyle, *An Essay About the Origin and Virtues of Gems. Wherein are proposed and historically illustrated some Conjectures about the Consistence of the Matter of Precious Stones, and the Subjects wherein their chiefest Virtue resides*, in: *The Works of the Honourable Robert Boyle. In Six Volumes. To which is prefixed The Life of the Author. Volume the Third A New Edition*, London, 1762, 512–61 (republished in facsimile with an introduction by Douglas McKie, Hildesheim, 1966). The *Essay* was first published in London in 1672.

[111] N. Steno, *Nicolai Stenonis de Solido intra solidum naturaliter contento dissertationis prodromus. Ad serenissimum Ferdinandum II. Magnum Etruriae Ducem*, Florence, 1669. English translation by J. G. Winter (*Michigan Humanistic Studies*, **11**, pt. 2, 1916, reprinted in facsimile, New York and London, 1968: *Nicolaus Steno (1631–1686) The Prodromus of Nicolaus Steno's Dissertation Concerning a Solid Body Enclosed by Process of Nature Within a Solid . . .*). See also: G. Scherz (ed.) and A. J. Pollock (translator), *Steno's Geological Papers*, Odense, 1969.

[112] *Ibid.* (1969), 175.

interest in the general problem of the formation of solids within solids, resulted from an attempt to clarify the role of the "matrix" in mineral formation. The change in the mode of explanation was not immediate, of course, and as Adams noted,[113] evidence for the persistence of the organic simile in mineralogy may be found in the eighteenth century, and possibly even in the nineteenth century. Thus Buffon[114] believed that the regular external features of crystalline minerals were to be accounted for by the action of "organic molecules" on the outer edges of the mineral matter. Being unable to penetrate to the interior of the hard mineral matter they could only bring about a superficial organization, whereas a penetration to the soft interiors occurred, with consequent internal organization, when animals or plants were generated. Crystalline minerals were viewed as the very lowest links in the "Great Chain" of organized beings. And at the beginning of the nineteenth century, we find Faujas de St. Fond toying with the idea that the rapid formation of deposits of peat might in some way be associated with the organic growth of metallic iron.[115] Nevertheless, that there was a decline and eventual demise of such mineral theories after the seventeenth century is a matter not seriously in question, and it seems reasonable to see this as but one aspect of what is customarily referred to as the "Mechanization of the World Picture".[116]

It remains for us to consider the question of whether or not the theories which we have been considering should be dignified with the term "scientific", and whether the explanations that these theories offered should be judged in any way adequate. It is clear that they formed an important part of the background from which the science of mineralogy gradually emerged and came into focus in the latter half of the eighteenth century. These hylozoist theories parallel the alchemical ideas from which chemistry eventually emerged, and indeed, as we have seen, they are themselves interwoven with the alchemical doctrines. It is also clear that for the most part the theories purported to explain, but not to predict, phenomena. Some would say that this in itself is sufficient to deny these theories any kind of scientific status, though the general feeling today is that this is an unwarranted limitation of the domain of scientific explanation. (Does not the Darwinian theory itself explain, rather

[113] Adams (2), 296.

[114] *Oeuvres Complètes de Buffon, avec des extraits de Daubenton, et la classification de Cuvier*, 6 vols, Paris, 1838, vol. II, 1–4.

[115] B. Faujas de Saint-Fond, *Essai de Geologie, ou mémoires pour servir a l'histoire naturelle du globe*, 2 vols, Paris, 1809, vol. II, 343.

[116] *Cf.* E. J. Dijksterhuis (translated from the Dutch by C. Dikshoorn), *The Mechanization of the World Picture*, Oxford, 1961 (*De Mechanisering van het Wereldbeeld*, Amsterdam, 1950).

than predict?[117]) These theories, then, attempted to weave a fabric of explanation round the complex phenomena of the mineral kingdom. And in this they were not wholly unsuccessful. The organic explanation of the regularity of the crystalline structure of minerals was by no means an unreasonable doctrine in the days long before (say) Haüy's important breakthrough in crystal theory. But although the theories explained mineral phenomena well enough (once certain premises had been accepted), their lack of predictive power was a serious practical limitation. They did not suggest appropriate experimental tests, and they did not (in themselves) encourage the making of new observations. They were not particularly helpful to the practical metallurgist. They did not give him any reliable information about where to look for ore deposits, or how to extract metals from their ores in an economical fashion.

It has been suggested above that the doctrine of the seeds of minerals may be construed as a specimen of essentialist thinking in science. Was it this feature of the theory which made it an inappropriate form of explanation, which had to be put aside before mineralogy could emerge as a mature science? It appears that the situation is not wholly comparable to the situations in biology and chemistry. I suggest that thinking about the mineral kingdom went through three broad though fairly distinct phases prior to 1800.

(1) The period prior to the middle years of the seventeenth century, when "organic" explanations of mineral substances prevailed;
(2) The period from the middle of the seventeenth century to the latter part of the eighteenth century, during which time explanations were frequently given in terms of the mechanical philosophy, but with inadequate systems of chemical analysis, so that schemes of classification still had to be based largely on external characteristics;
(3) The period in the latter part of the eighteenth century when, with notable advances in the chemical analysis of minerals[118] (stemming particularly from the work of Bergman), chemical systems of mineral classification were devised, while men remained to a large extent agnostic about the nature of the ultimate constituents of matter.

The last of these three phases does seem to be associated with a movement in chemistry from an earlier essentialist position to a more positivist approach.[119] Thus here, mineralogy apparently benefits from a movement away from the

[117] *Cf.* M. E. Scriven, "Explanation and Prediction in Evolutionary Theory", *Science*, **130**, 477–82, 1959.

[118] *Cf.* W. A. Campbell and C. E. Mallen, "The Development of Qualitative Analysis from 1750 to 1850 Part 1", *Proceedings of the University of Durham Philosophical Society*, **13** (Series A, Science, No. 12), 108–18, 1959. I have also considered this topic in a brief paper, "Some Eighteenth Century Methods for the Chemical Analysis of Minerals", *Journal of Chemical Education*, **50**, 337–40, 1973.

[119] See note 72, above.

essentialist mode of thinking. We may feel that the earlier change from organic to mechanical explanations was likewise a progressive step. With the advantage of hindsight, we see that the mechanical philosophy provided a favourable forward route. Yet it can be argued that the Cartesian type mechanism of (say) Steno has essentialist features. He attempts to explain the formation of crystals in terms of imperceptible particles, but these explanatory entities are not observed directly—they are hypothetical for Steno. But given these particles, he hopes to be able to make the formation of crystals intelligible to himself and to his readers. It would seem, then, that there are ways forward in science besides the removal of essentialist "barriers".

We return to the question of whether or not the "organic" mineral explanations should be deemed in any way adequate. Science cannot operate in a total theoretical vacuum, and the early mineralogical notions that have been described in this paper, despite the fact that they are so alien to our present thinking, provided the first pattern of mineralogical explanation, coordinating mineral phenomena with some degree of success. We believe that such ideas were mistaken, but they should not be ignored; I consider that they should be examined with some sympathy as we try to build up a picture of the development of men's ideas about the mineral kingdom.

*Acknowledgements.* I am much indebted to my colleagues Professor G. Seddon, Dr. R. M. Gascoigne and Dr. W. H. Leatherdale for their comments on this paper during its preparation and also for the advice given by the Editor's anonymous reviewer. I have been pleased to adopt a number of their suggestions in the final version.

# III

## MECHANICAL MINERALOGY

In another paper[1] I have suggested that it is convenient to analyse the pre-nineteenth-century ideas concerning the mineral kingdom into three broad stages. Before the middle of the seventeenth century, accounts of the formation and properties of minerals were customarily given in what we may call "organic" terms: minerals were believed to grow in some manner within the body of the Earth, and frequently their generation was ascribed to some kind of sexual process. From the middle years of the seventeenth century one finds a decline in the use of this "organic" metaphor, and its replacement by attempted accounts of mineral phenomena in "mechanical" terms. This mode of mineralogical explanation received an initial impetus from the detailed cosmological scheme of Descartes's *Principia*,[2] which contained a mechanical mineralogical-cum-chemical system, and achieved its maximum development about 1720 in the work of writers such as Réaumur. Thereafter, there was a decline in attempts to account for mineral phenomena in purely mechanical terms, and an increased emphasis on more strictly chemical modes of explanation, and this seems to have paralleled the developments that were occurring in chemistry at the same period.[3] However, the assistance that chemistry could offer to mineralogy was limited, until viable methods of chemical analysis were developed by Bergman and his disciples, towards the end of the eighteenth century. As is well known, this rise in the application of chemical analysis to mineral substances was paralleled almost simultaneously by developments in crystallography, through the work of Bergman, Romé de l'Isle and Haüy,[4]

[1] D. R. Oldroyd, "Some Neo-Platonic and Stoic Influences on Mineralogy in the Sixteenth and Seventeenth Centuries", *Ambix*, **21**, 128–56, 1974.

[2] R. Descartes, *Renati Des-Cartes Principia philosophiae*, Amsterdam, 1644. I have used the French translation as revised and edited by Clerselier: *Les Principes de la Philosophie de René Descartes, Quatrieme Edition. Revuë & corrigée fort exactement par Monsieur CLR. Avec des Figures dans le corps du Livre; Et celles en taille-douce, de la premiere Edition, mises a la fin du Livre*, Paris, 1681 (*Oeuvres de Descartes Publiées par Charles Adam & Paul Tannery sous les auspices du Ministere de l'Instruction Publique*, 12 vols, Paris, 1896–1910, IX).

[3] The increasing emphasis on chemical, as opposed to physical, modes of explanation, in France during this period, has been emphasized by C. C. Gillispie in his paper, "The *Encyclopédie* and the Jacobin philosophy of science: a study in ideas and consequences", in M. Clagett (ed.), *Critical Problems in the History of Science*, Madison, 1962, 255–89.

[4] For an account of the crystal theories of Bergman, Romé de l'Isle and Haüy, see: J. G. Burke, *Origins of the Science of Crystals*, Berkeley and Los Angeles, 1966; and H. Metzger, *La Genèse de la Science des Cristaux*, Paris, 1918.

and it was the last three decades of the eighteenth century that saw the establishment of mineralogy in its own right.

In my paper referred to above, I have attempted to give some account of the ways in which the "organic" metaphor was applied to the mineral kingdom; in the present essay I aim to outline the way in which the "mechanical philosophy" was utilized in attempted interpretations of mineral phenomena. These two different modes of explanation, I suggest, provided the two principal "world-views" influencing accounts of mineral substances, before the achievement of successful methods of chemical analysis of mineral substances, and the crystallographical developments of Haüy and Romé de l'Isle. Needless to say, no rigid line of demarcation can be drawn between the period of decline of "organic" explanations, and the period in which the use of mechanical models first became prominent. Indeed, we find the two co-existing in the writings of some authors, and it is possible to find a signal example of the use of the organic mode of explanation as late as 1761.[5] Nevertheless, the change in emphasis from organic to mechanical ways of viewing mineral phenomena may be discerned clearly enough.

The manner in which the mechanical philosophy influenced the physical and biological sciences has been discussed in the historical literature for many years, but relatively little attention has been given to the influence of this mechanical mode of thinking on ideas concerning the mineral kingdom,[6] and consequently it is the intention of this present brief enquiry to do something towards opening up this field.

Before proceeding, it may be helpful to state what I understand the chief characteristics of the use of the mechanical philosophy in explanations of mineral phenomena to have been. I suggest that any scheme that attempted to account for such phenomena in terms of the shapes, and interactions by direct contact, of hypothetical corpuscular components of matter, without necessarily invoking any additional chemical explication, offers us an example of the influence of the mechanical philosophy on mineralogy.

It is customary to think of the rise of the mechanical philosophy as an event in the history of ideas that occurred in the seventeenth century. Nevertheless, some intimations of quasi-mechanical modes of accounting for mineral

[5] J. B. Robinet, *De La Nature* Amsterdam, 1761.

[6] Thus, important accounts of the influence of the mechanical philosophy in science, such as E. J. Dijksterhuis's *The Mechanization of the World Picture*, Amsterdam, 1950 (in Dutch), Oxford, 1961, or R. S. Westfall's *The Construction of Modern Science: Mechanisms and Mechanics*, New York and Toronto, 1971, do not consider the relationship between this philosophy and mineralogical thinking. On the other hand, the topic discussed in the present paper was given serious consideration in Metzger's book, referred to above (note 4), pp. 212–19 and *passim*.

phenomena may be discerned in the sixteenth century, in the work of Agricola,[7] though he did not, one must suppose, put forward such views because of a commitment to any particular philosophical world-view, and I shall not, in fact, attend to the development of crystal or mineral theories in the period prior to the seventeenth century.

Considering, then, the influence in the seventeenth century of mechanical ideas on theories of the mineral kingdom, it should be noted that it mattered little whether such corpuscles were held to be divisible or otherwise. Both atomic and "corpuscularian" hypotheses could offer mineralogical explication, and we find, for example, that both Gassendi and Descartes put forward accounts of the mineral kingdom that were based on the principles of their respective matter theories. Thus, though Gassendi's opinions were in a sense traditional, in that he subscribed to the doctrine of mineral and metallic seeds,[8] he believed that the structures of the seeds themselves were to be thought of in terms of the refurbished atomic hypothesis:

> Je passe sous silence qu'il faut aller reprendre & chercher ce Principe ou semence jusque dans la source, je veux dire dans les Atomes mesmes, qui estant d'une figure & d'une tissure particuliere, font la propre structure & la configuration de ce germe, & qui penetrant ensuite dans la Matiere par leur mobilité naturelle, remüent, transposent, & disposent d'une telle maniere ses Atomes, qu'elle prend la forme, & la figure d'une telle espece de Metail.[9]

However, when attempting to account for the external forms of crystals, Gassendi suggested that they might be due to the shapes of the constituent particles, and he did not attempt to invoke the organic model. Thus amethysts and diamonds, for example, were supposedly constructed of hexahedral and octahedral units respectively.[10] Here his approach seems to have been essentially the same as that of the corpuscularian philosophers of the Cartesian school—and this was evidently a fairly general position. Thus it was primarily through mechanical corpuscularianism, rather than explanations in terms of

[7] See, for example, *Georgii Agricolae De Ortu et Causis Subterraneorum*, Basle, 1546, 13. Here, in his discussion of the origin of his *succus lapidescens*, Agricola includes, as one possible cause, the abrasion (literally the "licking") of rocks by water. Subsequently, he accounts for the petrification of wood by the hypothesis that deposition of stony matter in the interstices of the wood occurs when the *succus* penetrates therein and its motion ceases (56). For the most part, however, Agricola attempts to account for the formation of mineral substances by the various actions of heat and cold, and by sundry processes of "coction". This gives his work a strongly Aristotelian flavour.

[8] P Gassendi, *Abrege de la Philosophie de Gassendi En VIII Tomes. Par F. Bernier Docteur en Medecine de la Faculte de Montpelier*, Lyon, 1678, V, 85, 91, 107, and *passim*.

[9] *Ibid.*, 321.

[10] *Ibid.*, 268.

the configurations of constituent atoms, that men attempted to make mineral phenomena intelligible to themselves. It is true that in the work of Hooke,[11] we find a significant seventeenth-century attempt to account for the structures of crystals in terms of the *configurations* of constituent corpuscles.[12] Nevertheless, it appears that this approach did not prove to be as attractive or as influential as the ultra-mechanical corpuscularianism of Descartes's *Principia*, so far as the provision of a model for mineralogical explanation was concerned. Thus we shall concern ourselves here primarily with the mineral theories of Descartes and his followers.

It might be thought that there is today little of significance remaining to be said about Cartesian matter theory. Yet, in fact, historians have given little attention to Descartes's geological and mineralogical asseverations,[13] so that it may be worthwhile to give some consideration to these less well known aspects of his total cosmological scheme. Certainly they are of paramount importance so far as the present inquiry is concerned.

It will be recalled that in his *Principia Philosophiae*[14] Descartes followed a venerable tradition by attempting to give an account of the origin and geological history of the globe. Beginning with a few ostensibly self-evident principles, he claimed to give a deductively valid account of the construction and history of our planet. It is true that he admitted that there were various possible ways in which the deductive steps might be carried out, so that, in theory there could be a variety of "possible" worlds, it being the task of the empirical investigator to determine which of these "possible" worlds we happen to inhabit. Nevertheless, it seems pretty clear that Descartes believed that he was giving an account of the actual (rather than a possible) history of the

[11] R. Hooke, *Micrographia: or some Physiological Descriptions of Minute Bodies Made By Magnifying Glasses. With Observations and Inquiries thereupon . . .*, London, 1665, 85–6 (facsimile reprint, Toronto and London, 1961). Earlier, Kepler had thought of crystals being composed of regularly packed spherical corpuscles, but he also invoked a *facultas formatrix*, which appears characteristic of Neo-Platonic, rather than "mechanical", theorizing. (*Ioannis Kepleris. D. Maiest. Mathematici Strena Seu De Nive Sexangula*, Frankfurt on Main, 1611; English translation by C. Hardie, and introduction by L. L. Whyte, Oxford, 1966: J. Kepler, *The Six-Cornered Snowflake*. See also: C. J. Schneer, "Kepler's New Year's Gift of a Snowflake", *Isis*, **51**, 531–45, 1960.)

[12] See J. G. Burke (4), 38–40, for a discussion of Hooke's work, and also (41–2) for an exposition of the somewhat similar crystal theory of Huygens. I shall not enlarge upon Burke's account here.

[13] A paper with a promising title (G. A. Daubrée, "Descartes, l'un des créateurs de la cosmologie et de la géologie", *Journal des Savants*, 165–75 and 208–21, 1880) proves disappointing on examination, being primarily concerned with developments long after the time of Descartes.

[14] See note 2, above. The account given below is extracted from Part 4 of Descartes's book.

globe. Only one scheme was worked out for our consideration, and when the tortuous account of the evolution of the globe was completed in theoretical terms, Descartes triumphantly identified the theoretical entities of his scheme with the atmosphere, the seas, and the crust of the Earth of our every-day experience.

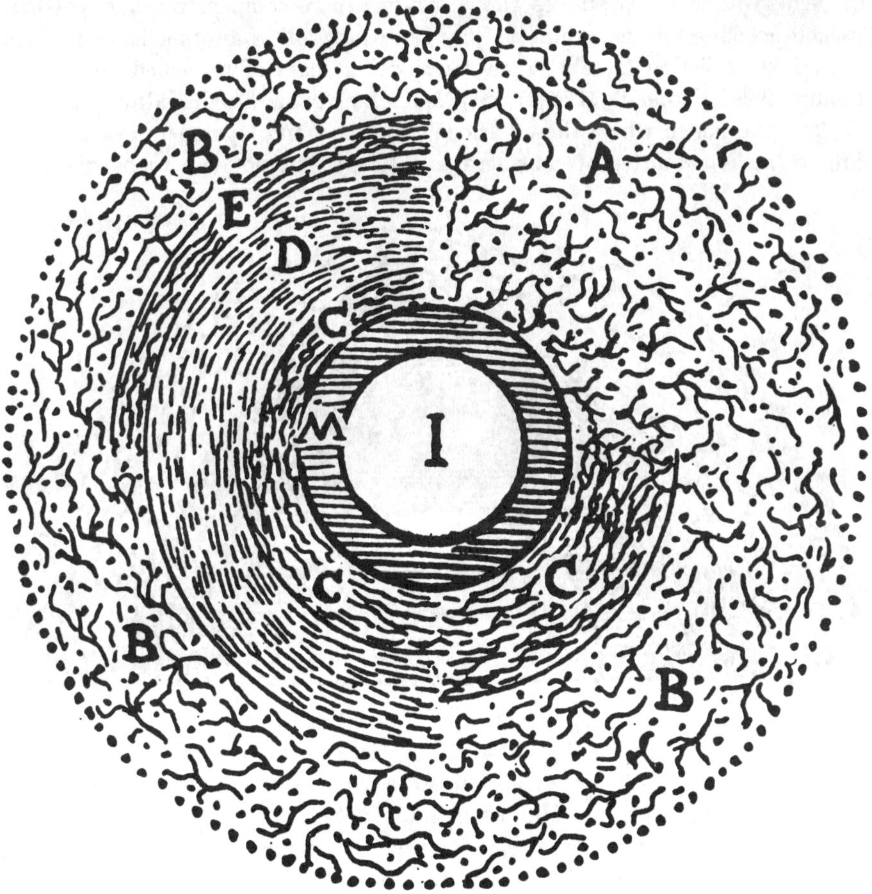


Let us consider Descartes's account of the evolution of minerals within the body of the Earth, commencing from the moment when the various layers of the planet have formed, according to his well-known cosmological scheme.[15] This stage may be represented by a reproduction of Descartes's own diagram (Figure 1).

[15] For a general discussion of the Cartesian cosmological theory, see E. J. Aiton, *The Vortex Theory of Planetary Motions*, London and New York, 1972.

Here, in the upper left-hand quadrant of this diagram (which illustrates four stages in the evolution of Descartes's Earth), region **I** contains the fiery particles of his first element; **M** is a compact and opaque region (formed by the coalescence of sun-spot material); **C** is a layer of very irregular, branched corpuscles; **B** is a liquid layer (later to form the water of the oceans) consisting of flexible aqueous corpuscles and also some more compact rod-like particles (which account for the salinity of the sea-water); **E** is another layer of solid material, consisting largely of an interlocking mass of branched corpuscles; finally, **B** is another fluid layer, corresponding to the Earth's atmosphere.

The manner in which these various corpuscles and layers supposedly came into being is a matter of considerable interest, but not one to be discussed

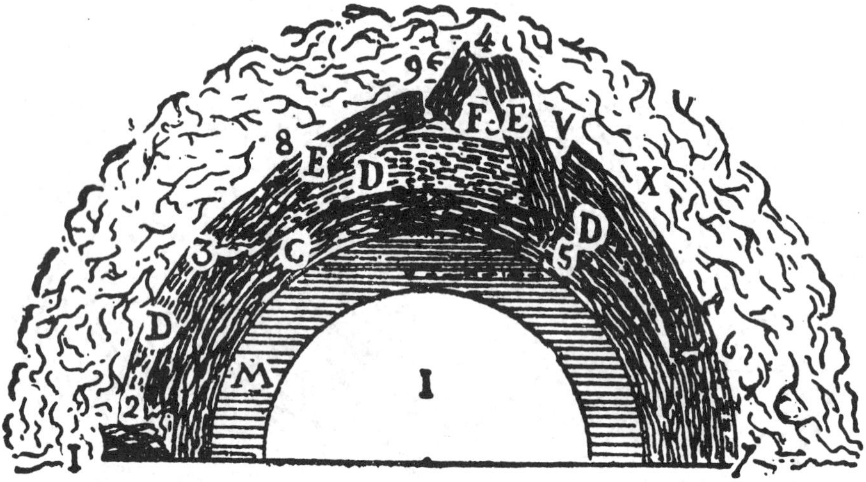


in a paper on the history of mineralogy.[16] Here we confine ourselves to a brief discussion of the subsequent evolution of the Cartesian Earth, before considering Descartes's mineralogical opinions in more detail.

Descartes suggested that in hot weather the corpuscles of the liquid layer, **D**, might pass up through the pores in **E**, and, on combining with aerial particles of region **B**, might form further material for layer **E**, and consequently be unable to return to layer **D**, whence they came. The process being repeated over a number of years, "spaces", such as **F**, might develop below **E**. (See Figure 2.) Thus, various subterranean cavities might be formed, and this

[16] For an exposition of the earlier stages of the evolution of the earth, according to the Cartesian scheme, see J. F. Scott, *The Scientific Work of René Descartes 1596–1650*, London, 1952, chapters 11 and 12.

might eventually lead to a fragmentation of the crust, **E**, with collapse of some of the fragments into the oceans, as indicated in the diagram, which shows the final stage in the evolution of the Cartesian Earth.

It is not necessary to emphasize that Descartes was no mineralogist, but he was familiar with the substances commonly found within the Earth, and he attempted to give his own "mechanical" account of their formation. As we shall see, this provided a pattern for mineralogical explanation (in France at least) for nearly a hundred years, so I shall proceed to an exposition of Descartes's ideas on these matters, despite the fact that they have been described previously as being of "little interest".[17]

Region **C** consisted largely of corpuscles of the third element, supposedly in motion because of the action of the other two elements within their interstices. Some of the pores were occupied by corpuscles of salt and water from the overlying oceans (**D**). The salt particles might be so buffeted that their previous baton-like shapes might be altered so as to produce flat and pliable corpuscles with sharp points (just as a piece of iron may be beaten into a blade); thereby various corrosive juices might be formed. On the other hand, the aqueous particles might be divided into several fine and flexible branches, thus giving rise to fatty and oily substances. Particles in **C** that were initially somewhat rounded might in time be worn down to smooth, round and slippery, but rather large and heavy, corpuscles. Thus quicksilver might be formed. The reader is invited to imagine that the branched corpuscles of **C** waved back and forth (owing to thermal agitation) like the branches of trees in the wind. Thus the intervening pores were constantly fluctuating in shape and size, so that particles that happened to get trapped in these pores might get beaten into the shapes described above.

Somehow, these three kinds of corpuscle (saline, mercurial and sulphureous) ascended from **C** towards **E**, though the motive power for the ascent is not made entirely clear. The corrosive juices interacted with the particles of **E** to form saline materials, such as vitriols, alum and other minerals. The oily particles formed bitumens and sulphurs when they reached **E**. If some of the corrosive juices interacted with the oily particles, then metals were formed, and might also be raised (again in some not clearly specified manner, but with the aid of the mercurial corpuscles) from **C** to **E**. Various shapes and sizes of corpuscle accounted for the different metallic species. It appears, however, that Descartes was not entirely happy with this account, and he stated that he would say more about it if he had sufficient leisure to carry out all the experiments needed to verify the theory. Quickly, he hurried on to the next topic: springs and rivers. Nevertheless, the problem of the origin of metals was taken up again a few pages later in the book where Descartes seems to

[17] *Ibid.*, 188.

be attempting to reconcile his theory with the old Aristotelian theory of exhalations,[18] or such versions of this doctrine as had come to be accepted in seventeenth-century theories of the Earth.

Descartes speaks of corrosive spirits and volatile salts arising from the interior layer of the Earth (**C**). He uses the term "exhalation" for the oily vapours also supposedly arising from the layer **C**, and refers to the particles of these as flying freely. The vaporized spirits wedged into the terrestrial particles of **E**, to form various (unspecified) opaque minerals. They might also combine together in various (again unspecified) ways, within cracks in the crust (**E**), to form transparent materials such as diamonds, agates, or rock crystal. Quicksilver vapours might bear up particles of metal, to lodge in layer **E**, and this allowed some quicksilver to be found there also, though combination with the oily exhalations might also produce vermilion, or cinnabar. Likewise, metals might be borne up by oily exhalations or saline spirits. It appears that Descartes was trying to account for the known association of sulphur and vitriols with metallic ore deposits. He believed that metals are found most commonly on the sunny sides of mountains where the solar heat will assist the metalline distillations within the Earth's interior most effectively.

Perhaps, unexpectedly, Descartes did not think of brimstone as a direct product of the oily exhalations. Rather, it was presumed to arise from some combination with saline, spiritual particles. If similar juices from the external layers of the Earth (**E**) were also incorporated, then inflammables such as bitumen and naphtha were formed. If there was a significant portion of earthy matter, then a clay resulted. A general agglomeration of such particles gave rise to petroleum.

Of all the mechanical mineralogical schemes propounded in the seventeenth and eighteenth centuries, this was undoubtedly the most detailed, and also the most influential. It is easy to understand its appeal. Earlier schemes, or theories of the Earth, had never offered anything like such a comprehensive treatment. The corpuscles, of which Descartes wrote so confidently, were readily visualized in the mind's eye. As scaled-down versions of macro objects, they were conceptually so simple, so easy to envisage, that the student might readily imagine that he had at last, comprehended the true essence of mineralogical phenomena. Moreover, the whole Cartesian scheme was ostensibly guaranteed by certain self-evident principles. Little wonder, then, that Descartes's system found such slavish imitators,[19] or that Fontenelle (using his

[18] Aristotle, *Meteorologica*, translated by H. D. P. Lee, London and Cambridge (Mass.), 1952, 287 and 289.

[19] For a system that was an almost exact replica of the original Cartesian scheme, see A. Le Grand, *An Entire Body of Philosophy, according to the principles of the famous Renate DesCartes* . . ., 2 vols, London, 1694. Somewhat more independent were: P. S. Regis,

charming simile of the mechanical operations of the theatre[20]) could imply that, at long last, the inward essence of things had been grasped, in all its mechanical simplicity. As his Marchioness put it, so aptly: "I perceive . . . Philosophy is now become very mechanical".[21]

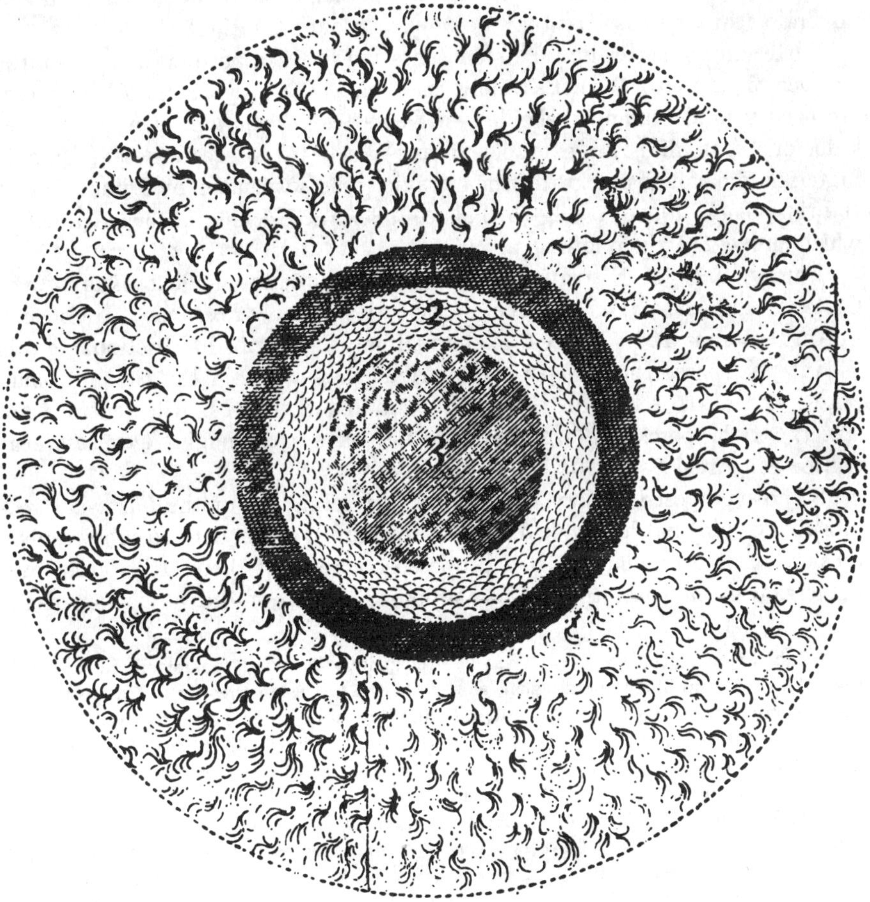


*Systême de Philosophie, Contenant La Logique, La Metaphysique, et la Morale*, 2 vols., Paris, 1690; and J. Rohault, *Rohault's System of Natural Philosophy; illustrated with Dr. S. Clarke's notes, taken mostly out of Sir Isaac Newton's Philosophy. With additions. Done into English by John Clarke*, 2 vols., London, 1723. This latter is discussed below, pp. 168–70.

[20] B. le B. de Fontenelle, *Conversations with a Lady, on the Plurality of Worlds. . . Translated by Mr. Glanville. The Fourth Edition. With the Addition of a Sixth Conversation . . .*, London, 1719, 8–10, and the figure facing page 1.

[21] *Ibid.*, 10.

That there were many who opposed Descartes's simplistic scheme is well known,[22] but equally it acquired numerous disciples, and it is apparent that the scheme (or, at least, the mode of thinking that it typified) ) gradually began to exert a considerable influence on both geological and mineralogical theory, and particularly so in France. It is to an examination of these geological and mineralogical repercussions that we now turn our attention.

Cartesian influences may be seen operating in ways both crude and subtle. The very diagrams of Burnet's widely-read *Sacred Theory of the Earth*[23] (Figure 3) are clearly (and crudely) beholden to Descartes's *Principia*. On the subtler side, one may find, for example, John Woodward suggesting that organic materials did not disintegrate when gravity was suspended by Divine fiat at the time of the Flood, owing to their complex interlocking fibrous structures, which prevented their falling asunder.[24]

The writings of Nicolaus Steno reveal a number of clear examples of Cartesian influence. For example, in the early work, *De Thermis* . . ., published in Amsterdam in 1660, the heat of thermal baths is ascribed to frictional effects as water squeezes through pores of various shapes within the Earth. Small pores, and sharp edges on the rocks bounding the pores, will generate hot water. If the pores are larger, and do not have such rough edges, less heat will be generated.[25]

Steno's proposed history of Tuscany is reminiscent of the Cartesian geomorphology, in that he invoked *collapse* structures, subsequent to subterranean erosion. Descartes likewise had proposed collapses of the Earth's crust into subterranean cavities, though the supposed causes of the formation of the

[22] Controversy, needless to say, was centred primarily round the astronomical, rather than the geological and mineralogical, aspects of the Cartesian theories. The portions of the *Principia* that related to the mineral kingdom were fairly evidently conjectural in character, and were not adopted widely as they stood. On the other hand, I contend that the general mode of explanation that they provided was of considerable influence in the history of theories of the mineral kingdom.

[23] T. Burnet, *The Theory of the Earth: Containing an Account of the Original of the Earth, and of all the General Changes Which it hath already undergone, or is to undergo Till the Consummation of all Things. The Two First Books Concerning The Deluge, and Concerning Paradise*, 2nd ed. London, 1691 (republished London and Fontwell, 1965).

[24] J. Woodward, *The Natural History of the Earth, . . . Written originaly* [sic] *in Latin: and now first made English by Benj. Holloway, L.L.B. . . . To which are added, Physical Proofs of the Existence of God, his actual incessant Concurrence to the Support Of The Universe, and of all Organical Bodyes, Vegetables, and Animals, particularly Man; . . .*, London, 1726, 160–2. This work was a translation of Woodward's reply to criticisms of his theory of the Earth, made by Camerarius.

[25] For a discussion of this work, with a facsimile of the Latin original, and accompanying English translation, see G. Scherz, "Niels Stensen's First Dissertation", *Journal of the History of Medicine*, **15**, 247–64, 1960.

caverns were quite dissimilar. Steno[26] refers to three causes of motion: natural, artificial and divine, the first arising from the motion of the subtle fluid penetrating all bodies, the second referring to the motion of animals, and the third being due to the "first and unknown cause". The subtle fluid is reminiscent of the first element of Descartes. Moreover, in one significant passage, concerned with the layered strata of the Earth, the account of Descartes is referred to with specific approval.[27]

It is, however, in Steno's crystal theory that we find the most prominent application of the new way of visualizing phenomena. Crystals, he maintained, do not grow vegetatively; they increase by the addition of crystalline matter "to the external planes of already delineated Chrystal".[28] Perhaps a subtle fluid flows out of the crystal, and in some manner orders the fresh material in such a way as to promote the continuous production and maintenance of the crystalline form:

> in the increase of Chrystal . . . [there is] a double motion: *one* which maketh the Chrystallin matter to be joyned to these and not to other places of the Chrystal; which motion, I guess, is to be adscribed to the *Permeating* subtil fluid; to be illustrated by the lately alledged Example of the Magnet: The *other*, whereby the new Chrystallin matter, joyned to the Chrystal, is spread over the Plain, which motion is to be derived from the *Ambient* Fluid.[29]

Thus it is the accretion of corpuscles[30] of "Chrystallin matter" which is to account for the growth of the crystal, and, as a somewhat pious hope, the motion of the ambient fluid is to maintain the regular growth of the crystal faces:

> Thus, when the Iron-Threds rise up upon a Loadstone, what by the motion of the Air is struck off from one cometh to the other. To this motion of the Ambient I should attribute, that not only in Chrystal, but also in many other Angular Bodies the opposite planes are parallel to one another.[31]

[26] N. Steno, *The Prodromus to a Dissertation Concerning Solids Naturally Contained within Solids. Laying a Foundation for the Rendering Rational Account both of the Frame and the several Changes of the Masse of the Earth, as also of the various Productions in the same. By Nicolaus Steno. English'd by H.*[*enry*] *O.*[*ldenburg*], London, 1671, 16. For more recent English editions, see the translation by J. G. Winter, *Michigan Humanistic Studies*, **11**, pt. 2, 1916—reprinted in facsimile, New York and London, 1968); and G. Scherz (ed.) and A. J. Pollock (translator), *Steno Geological Papers*, Odense, 1969.

[27] *Ibid.* (1671), 39.

[28] *Ibid.*, 55.

[29] *Ibid.*, 61–2.

[30] Earlier in the *Prodromus* (*Ibid.*, 15). Steno had explicitly stated that he took a corpuscular view of matter.

[31] *Ibid.*, 62.

It is clear, then, that the seventeenth-century mechanical philosopher sought to explain the mineral phenomena that had hitherto been accounted for by a variety of "organic" metaphors in terms of mechanical interactions of corpuscles, or corpuscularly constituted fluids of various kinds.

More extreme examples of the mechanical mode of mineralogical thinking are to be found in the work of N. Lemery, best known today as one of the more extravagant exponents of mechanical chemistry. His well-known doctrine of acid and alkaline corpuscles, slotting into each other to give rise to neutral products,[32] was invoked to account for the formation of pebbles. Sharp acid particles could meet with earthy matter, and from the coagulum resulting a pebble might harden, either by heat or by cold.[33] An *union estroite* of the two components might yield gems. This, despite the fact that (as Baron subsequently pointed out[34]) the customary product of an acid/base reaction is saline, not earthy.

We should also note Lemery's ingenious account of the cause of the heat of the interior of the globe, which writers in the Neo-Platonic tradition had ascribed to internal ferments. On the basis of his well-known experiment, in which a 50-lb mixture of moistened iron and sulphur was buried under soil and allowed to react, thus simulating a volcano,[35] Lemery argued that the internal heat of the globe might be attributed to the violent rubbing of pointed acid particles of sulphur against the corpuscles of iron. Earthquakes, thunder and lightning, and other geophysical and meteorological phenomena, might have a similar origin.

Such largely *a priori* accounts of mineral phenomena may appear incongruous and inherently unsatisfactory today, and the Cartesian theories were indeed criticized on a number of important grounds by the later Stahlian chemists and mineralogists, as well as the Newtonian philosophers. Nevertheless, I would argue that the swing to the "mechanical" mode of explanation was a progressive movement, for the new ways of visualizing phenomena were seen by their advocates as being incompatible with the "old wives' tales" that still formed the background to many seventeenth-century opinions concerning the mineral kingdom. The arch-Cartesian, Rohault, for example, maintained that:

[32] N. Lemery, *Cours de Chymie, Contenant La Maniere De Faire les Operations qui sont en usage dans la Medecine, par une Methode facile. Avec des Raisonnements sur chaque Operation pour l'Instruction de ceux qui veulent s'appliquer à cette Science*, 2nd ed., Paris, 1677, 25.

[33] *Ibid.*, 279.

[34] N. Lemery, *Cours de Chymie . . . Nouvelle Edition, . . . par M. Baron*, Paris, 1756, 410.

[35] N. Lemery, "Explication Physique & Chymique des Feux souterrains, des Tremblemens de terre, des Ouragans, des Eclairs & du Tonnere", *Histoire de l'Academie Royale des Sciences. Année M.DCC. Avec les Memoires de Mathematique & Physique, pour la même Année*, 2nd ed., Paris, 1719, 101–10.

> From the Nature of Stones both precious and common which we have laid down, I don't see how we can deduce certain Properties which are mentioned by the Writers of Natural History: As, for Instance, that a Blood-Stone, worn by the Person who has the *Bloody-Flux,* will stop the Distemper, and that other Stones will cure other Distempers. And therefore we find by repeated Experiments, that these Sorts of Properties are falsely ascribed to the greatest Part of these Stones.[36]

Certainly, the argument was based on empirical grounds, but also Rohault saw the theories that he espoused as being incompatible with the fabulous properties of gems and minerals to which many of his contemporaries still subscribed. What were his ideas on the "Nature of Stones both precious and common"? Petrifying springs were held to run with water containing a deal of "terrestrial Matter" which could enter the minute cavities of porous substances, thereby causing petrification.

His account of the formation of grains of sand is engaging, though not entirely clear:

> Though those Places in the Earth where Metals are formed are very much pressed upon by the Weight of all that terrestrial Matter which is betwixt these Places and the Surface of it, yet those Parts which are near the Surface are so little pressed upon, that they are separated from one another by an infinite Number of Chops and Chinks, which are open every Way, and which give a free Passage to Vapours and Exhalations, and to a great many other Parts of Matter which are put in Agitation, by the Heat which is in the Bowels of the Earth: And because it is the Property of Exhalations to mix themselves very easily with those very fine terrestrial Particles which they themselves loosen, they must compose a great many little Heaps, the Parts of which, after having been differently agitated amongst each other, will agree to move all the same Way at last, which will cause them to be at rest with respect to one another. After this, the Body which is thus composed, having a Force sufficient to put the Matter which surrounds it in Agitation, will by Degrees transfer all its Motion to this Matter, and at last be at Rest, being formed into a Figure very near round. And this, in my Opinion is the Method in which *a Grain* of Sand is formed, and in the same Manner innumerable Grains may also be formed.[37]

The six-sidedness of crystals (which were thought to have been in an initially fluid condition) was ascribed to their having consolidated from plastic spheres pressed together: "For as every Piece of Crystal, is surrounded and compressed betwixt six others, so it is squeezed into a Body with six Sides very nearly

[36] J. Rohault (19), II, 162.
[37] *Ibid.*, 159–60.

equal".[38] Once again, mechanical action alone was believed to be competent to account for the observed shapes of crystalline minerals.

In an interesting development, the theory of the lapidifying juice of Agricola was taken over by the French mechanical philosophers, being given considerable attention by E. F. Geoffroy, in a paper presented to the *Académie Royale des Sciences* in 1716, entitled "Sur l'Origine des Pierres".[39] Geoffroy proposed the existence of two kinds of primitive corpuscle for the formation of stones: (*a*) uniformly fine and thin plates; and (*b*) irregular and varied corpuscles, water being the vehicle for the transport of both kinds of particle. Crystals might be formed from a pure combination of the first kind of corpuscle. An admixture of the second kind produced various stones. Thus a kind of scale was envisaged from the purest rock crystal, on the one hand, to a coarse chalk or bole, made almost exclusively of corpuscles of the second kind, on the other. The presence of particles of the first kind was held to be an aid to fusion, for if the particles of the first kind could form a continuous uninterrupted whole they would form a proper fluid when agitated by heat. By contrast, a predominance of earthy corpuscles would mean that mere calcination would occur on heating. Clearly, an insufficient supply of the *suc crystallin* would result in a body with perhaps a transparent interior and an opaque earthy outer crust. What we should call weathered agates or flints would be of this nature. Water being the vehicle for the transport of the hypothetical particles, by its slow evaporation it could permit the coming together of the corpuscles in orderly array to form a crystalline arrangement. Again, I suggest, we have a clear manifestiation of the mechanical mode of thinking providing guidelines for the development of a simple mineralogical theory.

However, the most detailed attempt (after Descartes) to offer a mineral theory based primarily on "mechanical" principles was provided by Geoffroy's colleague, R. A. F. de Réaumur, in a series of three papers presented to the *Académie* in 1721, 1723 and 1730.[40] Réaumur accepted the doctrine of the lapidifying juice, but gave more attention to the nature of the matter that was supposedly able to form crystals by the accretion of its particles. In his first paper, Réaumur did not advance much beyond the position reached by

[38] *Ibid.*, 161.

[39] E. F. Geoffroy "Sur l'Origine des Pierres", *Histoire de l'Academie Royale des Sciences. Année M.DCC XVI*, Paris, 1718, 8–16.

[40] R. A. F. de Réaumur, "Sur la Nature et la Formation des Cailloux", *Histoire de l'Academie Royale des Sciences. Année M.DCC XXI*, Paris, 1723, 255–76; "Sur la Rondeur que Semblent affecter certaines especes de Pierres, & entr'autres sur celle qu'affectent les Cailloux", *Histoire de l'Academie Royale des Sciences. Année M.DCCCXXIII . . .*, Paris, 1725, 273–84; "De la Nature de la Terre en général, et du caractere des différentes especes", *Histoire de l'Academie Royale des Sciences. Année M.DCC XXX . . .*, Paris, 1732, 243–83.

Geoffroy, but argued that the crystalline matter was like exceedingly fine particles of sand, that had got into solution by a mechanical rubbing action of the water on the stony matter over or through which it had passed. Thus Réaumur did not here envisage any kind of chemical reaction between water and the substrate. It was a mechanical abrasion or attrition of stony matter that gave rise to the lapidifying juice, the process being analogous to that by which colloidal gold is formed.[41] Subsequently, porous stones might be converted into hard pebbles by the deposition of crystalline matter from percolating ground-waters. Alternatively, a hooking together of the corpuscles in the appropriate way might give rise to well-formed crystalline growths within cavities beneath the earth's surface. It may be noticed that Réaumur did not attempt to make any real distinction between the various types of crystalline stones. His outlook was primarily physical (or mechanical), rather than chemical and mineralogical.

In his second paper, of 1723, Réaumur took up the problem again, and considered the problem of why so many pebbles have rounded surfaces. He seems to have been attempting to account for the nature of what we are accustomed to call concretionary nodules, and this was done within the framework of the mechanical mode of explanation. Thus, Réaumur suggested that the deposition of stony matter might have occurred by a kind of sieving of the lapidifying juice. The reader was invited to imagine that some portions of a bed of rock were fortuitously more earthy and less permeable than other portions. Consequently, there might be pores so fine that they could only be penetrated by aqueous corpuscles, so that the passage of the lapidifying juice would result in the deposition of *matière pierreuse*. Perhaps the deposition of minerals and metals might be accounted for in similar fashion—though Réaumur made no attempt to inform us how the different metallic or mineral corpuscles were formed in the first place, nor how they differed in shape or size from stony corpuscles.

In his third memoir, Réaumur took up the problem of the nature of the earth that supposedly acted as a basis for the deposition of the material from the lapidifying juice. He also endeavoured to find out something about what actually happened when the deposition of the stony material occurred. Here a genuine attempt was made to treat the problem experimentally, by contrast with the variety of largely speculative approaches that we have encountered hitherto.

Earth, Réaumur argued, could not merely be finely divided sand, for there was no swelling when water was added to sand, whereas there was an increase in volume when water was added to earth. Again, earth settling in water did

[41] R. A. F. de Réaumur (40) (*Année M.DCC XXI*), 258.

so in a flocculent fashion, but this was not the case with very finely powdered sand. Such sand exhibited little cohesion when moistened with water, whereas earth behaved like a tenacious paste. These experiments, which offer some of the first published descriptions of experimental investigations in the field of sedimentary petrology, suggested to Réaumur that the corpuscles of earth were porous (being apparently distensible by the penetration of aqueous corpuscles), whereas the sandy particles, having little tendency to adhere with each other, must be solid and angular. The addition of water to a dry earth might bring about an interlocking of corpuscles, accounting for the cohesive properties of moist earth. Nevertheless, the corpuscles were probably in some measure flexible, as suggested by the ductility of moist clay. Réaumur also entertained the possibility that the mutual adhesion of earthy corpuscles might be due to the existence of certain flat polished surfaces. He referred to the mutual adhesion of flat polished surfaces of marble that apparently cling together. Clearly, no definite decision could be reached concerning the nature of the ultimate particles of earth on the basis of his experiments, but it is significant that he was making some effort to carry out an experimental enquiry into the nature of the hypothetical earthy particles.

Réaumur also made some attempt to determine experimentally whether any permanent change to the earthy corpuscles was brought about by the lapidificatory process. He showed that when a soft stone was crushed and treated with water, the product behaved very much like ordinary earth. By contrast, the finely ground particles of a flint had properties very different from those of ordinary earth. Hence, it seemed that the properties of the earthy corpuscles were permanently changed when the process of lapidification occurred. Réaumur suggested that the crystalline matter might have penetrated the earthy pores, rather like the acid spicules of Lemery or Hartsoeker[42] penetrating their porous alkaline counterparts. It should be remembered, of course, that for the members of the Lemery (or Cartesian) school, this was the way in which *chemical* reactions occurred, and thus the process of lapidification should, for Réaumur, have had certain chemical features.

For the chemist of the early eighteenth century, fermentation was one of the chief kinds of chemical reaction, and the effervescence that occurred between acids and alkalis was a prime example of a fermentatory process. Fermentation was generally accompanied by the production of some odorous substance, and hence the production of an odour was an indication that some kind of chemical reaction was taking place. Réaumur was familiar with the odour evolved

[42] Lemery (32), and also N. Hartsoeker, *Conjectures Physiques*, Amsterdam, 1706, 102. Réaumur's suggestion is made in his *Année M.DCC XXI*, 259.

when rain falls on earth after a long dry spell,[43] and this suggested to him that the interposition of aqueous corpuscles within porous earthy particles was a kind of chemical change. By analogy, then, it might be said that the lapidification of an earth was a chemical as well as a purely mechanical process.

It will be seen from the foregoing account that it was scarcely possible, in the early eighteenth century, to make quantitative applications of the principles of mechanics to mineral substances, and I have found virtually no examples of such attempted applications. A passage by the Swiss naturalist, L. Bourguet, does, however, give a slightly more quantitative account of the process of formation and action of the lapidifying juice of Geoffroy and Réaumur, and the subsequent precipitation of mineral matter:

> Les particules de tous ces Corps solides [minerals and metals] se séparent alors aisément, & leur pésanteur décroit en Raison triplée de la diminution de leur Diametre, à mesure que leur superficie ne décroit qu'en Raison doublée, ce qui fait que des Menstrues soutiennent toutes ces particules & ne composent qu'une même masse liquide. Mais cet état n'est point permanent. Les Molecules qui ne sauroient perdre entierement leur figure, se rallient peu à peu dans le fluide même, lorsque son mouvement se rallentit, ou qu'il s'en exhale une partie, & leur pésanteur augmente dans les mêmes proportions. Elles se précipi tent alors, ou en forme de poudre, ou elles se crystallisent & tombent ensuite au fond ou s'attachent aux côtés du vase qui les contient.[44]

Yet this did not amount to anything approaching a formal mathematical treatment of mineral formation, and one must wait for the work of Haüy at the end of the eighteenth century before one finds a reasonably adequate mathematical account of crystalline minerals.

It may be noted that Hélène Metzger placed Bourguet in a Newtonian context[45] (and indeed the passage just cited can more easily be reconciled with Newtonian than with Cartesian ways of thinking), but it is interesting to find that he also showed some attachment to the earlier "organic" world-view. Thus while he invoked particles of different shapes to act as building-blocks

[43] R. A. F. de Réaumur (40) (*Année M.DCC XXX*), 280. It is noteworthy that the odour that is evolved from the earth when rain falls after a long dry spell has only recently seriously attracted the attention of chemists, armed with their modern analytical techniques, although it has for long been considered by poets to be a phenomenon of significance. See Magnus Pyke, "Smells Like Rain", *The Listener*, 4 November, 1971.

[44] L. Bourguet, *Lettres Philosophiques sur la formation des sels et des crystaux. Et sur La Génération & le Méchanisme Organique des Plantes et des Animaux; A l'Occasion De la Pierre Belemnite et de la Pierre Lenticulaire. Avec un Memoire sur la Theorie de la Terre. Par Mr. Bourguet*, Amsterdam, 1729, 41.

[45] H. Metzger (4), 103–4. However, Bourguet's cosmological system was, I think, essentially Cartesian in character. (See below, note 53.)

for crystals of various forms (triangular prismatic molecules for rock-crystal, cubes for rock-salt, parallelipipeda for vitriols, *etc.*) he also maintained that: "Ces corpuscles doüés chacun d'une activité vitale convenable à sa figure, entrent dans la composition de tous les Amas qu'il y a dans le Monde materiel".[46] This position seems to have certain hylozoist features, reminiscent of the mineral doctrines to be found in the period of Renaissance naturalism, and it is indeed noteworthy that such ideas are to be found in association with what in other respects appears to be a strongly mechanical standpoint. It should be noted that Bourguet was apparently an exponent of the doctrine of the Great Chain of Being,[47] so that it is not altogether surprising to find evidence for a belief in organic features even among the lowly members of the mineral kingdom, and, as I have pointed out previously,[48] one may find evidence for this approach being followed until quite late in the eighteenth century. Nevertheless, it was an attitude of decreasing popularity.

The "mechanical" world-view can be found permeating most mineralogical writings of the early eighteenth century (other than the straight "natural histories" of minerals of that period), but it came under strong attack from the chemists of the Stahlian school, indeed from Stahl himself in the preface of his influential *Fundamenta Chymiae*,[49] though in fact Stahl's own doctrines reveal, on inspection, certain "mechanical" features,[50] with his very phlogiston having its own characteristic (gyratory) motion.[51] So far as I am able to discern, however, these mechanical features of Stahl's theories did not have much influence on the subsequent development of mineralogy. The Stahlian mineralogist, J. F. Henckel, made a special point of disavowing the use of

[46] L. Bourguet (44), 66.

[47] *Ibid.*, 58.

[48] D. R. Oldroyd (1).

[49] G. E. Stahl, *Fundamenta Chymiae Dogmaticae & Experimentalis* . . ., Nurenberg, 1723, preface (pages not numbered). For example, Stahl wrote: "Sed . . . Mechanica Philosophia, quamvis omnia summa se perspicuitate explicare glorietur, audacius sese applicuit ad Chymico-physicam rerum contemplationem. . . . Haerent plerumque dubia, intactoque nucleo cortices rerum & superficiem lambit, ex particularum figura & motu phaenomenorum rationes abstractas sane & generalissimas deducere contenta, quid mixtum sit quid compositum & aggregatum corpus, quaeque eorum natura, proprietates & discrimina sint, incuriosa. Atque hinc quidem tot infelices phantasiae chimaeras, tot vanas mutilasque applicationes in Chymia exstitisse apparet".

[50] I have discussed this aspect of Stahl's theories in a recent paper: "An Examination of G. E. Stahl's *Philosophical Principles of Universal Chemistry*", *Ambix*, **20**, 36–52, 1973.

[51] *Georgii Ernesti Stahlii, Experimenta, Observationes, Animadversiones, CCC Numero, Chemicae et Physicae* . . ., Berlin, 1731, No. CXLIII, 196–7. ("*Forma* haec, nihil aliud est, quam peculiaris plane, species *Motus*; tenerrimi, citatissimi, ultra propemodum, quam conceptu assequi licet: *Verticillaris*, seu circa Centrum materialium corpusculorum, illa rapidissime pellentis, & circumagentis").

hooked or rounded particles, or subtle aetherial corpuscles, in his theoretical schemes.[52]

As a result of such pressures, we find that by the middle of the eighteenth century, simplistic mechanism of the Cartesian variety was in a state of retreat, and the last directly Cartesian theory of the Earth with which I am familiar was put forward by Bourguet in 1742.[53] Readers will recall the well-known repudiation of the naïve application of the mechanical philosophy to chemistry, given by Venel in the *Encyclopédie*.[54] Similarly, mineralogists realized that mechanism alone was unequal to the task of providing a satisfactory mineral theory. The anti-mechanical Stahlian chemistry originated in Germany and spread to Sweden, France, Britain, and other countries. It was a chemistry of principles. For a Stahlian chemist (or mineralogist), the properties of salts were not to be explained in terms of corpuscles of this shape or that. But the assertion that a salt was composed of aqueous and earthy components was readily understood, and apparently confirmed by experience. The natures of the components would account for the properties of the substances of which they were constituents. And mineralogy, like chemistry, should be based on observations, and chemical principles, not on idle fancies about the invisible corpuscular structure of minerals. In mineral analysis, one should look for the proper chemical constituents of bodies, extracted in the laboratory, and one should eschew metaphysical corpuscular abstractions.

I will not here chart the rise of chemical mineralogy among the German and Swedish (and later the French) phlogistic mineralogists, for I have treated this topic elsewhere.[55] But it was from this school in particular, despite its somewhat unpromising beginnings,[56] that mineralogy grew to become a proper science in the last quarter of the eighteenth century. It was in this Stahlian school that chemists such as A. S. Marggraf worked, laying the basis for chemical

[52] J. F. Henckel, *Flora Saturnis, ou Preuves de l'Alliance qui existe entre le Regne Végétale: et le Regne Mineral: Tirés de l'Histoire Naturelle et de la Chymie . . . Par Jean-Frederick Henckel, Docteur en Medecine & Conseiller des Mines de S.M. le Roi de Pologne, Electeur de Saxe. Traduit de l'Allemand par M. Charas. Apothicaire de Paris*, Paris, 1760, 9.

[53] L. Bourguet and P. Cartier, *Traité des Petrifications. Avec Figures*, Paris, 1742.

[54] D. Diderot and J. L. D'Alembert (eds.), *Encyclopédie, ou Dictionnaire Raisonné des Sciences, des Arts et des Métiers, par un Societé de Gens de Lettres*, 17 vols. (with supplements and tables), Paris 1751–72, III (1753), 415–6.

[55] D. R. Oldroyd, "Some Phlogistic Mineralogical Schemes, Illustrative of the Evolution of the Concept of 'Earth' in the Seventeenth and Eighteenth Centuries", *Annals of Science*, in press.

[56] The Phlogistic mineralogy, referred to in the previous note, grew from the primitive chemical ideas to be found in Becher's *Physica Subterraneae*, a leading feature of which was a highly speculative account of the origin of our globe, along lines reminiscent of the Hermetic tradition.

methods of mineral analysis,[57] which later blossomed into the analytical work of Bergman, Kirwan and Klaproth towards the end of the century. By and large the Newtonian chemistry which proved so popular and influential in the eighteenth century was not of much value to mineralogists until it became unified with the Stahlian chemistry. This happened in the work of Bergman. As Thackray has pointed out,[58] Bergman was strongly influenced by the Newtonian chemical paradigm, and devoted much of his chemical career to elaborating and improving the standard affinity tables, which generally had a Newtonian theoretical basis. But to draw up his affinity tables, Bergman must have familiarized himself with an enormous number of chemical reactions, which were then concisely summarized by means of these tables. It is, I think, no coincidence that it was only a couple of years after the publication of Bergman's notable work on affinities[59] that he first published his general scheme of mineral analysis.[60] The clarification of ideas necessitated by the drawing up of affinity tables was, I think it is reasonable to suggest, a prerequisite for the working out of a general analytical method.

It will be noted that in this paper I have largely concentrated on ideas about the genesis of mineral substances, rather than their form and classification. But it was largely this aspect which attracted "mechanical" explanations. (It was also the focus of attention for the earlier "organic" explanations.) This is not to say that there was no progress in methods of mineral classification during the period that has been under discussion. There was, of course, and it is interesting to trace the development of schemes of mineral classification through the seventeenth and eighteenth centuries.[61] What we find, in fact, when such an exercise is carried out, is that the schemes show virtually no trace of the influence of the mechanical philosophy. If we consider the early

[57] See A. S. Marggraf, *Opuscules Chymiques de M. Margraf* [*sic*] *de l'Académie de Berlin*, 2 vols., Paris, 1762. His account of the reactions of the aluminous earth, for example, laid out the chemistry of this substance in a masterly fashion, revealing reactions of precipitation and solution that could be utilized in later schemes of mineral analysis ("Expériences faites sur la terre de l'alun", II, 111–75.)

[58] A. Thackray, *Atoms and Powers An Essay on Newtonian Matter-Theory and the Development of Chemistry*, Cambridge (Mass.) and Oxford, 1970, 218–21.

[59] T. O. Bergman, "De Attractionibus electivis disquisitio", *Nova Acta Regiae Societatis Scientiarum Upsaliensis*, **2**, 159–248, 1775.

[60] This was contained in the paper: "Disquisitio Chemica de Terra Gemmarum", *ibid.*, **3**, 137–70, 1777.

[61] For an account of some of the more important of such systems see C. S. St. Clair, *The Classification of Minerals: Some Representative Mineral Systems from Agricola to Werner*, Ph.D. thesis, University of Oklahoma, 1965. See, also, a tabular representation of the schemes of mineral classification that were put forward in the middle years of the eighteenth century, which I have drawn up for my paper: "A Note on the Status of A. F. Cronstedt's Simple Earths and his Analytical Methods", forthcoming in *Isis*.

seventeenth-century systems, such as those of Boethius de Boodt,[62] we find classification based on a variety of external characteristics, utilizing a series of dichotomies, which gave a taxonomic system somewhat resembling a "key" of a modern biological handbook. De Boodt did not link his system with what one might recognize as mineralogical "natural kinds". Later schemes, such as that of Walter Charleton,[63] did begin to incorporate clearly recognizable "natural kinds", such as "metals", "bitumens", "salts", *etc.*,[64] and by the time of publication of Woodward's system in 1704,[65] we have a fairly elaborate "natural history" of the mineral kingdom. The publication of such schemes continued in England, and elsewhere, until the middle years of the eighteenth century.[66] They did not display much emphasis on chemical criteria.[67]

By contrast, in Sweden, a tripartite division of stones was adopted, according to their behaviour on heating,[68] and the categories of salts, sulphurs and metals (which stemmed from the work of Avicenna[69]) were also used. Variations on this theme were adopted by various writers in Germany and Sweden until the 1760s or even later.[70] Finally, with the rise of chemical methods of mineral analysis in the "humid" way, we get systems such as those of Monnet,[71]

[62] A. Boethius de Boodt, *Anselmi Boetii de Boodt Brugensis Belgae . . ., Gemmarum et Lapidum Historia . . .*, Hanover, 1609. Here fold-out tables were printed, giving succinct representations of de Boodt's systems of classification.

[63] Gualteri Charleton, *Onomasticon Zoicon, Plerorumque Animalium Differentias & Nomina Propria pluribus Linguis exponens. Cui accedunt Mantissa Anatomica; Et quaedam De Variis Fossilium Generibus*, London, 1668.

[64] The use of such categories may be traced back to the work of Avicenna in the eleventh century. See *Avicennae de Congelatione et Conglutinatione Lapidum being sections of the Kitâb al-Shifâ'. The Latin and Arabic texts edited with an English Translation of the latter and with critical notes by E. J. Holmyard . . . and D. C. Mandeville*, Paris, 1927, 33.

[65] This appeared in: J. Harris, *Lexicon technicum, or an Universal English Dictionary of Arts and Sciences . . .*, London, 1704, under the head "Fossils".

[66] See, for example: J. Hill, *A General Natural History, or New and Accurate Descriptions of the Animals, Vegetables, and Minerals of the Different Parts of the World; . . .*, Vol. I, *A History of Fossils*, London, 1748; E. Mendes da Costa, *A Natural History of Fossils*, London, 1757.

[67] Hill investigated minerals by treating them with water, and by heating them, but he was primarily concerned with external appearances.

[68] See M. von Bromell, *Mineralogia, eller Inledning til nödig kundskap at igenkianna och uppfinna Allahanda Berg-Arter/Mineralier, Metaller samt Fossilier, Och huru de måge til sin rätta nytta anwändas . . .*, Stockholm, 1730. Here among a number of other categories, we find listed: stones that are refractory, those that form a powder on heating, and those that are vitrescible. These correspond to the "*Apyri*", "*Calcarii*" and the "*Vitrescentes*" of Linnaeus's *Systema Naturae* of 1735.

[69] The *tria prima* of the Paracelsians was clearly an additional consideration.

[70] See note 61.

[71] A. G. Monnet, *Nouveau Système de Minéralogie, ou Essai d'une nouvelle exposition du Regne minéral, auquel on a joint un Supplément au Traité de la dissolution des Métaux, avec des observations relatives au Dictionnaire de Chymie*, Paris, 1779.

Bergman[72] and Kirwan,[73] which were based upon chemical composition, according to the simple earths that were present.

But the mechanical philosophy was not able to make any significant contributions to these developments. It did not answer the problems of the mineralogist, any more than it truly answered the difficulties of the chemist. On the other hand, it did succeed in displacing the "organic" view of the mineral kingdom, and in retrospect this appears to be a matter of importance, being a significant preliminary to the development of modern mineralogy. Thus we today, who still find the mechanical mode of visualizing and interpreting phenomena congenial, can recognize the significance of the "mechanization of the mineralogical world-picture", despite the fact that pure mechanism proved to be an insufficiently powerful theoretical mode of explanation, and was eventually superseded.

*Acknowledgements.* I am indebted to my colleague, Dr. R. M. Gascoigne, for his comments made on the first draft of this paper, and to the Editor's referee for comments made subsequently. I have incorporated a number of their suggestions in the final version of the paper. I am also indebted to Centaur Press Ltd., for permission to reproduce Figure 3 from their publication: *The Sacred Theory of the Earth By Thomas Burnet with an introduction by Basil Willey*, London and Fontwell, 1965, p. 58.

[72] T. O. Bergman, *Sciagraphia Regni Mineralis secundum principia proxima digesta* Leipzig and Dresden, 1782.

[73] R. Kirwan, *Elements of Mineralogy*, London, 1784.

# IV

## SOME PHLOGISTIC MINERALOGICAL SCHEMES, ILLUSTRATIVE OF THE EVOLUTION OF THE CONCEPT OF 'EARTH' IN THE 17th AND 18th CENTURIES

### 1. *Introduction*

In previous papers, I have discussed the influences of the 'organic' and the 'mechanical' world-views on the development of ideas concerning the mineral kingdom,[1] and have suggested that the change from 'organic' to 'mechanical' modes of explanation was most beneficial for mineralogical enquiries. Nevertheless, despite the advances made as this important theoretical shift occurred, it appears that mineralogy did not make really substantial progress until the last three decades of the eighteenth century, when chemical methods of mineral analysis (in the wet way), and the crystallographic schemes of Bergman, Romé de l'Isle and Haüy, were established. It does not follow from this, however, that no developments of any significance were taking place in mineralogical theory and practice in the middle years of the eighteenth century. On the contrary, important changes in emphasis were occurring during this period. The purely mechanical modes of explanation, stemming primarily from the Cartesianism of the seventeenth century, declined in favour, to be replaced by increasing emphasis on more strictly chemical interpretations, which came to the fore in Germany (and also flourished in Sweden) under the influence of the theories of Becher and Stahl. And, as is well known, these modes of explanation achieved considerable popularity in France in the middle years of the eighteenth century, providing the basis of chemical (and, we may add, of mineralogical) theory in that country, until they were eventually swept away by the revolutionary doctrines of Lavoisier and his colleagues.

Historians have given much attention to the evolution of chemistry in the middle years of the eighteenth century, and the main aspects of the phlogistonism of this period are known in considerable detail. On the other hand, less consideration has been given to the mineralogical theories

[1] D. R. Oldroyd, 'Some Neo-Platonic and Stoic Influences on Mineralogy in the Sixteenth and Seventeenth Centuries', forthcoming in *Ambix*; 'Mechanical Mineralogy', also forthcoming in *Ambix*.

of the phlogistonists. This, in a way, is somewhat surprising, for the phlogiston theory grew from the work of J. J. Becher, whose principal scientific work[2] was a characteristically seventeenth-century speculative 'theory of the Earth'. Moreover, most of the phlogistonists took a great deal of interest in the mineral kingdom, and were frequently associated with the mining industry. It is, therefore, a matter of some interest to examine the relationship between the mineralogical theories of our period and the chemical doctrines that were emerging at the same time. Thus, in the present paper I wish to look at the mineral theories of some of the leading phlogistonists in Germany and Sweden, and also to consider some of the mineralogical ideas that were being promulgated in France in the period immediately before the chemical revolution of Lavoisier. It will be a point of particular interest to consider the extent to which these phlogistic schemes display evidence of the influence of earlier 'organic' patterns of explanation. It will also be instructive to note examples of 'mechanical' thinking in the theories, in view of the fact that a number of phlogistonists strongly maintained that it was improper to invoke 'mechanical' explanations in chemistry or mineralogy. In a subsequent paper I hope to discuss the general effect of the Lavoisier revolution on the progress of mineralogy.

As examples of the 'phlogistic' school of mineralogy, I have selected for consideration the work of J. J. Becher himself (1635–1682), J. F. Henckel (1678–1744), J. H. Pott (1692–1777), and J. G. Lehmann (1719–1767) in Germany; J. G. Wallerius (1709–1785) and T. O. Bergman (1735–1794) in Sweden; and G. F. Rouelle (1703–1770) and A. Baumé (1728–1804) in France. Stahl was not primarily concerned with mineralogical problems or theories of the Earth,[3] so I shall say little about his work, assuming that his modifications and elaborations of the scheme initially advanced by Becher are familiar to the reader.

[2] J. J. Becher, *Actorum Laboratorii Chymici Monacensis, Seu Physicae Subterraneae Libri Duo, . . . Joannes Joachimus Becherus, . . .*, Frankfurt, 1669. This was republished by Stahl in 1703 with the following title: *Joh. Joachimi Beccheri, D. Spirensis Germani Sacr. Caes. Majest. Consil. & Med. Elect. Bav. Physica Subterranea Profundam Subterraneorum Genesin, è principiis hucusque ignotis, ostendens. Opus Sine Pari, Primum hactenus & Princeps, Editio Novissima. Praefatione utili praemissâ, Indice locupletissimo adornato, sensuumque & rerum distinctionibus. Libro tersius & curatius edendo, operam navavit & Specimen Beccherianum, Fundamentorum, Documentorum, Experimentorum, subjunxit Georg. Ernestus Stahl, D. Prof. Publ. Ordin. Hall.*, Leipzig, 1703. The last edition of this was published with the same title in Leipzig in 1738. I have used these editions of 1703 and 1738.

[3] He did, however, publish an essay on the origin of mineral veins: *De Ortu Venarum Metalliferarum*, Halle, 1700. An English version of this is to be found in a collection of three *Pyrotechnical Discourses* (London, 1705 and 1730), the title being: *A short Discourse on the Original of Metallick Veins; by George Ernest Stahl, M.D. which may serve as an Answer to Dr. Woodward's Theory of the Earth, and was a Forerunner to.* Here Stahl argues against the occurrence of substantial changes in mineral veins since their initial formation, which, so he believes, took place when the globe was first formed.

2. *J. J. Becher* (1635–1682)

Despite his important position in the history of science, Becher has received scant attention from English-speaking historians, apart from the considerable amount of information that is contained in the second volume of J. R. Partington's monumental history of chemistry.[4] Here, I offer an analysis of such sections of the *Physica Subterranea* as have a particular bearing on mineral theory. It is an eclectic document, and many strands from earlier ways of thinking are found interwoven in this bizarre work. Nevertheless, a considerable number of ideas that came into prominence in the later eighteenth-century mineralogical and chemical theories may be found here in embryonic form.

Like other cosmogonists of the seventeenth century, Becher boldly attempts to give an account of the actual creation of the globe. Following the tradition recounted in *Genesis*,[5] he describes the separation of the Earth from an initial chaos, designated a *Panspermia*.[6] From this chaos separates the central globe, with overlying aqueous and aerial layers. With characteristically scholastic logic, Becher, argues that a *single* substance cannot give rise to a variety of products, so that at least two basic components for mixtion are required—an argument repeated *ad nauseam* by his successors. EARTH[7] and water are selected as these fundamental entities, air being deemed to act as an *instrument* (or agent)

[4] J. R. Partington, *A History of Chemistry*, 4 vols., vol. ii, London, 1961, pp. 637–652. See also P. Duhem, *La chimie est-elle une science française*? Paris, 1916, ch. 5.

[5] J. J. Becher, *op. cit.*, (1738), p. 10. It is a moot point as to the weight that should be allotted to the Hermetic tradition. As Frances Yates has pointed out (*Giordano Bruno and the Hermetic Tradition*, London, 1964, ch. 2), the Mosaic and Hermetic accounts of the origin of the cosmos have much in common.

[6] J. J. Becher, *op. cit.* (1738), p. 11. The designation of the original chaos as a *Panspermia* suggests Stoic roots, but may also have been transmitted to Becher via the Hermetic tradition.

[7] Becher uses the work *terra* throughout the *Physica Subterranea* in three senses: as a hypothetical elementary principle (perhaps corresponding to our concept of the solid state) with three subdivisions; as a substance that one may dig from the garden or otherwise extract from the ground; and as the terrestrial globe. I have tried to keep these senses separate by using 'EARTH' for the elementary principle, 'earth' for common or garden earth (or later as an empirically identifiable earthy component extractable from a mineral substance), and 'globe' for the planet Earth. When this is done, the evolution of the concept 'earth', through the writings of the mineral theorists that are considered in this paper, becomes clearly evident. It does not seem to be necessary to make such distinctions for 'water'. The original notion of water as a 'principle' did not suffer the same fate as 'earth' and 'air', each of which gradually came to be divided into a number of distinct, empirically identifiable substances during the course of the eighteenth century.

of chemical change, but not as a chemical component.[8] 'Heaven' (*coelum*) and fire are also stated to be agents (or instruments) of mixtion. The 'heaven' acts on the air, air acts on water, and water acts on EARTH, in a step-wise process of motion and mixtion, with gradual change from a state of subtlety to crassitude. The process is reminiscent (and doubtless ultimately derived from) the Stoic account of the evolution of the Cosmos, with a gradual successive separation of air, water and EARTH from fire, with accompanying decrease in tension.[9] For Becher, only the water and EARTH, at the completion of the process of separation from the initial chaos, can form material mixts.

Following the period of creation, Becher envisages a stage of '*productio*'. The most suitable word for the translation of this is not immediately obvious. 'Production', 'extension', or 'development' are possibilities. Alternatively, more overtly 'organic' terms may be suggested, such as 'procreation', 'generation', or 'bringing up' (in the sense of bringing up a family). It is perhaps safest to leave the word as '*productio*' in our discussions, not forgetting its possible 'organic' overtones.

The mixts formed at the time of '*productio*' can be of three kinds:

1. EARTH + Water —which supposedly give rise to stars, animals and plants, and also a number of minerals;
2. Water + Water —which supposedly give rise to meteorological products such as snow, frost, dew, hail and ice; and
3. EARTH + EARTH—which supposedly give rise to earths, stones and metals.

As is well known, however, three types of EARTH are postulated: vitrescible, fatty, and mercurial, the first two being found in stones, while all three occur in metals. Here, then, we have the germinal idea of Stahl's phlogiston theory.[10] Supposedly, the first EARTH is the 'mother' of stones, the second the 'mother' of earths, and the third

[8] J. J. Becher, *op. cit.*, (1738), p. 19. This is clearly a precursor of the element/instrument theory of Stahl, in which EARTH, water, air and fire could all function as physical agents, whereas only EARTH and water were deemed to be capable of functioning as components of mixts. For G. F. Rouelle (see below, pp. 40–46), all four elements could play both rôles. (See R. Rappaport, 'Rouelle and Stahl—the Phlogistic Revolution in France', *Chymia* 1961, **7**, 73–102).

[9] Cf. E. V. Arnold, *Roman Stoicism* . . . , Cambridge, 1911 (re-issued London, 1958), pp. 195–196.

[10] The word 'phlogiston' has, of course, much earlier antecedents. But the theory of the three EARTHS (though doubtless inspired by the *tria prima* of Paracelsus) is, so far as I am aware, Becher's original contribution—a new version of an old tradition. Stahl's theory is an obvious modification of Becher's.

the ' mother ' of metals. Moreover, the first ostensibly is distributed throughout the globe, while the other two arise at its centre and in the sea. If we tentatively identify the second and third EARTHS with descendants of the dry and moist exhalations of Aristotle's *Meteorologica*[11] (*viz*, the sulphureous and mercurial exhalations of the alchemists), then it is clear why Becher should have asserted that they arise at the centre of the globe. The reason for postulating the sea as a source is not so readily apparent, though it will be remembered that Paracelsus believed that metals were the fruits of the ' matrix ' water.[12]

Linking the theory more closely with laboratory experience, Becher suggests that the first EARTH is to be found in (or lies hidden with) spirit of vitriol, or even in alkali; the second occurs in spirit of nitre and certain oils; while the third occurs in spirit of salt, vinegar, soot and urine. It would appear that, so far as the acids are concerned, Becher is relating his hypothetical EARTHS to the three common mineral acids, according to their relative volatilities.[13]

As is well known, according to exponents of the chemistry of ' principles ', each ' principle ' supposedly conferred certain qualities upon the substance of which it was a component. Thus, says Becher, the first EARTH is responsibility for fusibility and solidity, diaphaneity and opacity; the second imparts colour, taste and odour; the third confers metallic form and the properties of malleability and ductility. It is subtle variations in these three principles that produce differences in the mixts of which they are constituents. Thus, the first principle may on occasions be vitreous or calcareous, and give rise to diaphanous glasses or opaque limestones accordingly, but we are not informed as to how or why these variations of the first EARTH occur as they do.[14] Apparently, there is less variation in the third principle than in the other two, for, in Becher's view, there is only one perfect metal—gold. Thus the different species of metals should be ascribed to modifications of the first and second EARTHS, rather than variations in the third.[15] Yet

[11] Aristotle, *Meteorologica,* translated by H. D. P. Lee, London and Cambridge (Mass.), 1952, pp. 29, 287 and 289.

[12] *E.g.,* A. E. Waite, *The Hermetic and Alchemical Writings of Aureolus Philippus Theophrastus Bombast, of Hohenheim, called Paracelsus the Great. Now for the first time faithfully translated into English,* 2 vols., London, 1894, vol. i, pp. 92–95 and 241.

[13] Clearly, we have here no direct association of the second EARTH with what we now term reducing agents. Later, Stahl, largely abandoning the third EARTH, saw soot and urine as sources of the second fatty EARTH, or *terra pinguis*—his phlogiston.

[14] Indeed, if Becher had embarked on such an account, he would soon have become entangled in an infinite regress of causal explanation.

[15] J. J. Becher, *op. cit.,* 1703, p. 536.

Becher's doctrine is not perhaps wholly consistent, for elsewhere[16] he seems to suggest that the observed diversity of earthy substances requires that there be more than one component principle—which may suggest that he believes that the diversity is due to the various *coalitions* of the principles, rather than variations in the natures of the *individual* constituent principles.

Becher suggests[17] that he is able to synthesize artificial stones and metals directly from his three EARTHY principles, so it is a matter of considerable interest to ascertain what he had in mind when writing of these three principles. The first is described as the mother and receptacle of the other two. Moreover, its presence is taken by miners to be an indication of a productive ore-body, or rich ore deposits. Commonly it is called '*zarter Kalch*' or '*Schi[e]fer*'.[18] Elsewhere,[19] Becher calls it '*Quartz*', and it may be that he envisages these various substances as being varieties of the first EARTH, with quartz being the closest available approximation to the pure principle. He maintains that this EARTH serves both as a womb (or matrix) and as an ingredient for the generation of stones and metals, and he asserts that the processes of mineral and metal formation continue to this day. With a scholastic style of logic-chopping, he argues that the material of stone cannot be stone itself, but only 'potential stone'. Mere juxtaposition cannot bring a stone into being, for if this were the case, all sands would promptly turn into stone. This discussion of potential stone scarcely seems to be reconcilable with the earlier treatment of quartz as a close approximation to the first EARTH—but Becher's writings have never been particularly noted for their internal consistency!

Elsewhere, I have drawn attention to the prevalence of Neo-Platonic notions in a number of seventeenth-century mineralogical writings.[20] Becher, too, seems to make use of such concepts. The '*idea*'[21] of the first EARTH may be expressed in the form of a most subtle and penetrating spirit, which, so it seems, may permeate the globe, nourishing and

[16] *Ibid.*, p. 115.

[17] *Ibid.*, pp. 117–118. This is a curious feature of Becher's theory. Clearly, it is his *ambition* to obtain his three EARTHS separate from each other, so that they can be used for synthetic purposes. In this respect, then, his aim is much the same as that of his successors. On the other hand, it is also clear, from his subsequent treatment of the question, that he has not succeeded in identifying any genuine elementary components of mineral substances.

[18] *Ibid.*, p. 121.

[19] *Ibid.*, pp. 123, 126 and 128.

[20] See note 1.

[21] J. J. Becher, *op. cit.*, 1703, p. 128. Here Becher transliterates the Greek ἰδέα directly into the Latin, as '*idea*'.

increasing ores and mineral veins. But Aristotle is not to be forgotten either. Two other exhalations are mentioned, which may become fixed in air and, adhere to the walls of mines.[22] In addition, certain Stoic ' seminal ' doctrines appear to be incorporated in Becher's theory, for we recall that in his account of the creation there is an initial separation of air, water and EARTH from the chaos. This is followed by the '*productio* ', in which individual substances are generated from EARTH and water. The ' most special ' principle for the formation of a body is deemed to be its specific semen, or seed, which is itself held to be constituted of EARTH and water.[23] The ' organic ' mode of thinking and the general electicism of Becher's theories are clearly apparent.

In his discussion of the second EARTH, Becher uses some curious arguments to justify its presence in minerals, or other terrestrial bodies. The first EARTH is held to be in plants, as well as in minerals. It can be recognized as the ash left after the calcination of plants, and is evidently a vitrescible principle, it being responsible for the fusibility of glasses. By considerations of symmetry, then, there should be a similar EARTH in animals—though it would be improper to publish the deeper mysteries of the EARTH of animals![24] Nevertheless, Becher assures us that man himself contains the vitrescible EARTH as his hypostasis. This point being established, the analogical argument could be put into reverse. By the ' analogy of nature ', the fat EARTH—the *terra pinguis*, clearly evident in animals and plants—should also be present in ' fossils ', a thesis supposedly confirmed by the inflammability of many mineral substances, metals themselves being calcinable in some instances.

The second EARTH is believed to be rarer and more humid than the first, and became so at the time of '*productio* '. It is analogous to the flesh and blood of an animal body. Thus, for Becher, the dry first EARTH of minerals, their hypostasis, is analogous to the dry bones of an animal; the second, more humid, EARTH may be compared with the covering of moist flesh and blood,[25] and it may also play the rôle of the soul of the body. The reader, however, must know that the second EARTH of ' fossils ' is only an analogue of the second

[22] One should not, perhaps, read too much into Becher's text, but one could put it this way: the first EARTH also has a related ' spirit ', expressing its ' idea '; likewise there are two other exhalations in the globe, corresponding to the dry (sulphureous) and moist (mercurial) exhalations of the alchemists, and (no doubt) expressing the ' ideas ' of Becher 's second and third EARTHS. This gives the whole scheme a nice symmetry, proper justice being done to several traditional theories.

[23] J. J. Becher, *op. cit.*, 1703, p. 114. EARTH and water are the primary principles. The seeds (themselves made of EARTH and water) are secondary principles (*principia principiata*).

[24] *Ibid.*, p. 132.

[25] *Ibid.*, p. 136.

EARTH of animals and plants. It is certainly *not* identical with it, being better able to endure fire. Thus the oil of minerals—their soul—is much more durable than the oil of animals or plants.

Becher's second EARTH comes in various grades, according to its purity and density. It is *not* sulphur, which is held to be a *mixt* of the ' universal acid ' (water plus the first EARTH) and the second EARTH. The second EARTH may, however, take on various ' guises ' (Lat., *larva* = mask). It may appear as a smoke (*fumus*), and may be found in substances such as sulphur, arsenic and realgar. It may be found in water, which it tinges with a cuticle of various colours. It may appear (in combination with the third EARTH) as fire-damp (*Schwaden*) in mines, and the heat of mines may be ascribed to it. It is the ' inhalation ', or ' *Einwitterung* ',[26] of the miners, and also (in a more or less fluid condition) their ' *Bergguhr* '.[27] It can penetrate stones, and give them tincture, and even produce sulphur or metal in their interiors.

As for the third EARTH, it supposedly gives metallic form to substances. Thus, returning to the analogue of the foetus in the womb, the first EARTH is analogous to the bones, the second to the flesh and

[26] *Ibid.*, p. 145. The etymology of the word ' *Einwitterung* ' is of considerable interest. *Prima facie*, one might imagine that it is directly related to the modern German terms ' *Einwitterung* ', ' *Auswitterung* ', or ' *Witterung* ' (' weathering down ', ' weathering away ' or ' efflorescence ', and ' weathering '). But, as F. D. Adams has pointed out (' Origin and Nature of Ore Deposits: An Historical Study ', *Bulletin of the Geological Society of America*, 1934, **45**, 375–424), in the seventeenth and eighteenth centuries the German miners used the term for hypothetical exhalations, supposedly to be seen on occasions rising from ore bodies, mines, or from the ground. This tradition doubtless derives ultimately from the Aristotelian doctrine of exhalations, though it may also have had an empirical basis in the phenomena of fire-damp or coruscations arising from the ground. It should be noted, however, that this miners' term, now obsolete, is not related to the modern notion of ' *Witterung* '. It is related to the largely forgotten English words ' withering ', ' widder ', and ' whyzt ', meaning ' wind ' or ' blast of wind '. These terms supposedly derive from the Old Norse word, ' *huitha* ', meaning ' squall '. Today, we remember it (in the northern dialect) from the title of Emily Bronte's novel—*Wuthering Heights*. The stated identity of Becher's ' *inhalationes* ' and ' *Einwitterungen* ' is thus readily accounted for.

The doctrine of inhalations and exhalations, as understood by the German miners of the seventeenth and eighteenth centuries is exemplified by the following passage from J. F. Henckel's *Pyritologia* (see note 33a, below): ' . . . and the matrix is not to be considered as a bare receptacle, but as a receptacle of proper materials too; that is, the inhalations, or impregnations, called *inweatherings*, cannot alone have their effect, nor their supplying their materials produce an ore, unless the matrix hold, and reciprocally furnish also, by means of an exhalation, called *outweathering*, its proper matter; that the passive may, by means of the agent, come into motion, to form the third body here intended ' (p. 79).

The theory of ore formation is evident from this passage, but one may suggest that the translator, in rendering literally ' *Einwitterung* ' and ' *Auswitterung* ' as ' *inweathering* ' and ' *outweathering*', has conflated two concepts, as we might easily do today.

[27] For a discussion of the meaning of the term ' *Bergguhr* ', see F. D. Adams, ' The Origin and Nature of Ore Deposits ', *Bulletin of the Geographical Society of America*, 1934, **45**, 375–424, and also my paper ' Some Neo-Platonic and Stoic Influences on Mineralogy in the Sixteenth and Seventeenth Centuries '. (See note 1.)

blood (and perhaps also the soul), while the third corresponds to the overall form. The third EARTH, however, primarily characterizes metals, and is responsible for their malleability and ductility. Possibly, the volatile aromatic distillates of plants are the vegetable analogues of this third EARTH; they provide the characteristic '*idea*' of the vegetable kingdom.[28] Similarly, the '*idea*' of animals may be detected by their destructive distillation, hartshorn (ammonium carbonate) providing a good example.

Some chemists call the third Becherian EARTH 'mercury', but in Becher's view this is a mistaken identification. Mercury itself is a mixt. Nevertheless, the third EARTH may properly be said to be mercurial. It has mercurial qualities, but is not itself mercury. It is exceedingly penetrating, and to be avoided in mines where it may cause fevers, phthisis and paralysis.[29] It is found in waters, and indeed the whole sea is full of it, it being the *primum ens* of salt. It is, perhaps, to be identified with the alkahest of the alchemists or philosophers.

Here, then, we have a sketch of the overall mineralogical and chemical theory of Becher. I will not list the ways in which he considered his four components of mixts (water and the three EARTHS) to have been combined in the various subterranean bodies. Some of these have been given by Rodwell[30] and have been published subsequently by Partington.[31] The eclecticism of Becher's scheme is noteworthy. Aristotelian, Hermetic, Alchemical, Neo-Platonic, Stoic, and Biblical threads are to be found ravelled in his argument. The doctrines of Paracelsus and Van Helmont likewise play a part in the overall scheme, and a place is found for the traditions of the German mine-workers. Nevertheless, despite the broad eclectic sweep of the theory, Becher makes little use of his complex conceptual scheme when he comes to develop his own system of mineral classification. For this, he uses external characters, in much the same way as his peers.[32]

The Stahlian scheme in many ways resembles that of Becher, except that less attention is given to the third EARTH, the first two being deemed sufficient to account for the properties of metals, while the second comes largely to play the rôle of a 'principle' of combustibility, though it also continues to confer colour and odour. For Stahl, the scheme is the basis of a chemical rather than a mineralogical theory; Becher's treatment is improved very considerably by the provision of a wealth of experimental support for the theory.

[28] J. J. Becher, *op. cit.*, p. 150.

[29] *Ibid.*, p. 155.

[30] G. F. Rodwell, 'On the Theory of Phlogiston', *Philosophical Magazine*, 1868, **35**, 1–32.

[31] J. R. Partington, *op. cit.*, (note 4), p. 646.

[32] J. J. Becher, *op. cit.*, (1738), pp. 238–252.

Let us now turn to an examination of some of the ways in which such doctrines were utilized by phlogistic mineralogists in the eighteenth century, remembering that they tended to view Becher's doctrines through the lens of Stahl's writings.

3. *J. F. Henckel* (1678–1744)

The work of the German mining engineer, J. F. Henckel,[33] provides a suitable example of the application of the phlogistic theories to mineralogical problems in the early years of the century. Although his initial training was in medicine (he was a disciple of Stahl at Halle), Henckel worked as a mining director at Freiberg in Saxony, the district that was subsequently to become celebrated as the home territory of Abraham Werner. Henckel was Councillor of mines to the King of Poland and Elector of Saxony, and he was also a member of the Berlin Academy. He spent about twenty years in Meissen, where he was active in the work of the porcelain industry.

Henckel is clearly a follower of the chemical principles of Becher and Stahl. He advocates a chemical approach to mineralogy,[34] follows Stahl in the rejection of the direct application of the mechanical philosophy,[35] and extols the virtues of empiricism.[36] Yet his account of the

[33] For Henckel's mineralogical opinions, I have used the following translations:
(*a*) *Pyritologie, ou Histoire Naturelle de la Pyrite, Ouvrage dans lequel on examine l'origine, la nature, les propriétés & les usages de ce Minéral important, & de la plûpart des autres Substances de même Regne: on y joint le Flora Saturnisans, où l'Auteur démontre l'alliance qui se trouve entre les Végétaux et les Minéraux; et les Opuscules Minéralogiques Qui comprennent un Traité de l'Appropriation, un Traité de l'Origine des Pierres, plusieurs Mémoires sur la Chymie & l'Histoire Naturelle, avec un Traité des Maladies des Mineurs & des Fondeurs. Par M. Jean-Frederic Henckel, . . . ,* Paris, 1760.
(*b*) *Introduction à la Minéralogie; ou Connoissance des Eaux, des sucs terrestres, des Sels, des Terres, des Pierres, des Minéraux, et des Métaux: Avec une Description abrégée des opérations de Métallurgie. Ouvrages Posthume de M. J. F. Henckel, publié sous le titre de Henckelius in Mineralogiâ redivus, & traduit de l'Allemand,* 2 vols., Paris, 1756.
(*c*) *Pyritologia: or, a History of the Pyrites, the Principal Body in the Mineral Kingdom. In which are considered its Names, Species, Beds, and Origin; its Iron, Copper, unmetallic Earth, Sulphur, Arsenic, Silver, Gold, original Particles, Vitriol, and Use in smelting. The whole compiled from a Collection of Samples; from visiting Mines; from an Intercourse and Correspondence with Naturalists and Miners; but chiefly from a Course of Chymical Enquiries.* . . . , London, 1762.
The *Pyritologia* was first published in Leipzig in 1725; the *Introduction* first appeared in Dresden in 1747; the *Flora Saturnizans* in Leipzig in 1722, and the *Opuscules* in Leipzig in 1744. The posthumous *Introduction* (*Henckelius in Mineralogiâ redivus*, 1747) contained a *précis* of his mineralogical lectures at Freiburg.

[34] J. F. Henckel, *op. cit.*, (note 33 b), vol. i, p. 4. Chemistry, he says, is the key to mineralogy.

[35] J. F. Henckel, *op. cit.*, (note 33 a), p. 9 (in the *Flora Saturnisans*).

[36] J. F. Henckel, *op. cit.*, (note 33 c), p. 5.

formation of pebbles has much in common with the mechanical schemes advocated by his French contemporaries,[37] and in following Becher, by attempting to trace the mineral history of the globe back to the time of the Creation,[38] Henckel far exceeds the constraints imposed by an empiricist methodology.

In his earliest mineralogical work, the *Flora Saturnizans* of 1722, the ' organic ' metaphor, often characteristic of mineralogical patterns of explanation since the Renaissance, is pressed so hard that it is, so to speak, pushed into reverse. Minerals and plants are held to be so similar in their constitutions that they may be thought of as ' brothers '. Thus, whereas earlier writers on the mineral kingdom had accounted for the formation and growth of minerals by the assumption that they were living organisms, Henckel believes that phenomena relating to the vegetable kingdom may be elucidated by an examination of the processes that occur in the very closely related mineral kingdom.

In his *Flora Saturnizans*, and again in his *Pyritologia*, Henckel (like Becher) attempts to trace the formation of minerals back to the period of the original creation of the globe:

> ' On the second day, the dry land, as Moses calls it, began to appear, out of the chaos, and acquired its form. Now from this, and particularly the third day, on which the earth was by an express *fiat*, made fit for the production of vegetables, it is scarcely to be doubted but that the earthy particles were hereby multiplied, and the tenderest of them formed, by means of the water, for vegetation; and the more dense, heavy particles, rendered in some measure fit for the production of ores and metals: from ever varying circumstances and causes, the several species of stones and ores were still more and more multiplied, and have appeared in the disorder and confusion we now find them.[39]

[37] J. F. Henckel, *op. cit.*, (note 33 a), pp. 414, 440 and *passim*. Cf., for example, E. F. Geoffroy, ' Sur l'Origine des Pierres ', *Histoire de l'Académie Royale des Sciences, Année M.DCC XVI. Avec les Mémoires de Mathématique & de Phisique, pour la même Année*, Paris, 1718, pp. 8–16. Both Henckel and Geoffroy envisage the formation of crystals or stones occurring when minute terrestrial corpuscles are brought together by the agency of water. On the other hand, Henckel's theory does seem to have considerably stronger chemical features. He makes much of the appearance of ' stones ' in urine that has been allowed to stand for some months; he considers that a ' calcareous earth ' is needed as a ' base ' for the formation of stones, serving to join together water and earth.

[38] See, for example, note 39 below, and: J. F. Henckel, *op. cit.*, (note 33 a), pp. 18–19.

[39] J. F. Henckel, *op. cit.*, (note 33 c), pp. 62–63. This passage is perhaps suggestive of ideas in Henckel's theories which might be construed as forerunners of the celebrated ' aqueous ' doctrine of A. G. Werner. Elsewhere (*op. cit.*, note 33 a, p. 438), Henckel refers to a thick, coagulated and gelatinous fluid, existing in a very soft state after the universal deluge—dense and very different from anything known today. This seems to be a likely forerunner of Werner's universal ocean, which should not, therefore, be thought of as being merely a scaled-up version of our modern oceans.

As at the period of Becher's '*productio*', the mixtion of water and EARTH occurs to form mineral and metallic matter. Henckel visualizes this as a kind of copulatory process:

> 'Earth is the mother and the passive; on the contrary, water, as being active and what must impregnate, may be called the father'.[40]

Yet there is some internal inconsistency in Henckel's account, for we are also told that 'minerals grow by a bare external accretion'.[41] And in a later work,[42] he gives a *cri du coeur*, bemoaning his ignorance concerning the origin of flints. Believing that pebbles are no longer being formed, so that he cannot investigate the actual processes of their formation, he is forced to push back their origination to the early period of '*productio*'. Here[43] he disclaims any intention of attempting to trace the history of the globe back to the Creation, and then blandly makes the conjecture that the Earth was originally soft and spongy after the separation of the waters, but subsequently became harder—when pebbles were formed. It would seem that Henckel's empiricism is at odds with his belief that the Mosaic account of Creation is pertinent to his enquiries.

So far as the actual use of the chemical theories of Becher and Stahl are concerned, we find Henckel following his predecessors quite closely. The hierarchy of principles, mixts, compounds, decompounds, and superdecompounds, is adopted, and an attempt is made to utilize the three Becherian EARTHS and water as principles.[44] For Henckel, as for Becher, the first EARTH is a hypostasis for the other two—it gives them body, and enables them to combine. The second EARTH, however, is not quite the same as Becher's, being closer to Stahl's version. It imparts inflammability, and metalleity. Like Stahl, Henckel does not have much to say about the third EARTH. It may be volatile, but is, perhaps, equivalent to the second sulphureous 'principle'. When lead calcines, one component EARTH forms a glass, another burns, and the third volatilizes and is dissipated. The proper relationship between these three supposed components and ordinary common or garden earths is, however, never really made clear. Thus, when discussing possible

[40] *Ibid.*, p. 63.

[41] *Ibid.*

[42] J. F. Henckel, *op. cit.*, (note 33 a), p. 410. (Originally published as: *Idea Generalis de Lapidum Origine* . . . , Dresden and Leipzig, 1734.)

[43] *Ibid.*, pp. 395–396.

[44] J. F. Henckel, *op. cit.*, (note 33 b), vol. i, pp. 36–37. It may be noted, however, that in the *Pyritologia* (p. 5) Henckel states: 'I attempt not to separate or resolve the *pyrites*, either into the four elements, the three, or other imaginary principles; as being things that can never be made objects of our senses . . .'.

methods for the classification of stones, Henckel considers that subdivisions might be made according to: the three kingdoms of nature, transparency, hardness, colour, commercial uses, medicinal uses, essential and intrinsic properties, and behaviour in the fire. He is aware that the action of the furnace does not yield a system of classification according to any recognizable 'natural kinds'. Evidently, he wishes to base his classification on essential and intrinsic properties, yet, in fact, the method actually adopted owes nothing to Becher's trichotomy of EARTHS. Thus he uses the effect of acids on stones as his main guide to their classification:[45]

(*a*) *Calcarei*

Limestones
Alabasters

(*b*) *Silicei*

Precious stones
Pebbles and sandstones

(*c*) *Calcareo-silicei*

Spath
Talc
*Glacies Mariae*
Mica
Stalactites
*Flos Martis* [a form of aragonite]

(*d*) *Limosi*

Slate
Gneiss
Marlstone

As might be expected, the phlogistic hypothesis is invoked to account for the processes of calcination and reduction, but it is clear that the matter theory of Becher and Stahl is of little real use in the classification of minerals and stones. Instead, simple chemical tests and external characters are used.

Perhaps Henckel's most interesting work, and certainly his most influential, was his *Pyritologia*, the first detailed memoir on a single class of mineral substances. Here, he inclines to the opinion that pyrites,

[45] J. F. Henckel, *op. cit.*, (note 33 b), pp. 65 ff.

like other minerals, are constantly being generated, though some deep-seated ores were doubtless created ' at the beginning '. The pyrites found in fossil shells (thought to have been emplaced by the Deluge) appeared to be good evidence for the formation of the substance long after the Creation.

It must have been clear to Henckel, from his knowledge of mining and smelting processes, that the constituents of pyrites are iron, sulphur, and possibly also arsenic, copper, silver and gold. How are these to be reconciled with the EARTH and water of Becher and Stahl? The answer is to relegate these latter to ' remote ' constituents, while sulphur, arsenic, etc., are to be brought forward as the 'proximate parts '. Yet Henckel clearly hankers after the pristine theoretical scheme of Becher, for we are also informed that sulphur is the ' fat earth ', arsenic the (volatile) 'mercurial earth ', while iron is the ' metallic earth '.[46] Henckel is impressed by the ease with which iron rusts, and suggests that it is the closest available approximation to the ' crude, undetermined earth itself '.[47] He would like to confirm his assertions by both analysis and synthesis, but this is only possible in a few cases, notably cinnabar.

Like many of his predecessors, Henckel makes use of the concept of an earthy matrix in which mineral substances may form. Becher's well-known experiment, in which iron was produced by heating clay with linseed oil,[48] suggests to Henckel that a loamy soil might be a suitable matrix in which pyrites might form.[49] But what are the substances which come together in this matrix? Here, Henckel has recourse to the ancient doctrine of exhalations, and, in part, retreats from the methodological maxim that analysis and synthesis should provide a proper understanding of the composition of a body. Analysis of pyrites reveals sulphur and metal, yet Henckel is not prepared to assert, without prevarication, that the mineralizing vapours that mingle in the matrices are ' straight ' sulphur and metal:

> ' The mineral damps, that lodge anywhere for mineralization, may happen to be of different *mixtions* and natures, yet, doubtless, in their particles, are still in an undetermined, and, as it were, *chaotical* state; in which, though caught and examined fresh, neither the *sulphur-earth* apart, nor the metal-earth; neither the dry, nor the *moist*, are separately distinguishable: either they are found in a kind of seminal state, wherein the parts of the body to be formed are not yet distinctly exhibited or elaborated, tho' they lodge *potentially* therein '.[50]

[46] J. F. Henckel, *op. cit.*, (note 33 c), p. 97.

[47] *Ibid.*, p. 100.

[48] J. J. Becher, *Experimentum Chymicum Novum, quo artificialis & instantanea Metallorum Generatio & Transmutatio ad oculorum demonstratur*, . . . , Frankfurt, 1671, pp. 36–37.

[49] J. F. Henckel, *op. cit.*, (note 33 c), p. 143.

[50] *Ibid.*, p. 145. (Cf. note 26.)

This reluctance to allow the proximate parts of pyrites to be common sulphur and iron doubtless arises from the failure to achieve a direct synthesis from these components in the laboratory. Moreover, the internal structure of nodules of pyrite (with their radiating or testaceous appearances) suggests to Henckel that they are produced by some process of coction, fermentation, or transmuting elaboration. In pyrites itself the sulphur may be compared to a ' juice ', and the iron-earth to a ' remaining dry body '.[51] Perhaps we have here a faint echo of Becher's second EARTH, that acted metaphorically as flesh and blood to the dry hypostasis of the first EARTH.

In his essay on the origin of stones, Henckel also makes some use of this doctrine of hypostasis and covering. Thus, the hypostatic matter of stones might be either a marl, a chalk, some intermediate between these, or a metallic substance. To these might be joined saline, oily, metallic, or sulphureous additions.[52]

Thus, I think it can be seen that although Henckel is not able to use the scheme of Becher and Stahl as a direct basis for his mineralogical theory, he draws on it for theoretical inspiration in a number of ways. It should be noted, however, that in the essay on the origin of stones (one of Henckel's later works, published in 1734 at a time when ' mechanical mineralogy ' was near its zenith) Henckel adopts quite a ' modern ' account of the process of crystallization, very similar to the views of his contemporaries in France; that is, an explanation in terms of the accretion of minute earthy corpuscles is advocated.[53]

4. *J. H. Pott* (1692–1777)

Next, I wish to examine some of the mineralogical doctrines of the German mineralogist, J. H. Pott, best remembered for his remarkable labours directed, at the instigation of Henckel, towards the discovery of the secret of the Meissen porcelain. His method was simply to heat together a selection of earths in various proportions, and to ascertain what happened at each trial. According to Partington,[54] some 30,000 experiments of this nature were attempted, but the crude empiricism of this approach did not solve the refractory problem. Partington has described the tables that Pott published at the end of his *Lithogeognosia*[55]

[51] *Ibid.*, p. 148.

[52] J. F. Henckel, *op. cit.*, (note 33 a), p. 430.

[53] Cf. note 37.

[54] J. R. Partington, *op. cit.*, (note 4), p. 718.

[55] J. H. Pott, . . . *Chymische Untersuchungen welche fürnehmlich von der Lithogeognosia oder Erkäntniss und Bearbeitung der gemeinen einfacheren Steine und Erden ingleichen von Feuer und Licht handeln*, vol. i, Potsdam, 1746, vol. ii, Potsdam and Berlin, 1751. I have used the French translation of Baron d'Holbach: *Lithogéognosie, ou Examen Chymique des Pierres et des Terres en Général,Et du Talc, de la Topaze, & de la Stéatite en particulier, Avec une Dissertation sur le Feu & sur la Lumière*, 2 vols., Paris, 1753.

(the work in which the results of this laborious series of experiments are described) as ' an early example of analysis tables ', but I confess that they do not appear to me in this light. If anything, they are ' tables of synthesis '. They do not seem to provide any scheme for the decompounding of a mineral body, though, to be sure, one might, by good fortune, chance on a substance that resembled the product of one of Pott's fusions.

Nevertheless, despite the crudity of Pott's basic mode of attack, his work is of considerable importance in that his fourfold classification of earths and stones is on a purely chemical basis, and includes perhaps the first clear recognition of silica as a distinct and special kind of earth:[56]

1. Alkaline and calcareous earths
2. Gypseous earths
3. Argillaceous earths
4. Siliceous earths.

Moreover, we see here a definite movement towards the recognition of common substances met in the field or in the laboratory as the fundamental earthy components of the terrestrial globe, despite the continuing official adherence to the doctrine of the three hypothetical EARTHY principles. Nevertheless, Pott was a convinced phlogistonist, and he followed his teacher Stahl in many respects, for example, by shifting the main cause of metalleity to the second of Becher's EARTHS. In his dissertation on the sulphurs of metals (an early work, published in Halle in 1716,[57] when he was under the direct influence of his celebrated mentor), Pott argues strongly for the presence of sulphur in metals. He claims authority for this opinion from the writings of many of his predecessors, including Paracelsus, Van Helmont, Sylvius, Becher, Hoffmann, and Stahl. But empirical proof may also be provided: silver is dissolved in *aqua fortis*, and spirit of salt is added. The resulting precipitate is then distilled with mercury, when, supposedly, cinnabar is formed, the sulphur ostensibly having been derived from the silver. It would appear that confusion arose through the identification of the red calx of mercury as cinnabar, but we observe the attempt to identify a ' principle ' with an empirically identifiable substance.

[56] *Ibid.*, vol. i, p. 7. (As we have seen above, Becher linked quartz with his first EARTH, though this was done in a tentative fashion.)

[57] Published in Latin, Berlin, 1738 (in *Exercitationes Chymicae*). It also appears in Demachy's translation of Pott's collected essays: *Dissertationes Chymiques de M. Pott, Professeur de Chymie, & Membre de l'Accadémie Royale des Sciences de Berlin* . . . , 4 vols., Paris, 1759, vol. i, pp. 1–132.

In a later work, Pott provides a detailed consideration of the phlogiston theory in an essay on the properties and effects of light and fire, given as an addendum to the *Lithgeognosia*.[58] The treatment is not entirely consistent. Supposedly, the inner forms of substances cannot be known—and there is particular difficulty for simple principles such as fire.[59] On the other hand, we are informed that the particles of phlogiston have an intestine, gyratory movement about their axes.[60] Light and fire are not one and the same. Light is perhaps an exceedingly fine and subtle matter. Its motion is progressive, rather than intestine. If the particles of fire are stationary, then no inflammation occurs. Thus coloured substances (such as metallic glasses) may contain phlogiston, without actually being inflammable. In passing from one substance to another, the fire particles become agitated, and consequently combustion occurs. Pott believes that water, rather than air, is a requisite for the production of a flame.[61] If an air-blast enhances a flame, this is simply due to the water mixed with the air. It is the water, escaping as vapour, and illuminated by the concomitantly escaping light, that produces the flame. If phlogiston is mingled with so much earth that an effective escape (and combustion) is not possible, then we will merely have a coloured stone, or coloured earth. Nevertheless, Pott believes that the escape of a little phlogiston renders the escape of further quantities easier (just as the escape of one prisoner may facilitate the get-away of others). In all this discussion, one sees a mingling of the pure Stahlian theory of combustion with what appear to be descendants of the Cartesian theories of matter, light and combustion. By contrast with much of his other work, Pott's essay on phlogiston does not have a very secure empirical base.

Phlogiston, for Pott, plays an important rôle in the mineral kingdom. It is present in coal, bitumen, sulphur jet, petroleum and amber. It is also present in metals, but is not very tightly bound in tin, which is why this metal becomes so hot on bending. Some bodies only melt on heating, rather than catching fire, because the phlogiston in them is particularly 'homogeneous'.[62] Phlogiston, in Pott's writings, is called upon to

[58] J. H. Pott, *op. cit.*, (note 55), 'Essai D'Observations Chymiques et Physiques, Sur les Propriétés et les Effets de la Lumière et du Feu', pp. 327–404.

[59] *Ibid.*, p. 328.

[60] *Ibid.*, p. 348. (This is a Stahlian doctrine. Cf. Stahl's *Experimenta, Observationes, Animadversiones, CCC Numero, Chymicae et Physicae*, Berlin, 1731, No. CXLIII, pp. 196–197.)

[61] *Ibid.*, p. 356. This doctrine is also to be found in the writings of Becher and Stahl (and even Lavoisier). See J. R. Partington, *op. cit.*, (note 4), pp. 648, 673.

[62] It is not immediately obvious why 'homogeneity' should prevent combustion. Presumably, Pott means that the phlogiston is particularly tightly bound in the bodies so that they resist combustion.

play innumerable rôles, supposedly accounting for metallic lustre, malleability, odour, colour, taste. It is the active principle in nature, but can both conserve and destroy. It can divide and attenuate, but also compose. As Lavoisier said, years later: ' C'est un véritable Protée '.[63] Yet all this theory is not well integrated with experimental investigations. By attempting to explain so many phenomena by means of the phlogistic principle, Pott ultimately explains very little. His massive empirical onslaught against the problem of porcelain could have been conducted just as effectively without the theoretical foundation supposedly supporting the investigations.

5. *J. G. Wallerius* (1709–1785)

The leading phlogistic mineralogist in Sweden in the middle years of the eighteenth century was J. G. Wallerius, professor of chemistry, metallurgy, and pharmacy at the University of Uppsala. In a detailed publication on the properties, classification, and theory of minerals, he made one of the first attempts to base a reasonably adequate system of mineral classification on chemical principles.[64] In a later work, he set out a speculative but interesting cosmogony, of a conservative hue, showing a considerable dependence on the type of thinking exemplified by Becher.[65]

Wallerius tends to follow Becher's account of the three EARTHS, rather than Stahl's, in that he maintains the earlier view that each metal is composed of a vitrescible EARTH, the sulphureous principle (or phlogiston), and a mercurial EARTH, supposedly responsible for conferring metallic properties. Moreover, in his discussions of the various metals, Wallerius gives detailed descriptions as to the manner in which the mercurial component may be extracted from each of them. The instructions make curious reading today. Thus, to extract the mercurial principle of silver, one should take ' horn silver ' (silver chloride), ' volatile alkaline salt ' (ammonium carbonate), white of egg, and spirit of wine, digest for three weeks, add black flux, and distil. Mercury supposedly appears in the distillate![66] For copper, one should distil the calx of copper with

[63] A. L. Lavoisier, *Oeuvres de Lavoisier* . . . , 6 vols., Paris, 1862–1893, vol. ii, p. 640.

[64] J. G. Wallerius, *Mineralogia, eller Mineral-Riket* . . . , Stockholm, 1747. I have used the French translation of 1753: *Mineralogie ou Description Générale des Substances du Regne Mineral*, 2 vols., Paris, 1753.

[65] J. G. Wallerius, *Meditationes physico-chemicae de origine mundi, imprimis geocosmi ejusdemque metamorphosi*, Stockholm and Uppsala, 1779 (not seen). I have used the French translation of J. B. Dubois: *De L'Origine du Monde, et de la Terre en particulier; Ouvrage dans lequel l'Auteur développe ses Principes de Chymie & de Minéralogie, & donne, en quelque manière, un abrégé de tous ces Ouvrages: Par M. Wallérius*, . . . Warsaw, 1780

[66] J. G. Wallerius, *op. cit.*, (note 64, 1753), vol. i, pp. 576–577.

'sal ammoniac' (ammonium chloride) and soap. Again, mercury is supposedly obtained in the distillate.[67] I cannot imagine that Wallerius could have obtained anything vaguely resembling mercury from such concoctions, and his judgement seems to have been prejudiced by his theoretical presuppositions. It is noteworthy that no such experiments for the extraction of the mercurial principle are described for the 'semi-metals' such as zinc and cobalt. These substances, discovered long after the six solid metals known from antiquity, were not immediately admitted into the august class of metals, supposedly because the semi-metals were less malleable and ductile than the true metals. So Wallerius would not have expected to find the mercurial EARTH among the semi-metals, and not surprisingly, therefore, he fails to record its discovery. Nevertheless, the attempts to identify an EARTHY principle with a particular, empirically detectable, substance should be noted. Similarly, it should be noted that Wallerius seems to identify the vitrescible EARTH of Becher directly with the various calces of the several metals. Thus, once again, we see Becher's single vitrescible EARTH being subdivided, this time into the assortment of metallic calces. All this is clearly part of the general eighteenth-century trend towards the subdivision of the 'ideal' EARTHS of earlier generations into a number of chemically related, but experimentally distinguishable, individual earths. Siegfried and Dobbs have seen this as an important step on the road to the 'chemical revolution',[68] an opinion that I fully endorse.

In extending the theory of the three EARTHS into the chemistry of non-metallic substances, Wallerius follows Becher by adopting the hypothesis that the three common mineral acids differ from each other in that each contains one of the three hypothetical EARTHY principles. Thus, 'vitriolic acid' supposedly contains the vitrifiable EARTH; 'nitrous acid' supposedly contains the inflammable EARTH; while the 'acid of sea salt' is believed to contain the mercurial EARTH.[69] The substance common to all three of these mineral acids is the 'universal acid' of Becher and Stahl.

If an acid reacts with different (solid) alkalis, then crystals of different shapes are obtained. This suggests to Wallerius that something rather similar occurs when mineral crystals are formed, except that some earthy or metallic base is involved. The opinion of Henckel—that 'marl' is the base of quartz—is quoted; Wallerius believes that a 'marly earth'

[67] *Ibid.*, p. 524.

[68] R. Siegfried and B. J. Dobbs, 'Composition, a Neglected Aspect of the Chemical Revolution', *Annals of Science*, 1968, **24**, 275–293.

[69] J. G. Wallerius, *op. cit.*, (note 64, 1753), vol. i, pp. 319–320.

is serving as a 'base'[70] for the formation of many non-metallic mineral substances. Even alum is believed to contain calcareous earth, whence slate also supposedly contains the same component. This should lead to the classification of slates with marbles and limestones, yet Wallerius is well aware of the obvious category difference between slate and limestone, the one being readily fusible, while the other forms a calx on strong heating. Thus his mineralogical good sense prevails over the erroneous chemical notions, and slates are classified with the vitrifiable stones. Wallerius's classificatory scheme may be given in outline here:[71]

| *Classes* | *Orders* |
|---|---|
| 1. Earths | 1. Powdery earths |
| | 2. Argillaceous earths |
| | 3. Mineral earths |
| | 4. Sands |
| 2. Stones | 1. Calcareous stones |
| | 2. Vitrifiable stones |
| | 3. Refractory stones |
| | 4. Compound stones or rocks |
| 3. Minerals | 1. Salts |
| | 2. Sulphurs |
| | 3. Semi-metals |
| | 4. Metals |
| 4. Concretions | 1. Porous substances |
| | 2. Petrifactions |
| | 3. Figured stones |
| | 4. Concretions. |

The neat four-by-four classification, suggestive of a rationalist mind, may be noted.[72] As remarked above, Wallerius's scheme is a mixed one, being based on both chemical and external characteristics. It is, perhaps, most successful in dealing with the metals, salts and sulphurs, and it is noteworthy that in these areas, in which the chemical methods of classification (albeit crude phlogistic methods) could be

[70] For a discussion of the use of the term 'base' in early chemistry, see my paper: 'Some Early Usages of Chemical Terms', *Journal of Chemical Education*, 1973, **50** (7), 450–454.

[71] J. G. Wallerius, *op. cit.*, (note 64, 1753), vol. i, p. xxiij.

[72] Cf. the neat three-by-three system of classification in the mineral system of Wallerius's rationalist countryman, C. Linnaeus: . . . *Systema Naturae* . . . , Leyden, 1735, unpaginated (facsimile edition, Nieuwkoop, 1964).

applied with some effect, the classificatory scheme comes closest to being based on what we might regard as 'natural kinds'.

In his later work, on the origin of the world,[73] Wallerius propounds a highly speculative theory of the genesis of the cosmos, which displays a number of similarities to Becher's *Physica Subterranea.*

Wallerius is well aware that the analysis of a body by fire might yield earthy, inflammable, saline, and aqueous principles, but since he believes that water and earth (*sic*) are interconvertible, and that salt can be resolved into aqueous and inflammable components, he thinks it necessary to consider two principles only: a fixed EARTHY principle, and an active fire principle (which is also the principle of light).[74] At the time of creation, these might be combined in various ways to give other important principles.

Thus (following the terminology adopted by Wallerius) if

a = particle of matter of light (active) } primary principles
b = particle of EARTH (passive) }

we have:

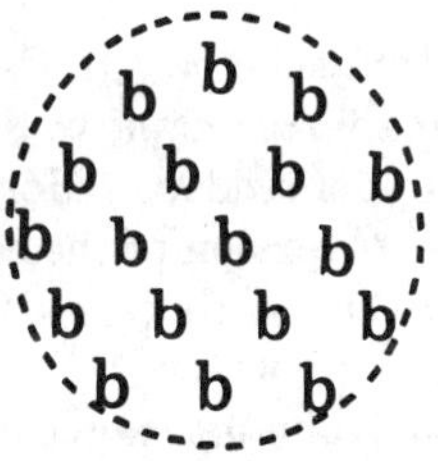

= EARTH

= matter of heat

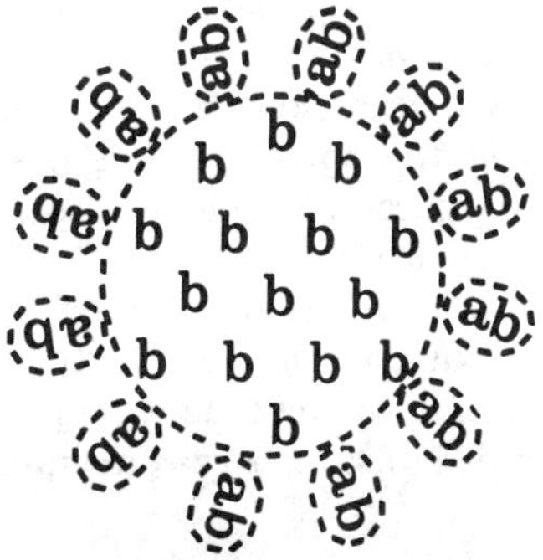

= water

} secondary principles

[73] See note 65.

[74] *Ibid.*, (1780), pp. 141–145. It should be noted that Wallerius's suggestions, which I have represented here in diagrammatic form, are explicitly stated to be conjectural.

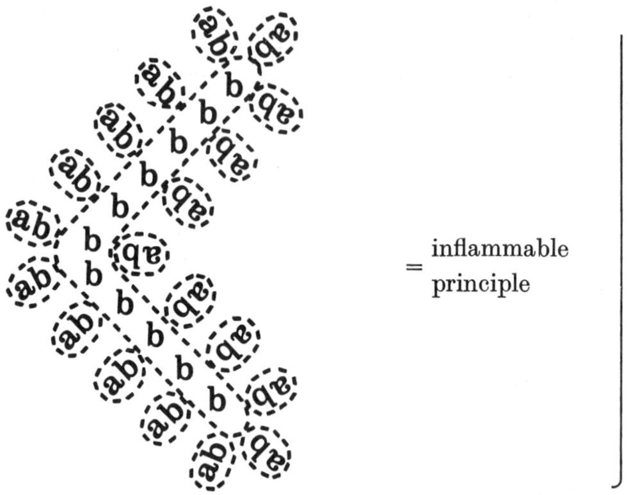


It may be noted that this scheme (its Cartesian overtones notwithstanding) allows the inflammable principle to be thought of as an EARTH, as the Becherian tradition demanded. Moreover, it appears that we have here a somewhat unusual version of the caloric theory of heat.

It is interesting to note that Wallerius also envisages the production of bodies from the corpuscles of water.[75] This gives him a place in the theoretical tradition of Van Helmont, not to mention the contemporary 'aqueous' theories being propounded by Werner at about the same time. (Indeed, for Wallerius, *all* the planets were initially composed of water.) Furthermore, it may be noted that in surmising that bodies are produced in water by a process of fermentation and effervescence, and that the whole process of formation of rocks and minerals is rather like that which occurs when a hen sits on her eggs,[76] Wallerius reveals his thinking to be influenced by the 'organic' tradition, traceable to Paracelsus and beyond. For Wallerius, it is the spirit of God that imparts the initial intrinsic motion to the corpuscles which results in the coalitions that form the bodies of our experience.[77]

Earths and rocks may be generated by coagulation or concretion. Wallerius gives the precipitation of a calcareous solution (in acid) by alkali, and the precipitation of a siliceous solution by acid, as examples of 'coagulation'. Pebbles, flints and even grains of sand may come into

[75] *Ibid.*, pp. 83, 86–93, 132 and *passim.*

[76] *Ibid.*, p. 161.

[77] *Ibid.*, p. 159.

being by such coagulations. By the force of affinity, adjacent particles may come together to form concretions. By gravity, the several coagulated (or concreted) masses may coalesce, and unequal precipitation of stony matters, or a drying out of such matter, leads to the formation of caverns within the central parts of the Earth, into which waters can withdraw. The passage of waters through the subterranean channels leads to the deposition of mineral veins. It can be seen that this theory also has much in common with the ' aqueous ' schemes being propounded in Germany at about the same period.

In Wallerius's lifetime there was a considerable increase in the knowledge of the variety of ' earths ' that may be found on our globe. Needless to say, he is well aware of these developments, but he claims that only two of these earths (calcareous and argillaceous) may be considered as candidates for the status of truly ' primitive ' earths[78] and he considers that these may be formed—in some not clearly specified manner—from water. Other earths (magnesia, silica, gypsum, *etc.*) are supposedly never found isolated from calcareous or argillaceous matter, so that their claims to independent status are rejected. Lime and magnesia, for example, merely differ in the degree of attenuation of their corpuscles. The calcareous and argillaceous earths are thought to have been derived from water by a process of transmutation.[79] Thus, calcareous earth appears to be deposited by waters in the formation of stalactites; argillaceous earth collects in the waters of cisterns. Rejecting the arguments of Pott and Lavoisier against the transmutation of water into earth, Wallerius maintains that water is made up of matter that is essentially the same as the matter of sand, differing only in the incomparably greater state of division of the particles. Thus, by evaporation, putrefaction, trituration, agitation, or coagulation, the corpuscles of water may come together to form earths. By analogy with the known precipitation of silica from ' liquor of pebbles ' by acid,[80] acids may cause the formation of earth from water. God probably used the elementary earth, derived from water, for the creation of Adam's body.[81] Numerous scientific, Biblical, and classical sources are cited in favour of the opinion that water and earth are interconvertible.

[78] *Ibid.*, p. 123.

[79] *Ibid.*, p. 124.

[80] *Ibid.*, p. 84.

[81] *Ibid.*, p. 118. It is a moot point whether one should represent Wallerius's intention by ' earth ' or ' EARTH '. He speaks of an ' elementary earth ', yet at the same time he seems to be trying to identify it with an empirically identifiable substance, referring to the products of the putrefaction or incineration of human bodies!

Water may also be transmuted into air, which explains why the formation of air is not mentioned in *Genesis*. Utilizing the phlogiston theory, Wallerius is not prepared to accept the new doctrines of Black concerning fixed air. A gas liberated in a reaction is a product, not an extract or educt. Thus the interaction of vitriolic acid with limestone is viewed as a process in which phlogiston (from the limestone) unites with the acid to form fixed air. Fixed air is phlogisticated vitriolic acid. It may also be taken to be the sulphur principle, and its presence may account for thunder and lightning in the atmosphere. In these discussions, Wallerius seems to be stretching the theory beyond its elastic limit, in his attempts to reconcile it with the traditional Stahlian doctrine that sulphur is composed of vitriolic acid and phlogiston.

It is a matter of some interest to attempt to evaluate these opinions. Is Wallerius an 'organicist', a 'mechanist', or a 'chemist'? We have discerned a few instances of the use of 'organic' models in his thinking, but the overall impression that one gains from reading Wallerius is that he is certainly not an exponent of 'Renaissance naturalism', and indeed it would be absurd to expect a man of the Enlightenment to be thinking in this way. Nevertheless, certain vestiges of the earlier 'organic' approaches are evident.[82] Also, as we have seen, Wallerius frequently invokes corpuscularian interpretations, and certain *quasi*-Cartesian mechanical modes of explanation can be discerned. Nevertheless, the general impression I gain is that Wallerius is thinking as a chemist, his favoured explanations being in Stahlian terms. Of particular interest, from our point of view, is the merging of the concept of hypothetical 'EARTHY' principles with the common earth-like substances of the laboratory. The greatest weakness in the scheme lies in the repeated attempts to offer an account of the original synthetic processes that supposedly occurred at the time of Creation. Here, Wallerius is following a venerable tradition,[83] but, clearly, one that transcends all empirical evidence. With the advantage of hindsight, it is apparent that mineralogists had to learn to eschew such grandiose speculative schemes, and

[82] In his second major exposition of his mineralogical system (*Systema Mineralogicum, Quo Corpora Mineralia in Classes, Ordines, Genera et Species, suis iam varietatibus divisa, describuntur, atque observationes, experimentis et figuris aeneis illustrantur, à Johan. Gotsch. Wallerio.* ..., 2 vols., Stockholm, 1772–1775), Wallerius insists on the continuing formation of minerals, but categorically denies that the increase is due to any kind of 'organic' process (vol. i, p. 3).

[83] Much of Wallerius's '*Origin of the World*' is based on the Mosaic account of Creation, or is devoted to attempts to reconcile this account with empirical observations. The belief in an early period during which the globe was in a fluid condition, and the assumption that all substances had an aqueous origin, are clearly beholden to the Mosaic tradition. Thus, I suggest, the celebrated 'aqueous' theory that culminated in the work of Werner, had important roots in the theories of the phlogistic mineralogists, themselves stemming in part from the tradition of *Genesis*. (See note 39, above).

concentrate attention on the details of the external characteristics of minerals, and their reactions with a multiplicity of chemical reagents.

6. *J. G. Lehmann* (1719–1767)

Let us now turn to a consideration of the ideas of J. G. Lehmann, physician and Councillor of mines in Berlin, and later professor of chemistry in St. Petersburg, perhaps best known today for his ' aqueous ' theory of the Earth, considered to be a forerunner of the better-known, but similar, theory of Werner.[84]

Lehmann gives us a detailed account of his views on the formation of metals—an account that is of interest in that (like Wallerius's work) it exhibits a blending of the matter theory of Becher with later, more mechanical, modes of explanation. At the same time we find some very early and primitive ideas outcropping. Following Becher, Lehmann adopts the tripartite division of EARTHS for the composition of metals.[85] He is inclined to identify the first vitrescible EARTH with some properly distinguishable earthy substance, and selects the vitrifiable earth of Pott's tetrad to serve in this way. The presence of such an earthy substance in metals is supposedly revealed by Becher's celebrated production of iron by the action of heat on a mixture of linseed oil and clay.[86] It imparts solidity and fixity to metals.

Lehmann does not believe that the second component of metals should be called an EARTH—the word ' phlogiston ' is preferred. It is made up of unctuous and saline corpuscles, with a small portion of ' subtle ' EARTH. It is perhaps a universal component of matter, and is driven off during calcination, leaving the other two EARTHS behind.[87] It is different combinations of phlogiston and the mercurial principle (rather than differences in the vitrescible components of metals) that account for the individuality of the different metals.[88] The phlogiston serves to join the particles of the first principle with those of the third.

[84] A. Geikie, *The Founders of Geology*, 2nd edn., London, 1905 (republished, New York, 1962), p. 233. Lehmann's principal geological theories are contained in his *Versuch einer Geschichte von Flötz-Gebürgen*, Berlin, 1756, which in translation forms the third part of the French edition of his major works: *Traité de physique, d'histoire naturelle, de minéralogie et de métallurgie*, 3 vols., Paris, 1759. The second of these volumes is the most important from our point of view. It bears the title: *Traité de la Formation des Métaux, et de leurs Matrices ou Minieres: Ouvrage fondé sur les principes de la Physique & de la Minéralogie, & confirmé par des Expériences Chymiques, Par M. Jean-Gotlob Lehmann, . . . Traduit de l'Allemand*, Paris, 1759.

[85] J. G. Lehmann, *op. cit.*, (*Traité de la Formation des Métaux*), p. 64.

[86] See note 48 above.

[87] J. G. Lehmann, *op. cit.*, p. 75.

[88] *Ibid.*, p. 85.

The third, mercurial, principle imparts density and other metallic characteristics. It is to be found in mineral exhalations and mineralizing vapours. It is *not* to be confused with metallic mercury.

How, we may ask, are metals formed from these component particles? Lehmann believes that it is sufficient to say that ' Nature ' brings the component corpuscles together to form mixts.[89] The inflammable particles, like acids, act on the particles of the first EARTH, dilating them, so that the mercurial particles may gain an entrance. But for all this to occur satisfactorily, the constituent particles must be in the correct proportions, and in the appropriate state of division—otherwise mere ' semi-metals ' will be formed. Antimony, for example, is thought to have plenty of the second and third EARTHS, but an insufficiency of the first component. Thus, by adjusting the relative quantities of the three EARTHS in a mixt, one might hope to convert ' semi-metals ' into proper metals. Lehmann clearly believes in the possibility of transmutation. Air and water, within the globe, may act as the agents that bring the constituents of metals together to form native metals, or ores. (He suggests that this may be why Aristotle believed that metals arose from moist exhalations!)

It might be thought from the foregoing discussion that Lehmann's treatment is very ' mechanical ', and by no means characteristic of a phlogistonist. This does indeed sometimes seem to be the case, yet he himself is most anxious to repudiate any such suggestion. Any attempts to reduce chemistry and mineralogy to mathematics (or more particularly to geometry) are to be rejected.[90] Yet, while ostensibly excluding mechanical or geometrical theories, he is, in fact, forced to draw on such modes of explanation in an effort to make the overall account intelligible.

Perhaps the most interesting portion of Lehmann's work, from the point of view of our present enquiry, is his section on the ' matrices ' of metals, since, *prima facie*, this is strongly suggestive of a continuation of the long ' organic ' tradition in mineralogical theory. According to Lehmann, a matrix is a body in which the ' germ ' of a metal is received, perfected, and conserved, until a superior force comes and releases it. Hoffmann's opinion—that matrices function as instruments that contribute to the perfection of metals—is mentioned with approval. This is a point of some interest, since it will be remembered that Paracelsus[91]

[89] *Ibid.*, p. 108. (This is also Stahl's opinion: *Philosophical Principles of Universal Chemistry . . . Drawn from the Collegium Jenense of Dr. George Ernest Stahl. By Peter Shaw M.D.*, London, 1730, pp. 8–9. For Lehmann, ' Nature ' is to be interpreted as the perpetual motion of bodies, the motion having been imparted by God.)

[90] *Ibid.*, p. 111.

[91] Paracelsus, *op. cit.*, (note 12), vol. i, p. 202.

considered earth, water, air, and fire served as *matrices* for the formation of bodies, while Becher and Stahl believed that the four elements played an *instrumental* rôle in chemical changes. Lehmann states that the ores of metal serve as their matrices, and so, too, do the veins, cracks, and faults with the Earth. Veins may be thought of as the branches of a large trunk within the interior of the Earth.[92] The elementary parts of metals are elaborated in the centre of the Earth, and rise towards the surface in a humid form, or as vapours, like sap rising in the branches of trees. Or, one may think of the process as being analogous to the collection of soot in a chimney. In certain instances, as for example when clay functions as a matrix, the matrices may act by filtering off the metallic matter from the percolating waters or exhalations.[93]

In these discussions of metallic ' germs ' and ' matrices ', and of aqueous ' trees ' within the interior of the Earth, we may feel that Lehmann is merely displaying manifestations of the organicism of earlier generations. I would suggest, however, that this picture should not be taken wholly at its face value. The *words* are there, to be sure, but they seem to be used largely in a metaphorical sense. Certainly, they reveal the tradition from which his ideas have developed, but it would, perhaps, be wrong to interpret Lehmann too literally. The overall impression I form from reading his work is that he thinks as a chemist, utilizing a number of ' mechanical ' ideas in his theories. The ' organic ' terms that are employed seem to be almost vestigial, and in his discussion of the classification of earths and stones (which perhaps forms the major *mineralogical* part of the work—though I have not discussed it much here) external characteristics are used as criteria for sorting, and the treatment does not have any strong theoretical commitment. Indeed, the same might perhaps be said for the earlier writers that we have been discussing. They devise (or utilize) grand speculative schemes concerning the origin of the cosmos, while simultaneously developing their knowledge of minerals according to their external characteristics and chemical behaviour. The practical and theoretical aspects of their work are not well meshed, though the later phlogistonists, such as Bergman, Scheele, or Kirwan, were increasingly successful in bringing about a satisfactory integration of theory and practice.

[92] J. G. Lehmann, *op. cit.*, p. 253.

[93] *Ibid.*, p. 374. This opinion is similar to that expressed by earlier French ' mechanical ' mineralogists such as R.A.F. de Réaumur. (' Sur la Rondeur que Semblent affecter certaines especes de Pierres, & entr'autres sur celle qu'affectent les Cailloux ', *Histoire de l'Académie Royale des Sciences. Année M.DCC XXIII. Avec les Mémoires de Mathématique & de Phisique, pour la même Année*, Paris, 1725, pp. 9–13 and 273–284).

7. *G. F. Rouelle* (1703–1770)

Let us now turn our attention to France and examine briefly the ways in which phlogistic ideas were utilized in mineralogical explanations in the period before the ' chemical revolution '. According to Rhoda Rappaport,[94] it was the work of G. F. Rouelle that was primarily responsible for the introduction of phlogistic chemistry into France. It is, therefore, a point of some interest to investigate the mineralogical theories that Rouelle taught in his famous course of lectures. For this, we must examine some copy of the Rouelle lecture notes, for the lectures themselves were never published. I have used the ' Clifton College ' manuscript, which is in what Rappaport calls the ' Diderot tradition '.[95]

By and large, Rouelle made little use of the Becherian/Stahlian scheme for mineralogical purposes, and there is an uneasy compromise between Becher's doctrine of EARTHS, the traditional Aristotelian EARTH, and plain common or garden earths. Thus we are informed that:

> ' La terre est un Corps dur, simple, friable—fusible au feu, et immiscible a l'eau, a l'esprit de Vin, et aux huiles. il est facile de démontrer la terre, dans le regne vegetal, et animal, cela n'est pas aussi aisé dans les métaux. l'eau distillée laisse une terre qui étant depouillée de toute espece de salure est tres simple, et tres subtile, on l'appelle *terre vierge* . . . c'est une grand question de savoir s'il n'y a qu'une terre dans la nature, ou s'il y en a plusieurs '.[96]

Earth imparts solidity and fixity to bodies, though there can be some very solid bodies containing much air and water, and only a little earth. Can these views by reconciled with Becher's doctrine of three EARTHS in metals? The notes inform us that:

> ' *Beccher* admettois trois sortes de terres, dans les métaux, une terre vitrifiable, . . . , une terre Colorante, parce qu'elle donne la Couleur aux métaux, c'est le phlogistique de *Staahl* [,] et une terre mercurielle, dont l'existence n'est pas aussy bien prouvée. il pretend qu'elle se trouve dans le sel marin, et qu'elle donne aux métaux la metallicité; il y a lieu de présumer qu'il a fait deux terres différentes du phlogistique, car les effets qu'il attribue a la terre colorante, et a la terre mercurielle [,] sont les mêmes que ceux du phlogistique, les terres qui ne sont pas vitrifiables par elles mêmes le sont souvent en y ajoutant l'alkali fixe.'[97]

[94] R. Rappaport, *op. cit.*, (note 8).

[95] R. Rappaport, ' G. F. Rouelle: An Eighteenth-Century Chemist and Teacher ', *Chymia*, 1960, **6**, 68–101.

[96] Clifton College MS: *Cours de Chimie, ou Leçons de Monsieur Rouelle, Recueillies pendant les années* 1754. 1755. *et rédigées en* 1756. *Revues, et corrigées, en* 1757. *etc.* 1758, p. 43.

[97] *Ibid.*, p. 44.

Certainly, the Stahlian phlogiston does play the rôle of both the second and the third EARTH of Becher, but this does not succeed in bringing about a reconciliation of the two traditions of matter theory, the Aristotelian (which is also manifest in the work of Macquer at the same period[98]) and the doctrine of the EARTHS of Becher and Stahl. Thus, like the German and Swedish mineralogists that we have considered above, Rouelle does not find the trichotomy of Becher particularly useful as a basis for mineralogical theory. Rather, he recognizes a number of different individual earths, which together provide him with quite a useful classificatory scheme, based upon chemical principles. It is noteworthy that this scheme (which, may have been earlier than that of the Swedish mineralogist, A. F. Cronstedt, who is frequently taken to have been the first to offer a mineral system based entirely on chemical principles[99]) envisages the possibility of transmutation of earths, so that the weaknesses of earlier beliefs have not been eliminated. Never-the less, Rouelle's whole scheme is argued on an empirical basis, and he offers an interesting set of empirical arguments for the interconversion of earthy substances. In his section on earths and stones, he claims the existence of six kinds of earths (or stones):[100]

1, Calcareous earth,

2, Gypsum,

3, Argillaceous earth,

4, Quartzose or siliceous earth,

5, Fusible spar (fluor spar),

and

6, Apyrous or refractory stones.

Yet some of these are held to be interconvertible. Thus, if '*liquor silicium*' (a solution of potassium silicate obtained by the action of potash on silica) is treated with acid, one obtains a precipitate of siliceous earth. But, if too much acid is added, then a [colloidal] solution is obtained. Rouelle believes that this occurs because the siliceous earth has in some way been converted into calcareous earth—which is known to be soluble in acid. He admits that this is but an isolated experiment, but believes that it throws much light on chemistry. It may indicate the possibility

[98] P. J. Macquer, *Elemens de Chymie Theorique*, Paris, 1749, p. 2.

[99] A. F. Cronsdedt, *Försök til Mineralogie, eller Mineral-Rikets upställning*, Stockholm, 1758. It should be noted that there is still considerable uncertainty as to the exact dating of the various parts of the Clifton College MS. The section on earths and stones occurs in the latter portion of the notes, which were, in any case, probably copied some years after the lectures were delivered.

[100] G. F. Rouelle, *op. cit.* (note 96), p. 910.

of the transmutation of the different kinds of earths, and suggests that they are all, at bottom, fundamentally the same. Repeated calcination of silex may convert it into a calcareous substance. He believes that a recognition of the interconvertibility of the several earths may prove helpful in accounting for a number of common observations, such as the frequent association of flint with chalk. However, as we now know (though doubtless it was not so apparent at the time), this was not a progressive doctrine. Rouelle hankers after a reduction of all earths to one fundamental, simple, EARTHY principle. Doubtless, he much regretted the failure to achieve this ambition. By contrast, we recognize the identification of a number of individual earths as a progressive step.

Rouelle recognizes that calcareous earth is produced in shells, or from the debris of shells, and he maintains that he was the first to make this suggestion. Regarding quartz, Rouelle believes that it is the principle of silex, jasper, rock crystal, agate, and precious stones. However, he also includes diamonds in his list, and states that quartz crystals are cubical. He speaks of quartz '*de l'ancienne terre*' and '*de la nouvelle terre*' [flint], which dichotomy may perhaps correspond to Lehmann's important subdivision of the globe's strata. Flints, when broken open, often reveal fossil shells, and this suggests to Rouelle that silica may indeed be formed from calcareous earth. Rock crystal is the primitive matter of stones, including precious stones. He proceeds to a description of the crystalline natures of the various precious stones, but the accounts are not very accurate, probably due to a paucity of good natural specimens for examination. There is no real experimental evidence for the claim that rock crystal is a component of other precious stones. Spar (fluorspar) is properly distinguished from quartz, but there is little account of its particular attributes. His separation of '*apyres*' from other types of stones was traditional, being found, for example, in the work of Linnaeus.[101] Despite their inclusion in Rouelle's scheme, he make little of the distinction and for all practical purposes the category was superfluous.

It seems clear that Rouelle had little that was really original in his mineralogical scheme. Its interest, for our purposes, lies in the way the phlogistic ideas are utilized for mineralogical purposes by a Frenchman who is an avowed phlogistonist. The traditional Becherian matter theory finds little place in Rouelle's mineralogy, and indeed the very phlogiston itself plays little part. What *is* important is the subdivision of the earths (the kind that one can dig from the ground) into several important categories, despite the fact that Rouelle favoured the reduction of all earthy substances to a few simple principles. This seems to be a direct response to the pressures of empirical discoveries. To be

[101] C. Linnaeus, *op. cit.*, (note 72).

sure, Rouelle doesn't really make very much of this important subdivision, though ultimately it provided the breakthrough for chemical mineralogy.

I find virtually no evidence of the old ' organicism ' of earlier phlogistonists in Rouelle. There are a few overtly ' mechanical ' explanations, but the whole scheme is not characterized by a ' mechanical ' approach. It is a chemist's mineralogical picture, though of a rudimentary kind. It would seem, then, that the French chemists extracted the valuable *chemical* aspects from the Becherian/Stahlian scheme, leaving as dross the residual ' organic ' features of their predecessors. A few vestiges of pristine Becher theory may be found, but it is not possible to assimilate them satisfactorily into the overall scheme, and they do not cohere well with the developing chemical empiricism. Rouelle, of course, makes no effort to detail the original syntheses of minerals at the time of the Earth's creation.

8. *Antoine Baumé* (1728–1804)

As a second French example of a treatment of mineralogical problems by a phlogistonist writing prior to the ' chemical revolution ', I consider the work of the French apothecary, and disciple of Macquer, Antoine Baumé.

Baumé's earlier work, of 1763,[102] is of particular interest in that it reveals the continuing struggle to establish the nature of the earthy constituents of mineral substances. Following Macquer, he returns to the use of EARTH, water, air, and fire as ' elements ', though he thinks that they should be treated as ' secondary principles ' or ' *principes principiés* '.[103] He mentions the water and three EARTHS of Becher and Stahl[104] with apparent approval, but fails to meld them satisfactorily with his chosen tetrad of elements. His chief concern is to find the true elementary EARTH, but he has to admit that, despite endless searching, chemists have failed to discover it. Perhaps diamond and rock crystal come the closest in their properties to the true EARTHY principle, or ideal EARTH.

As we have seen, previous chemists had considered a number of different earthy principles. Pott favoured four: calcareous, gypseous, argillaceous and vitrifiable. But the grand-master of chemistry, G. E. Stahl, had opted for two (if we disregard the somewhat dubious mercurial

[102] A. Baumé, *Manuel de Chymie, ou Exposé des Opérations et des Produits d'un Cours de Chymie. Ouvrage utile aux personnes qui veulent suivre un Cours de cette Science, ou qui ont dessein de se former un Cabinet de Chymie*, Paris, 1763.

[103] *Ibid.*, p. 16. This doctrine has Stahlian roots, as well as being found in Macquer's *Elemens*. It converts the four Aristotelian elements into mixts.

[104] *Ibid.*, p. 20.

EARTH). For this reason, Baumé considers it proper to reduce Pott's tetrad to two fundamental earths:[105]

> 'Ainsi nous nous en tenons au sentiment de Stahl, & nous ne reconnoissons, comme lui, que deux espèces de terre, scavoir la terre vitrifiable & la terre calcaire '.[106]

Two, therefore, is a good number to aim for. The search for a small number of hypothetical earthy principles has not been abandoned. We notice, however, that the chosen pair are, in fact, experimentally separable substances.

How does Baumé justify this reduction? He recognizes, very properly that the gypseous earth of Pott is merely a saline union of calcareous earth and vitriolic acid,[107] but in identifying the argillaceous and vitrifiable earth he is confused, like Rouelle, by that incubus, the reaction of '*liqueur des cailloux*', or '*liquor silicium*', with acid. He rightly takes the *liquor* to be a solution of vitrifiable earth, made possible by the intermediary of the fixed alkali. Thus, when acid is added, the alkali is 'seized' and the vitrifiable earth is precipitated. However, with excess acid the earth becomes soluble in acid [as colloidal silica], and the solution deposits crystals of alum on evaporation. To explain this result, we must presume that there was some aluminous earth in the original pebbles that were being investigated, or that some argillaceous matter was acquired from the crucibles used in the preparation of the '*liqueur des cailloux*', but Baumé's preferred explanation is that the vitrifiable earth, the argillaceous earth, and the earthy base of alum are all one and the same.[108]

Thus Baumé could reduce Pott's four earths to two by means of the equations:

Argillaceous earth = vitrifiable earth + vitriolic acid

Gypseous earth = calcareous earth + vitriolic acid.

In his later, much more detailed, exposition of his views,[109] Baumé turns backwards, for he decides to accept only *one* primitive EARTH,

[105] In making this reduction, Baumé was 'swimming against the tide'. The empirical investigations of his contemporaries were driving chemists to the acceptance of a considerable variety of fundamental earths.

[106] A. Baumé, *op. cit.*, (note 102), p. 65.

[107] *Ibid.*, p. 299.

[108] *Ibid.*, p. 100.

[109] A. Baumé, *Chymie Experimentale et Raisonée*, 3 vols., Paris, 1773.

the vitrifiable earth, calcareous earth being considered to be constituted of vitrifiable earth, phlogistic matter, air, and water.[110] Perhaps the interior of the Earth is made up of this pure, primitive EARTH. Phlogiston may be thought of as a compound of this pure, primitive EARTH with fire.[111]

We have seen, above, that Rouelle had offered the opinion that calcareous earth might be synthesized in sea water by marine animals. Now, in the last major French phlogistic treatise published before the 'chemical revolution', Baumé develops this idea in a most ingenious manner, with a return to something of the broad scope of the speculations that we have seen in Becher. Baumé suggests that the marine animals combine the earthy vegetable matter in their diet with water, binding it together with animal mucilage. Thus, living organisms are the instruments whereby the fire, arriving on the globe from the sun, is fixed, thus forming immense reservoirs of combustible matter on the globe, which can be incorporated in salts, sulphur, bitumens, or metallic minerals. One may say further, then, that organic creatures are the cause of volcanoes, earthquakes, and meteors! Baumé believes that without them our world would simply be a single pure crystal, or a mass of sand. Thus sea plants are, so to speak, a first combination of the primitive EARTH. They serve as food for animals, which synthesize calcareous earth from water, air, phlogiston and vitrifiable earth.[112] Then, in the immense basin of the sea, are formed the salts, sulphurs, metals, pyrites and bituminous substances. The decomposition of vegetables produces coal, which may be deposited on land during inundations by the sea. In all these operations, the sea becomes a rich reservoir of phlogiston, which, on meeting with vitriolic acid (in gypsum or alum) forms sulphur. The sea also contains the components necessary for the synthesis of metals. The process may be represented thus:[113]

$$\text{Calcareous earth} = \text{Vitrifiable earth} + \text{air} + \text{water} + \text{phlogiston}$$

$$\text{Calcareous earth} + \text{phlogiston} = \underbrace{\text{Vitrifiable earth} + \text{phlogiston}}_{\text{metal}} + \underbrace{\text{air} + \text{water}}_{\text{(escaping)}}$$

[110] *Ibid.*, vol. i, p. 162.

[111] *Ibid.*, p. 120. This is not unlike Wallerius's doctrine, discussed above, p. 31. Such a position goes some way towards retaining the traditional view of phlogiston as a kind of 'earthy' substance.

[112] *Ibid.*, vol. iii, p. 307.

[113] *Ibid.*, p. 321. (Supposedly, the phlogiston combines with calcareous earth more readily than with vitrifiable earth. Consequently, calcareous earth is a necessary intermediary in the formation of metals.)

Perhaps, also, a little saline matter is contained in the metal. This supposedly prevents the total removal of phlogiston from a metal during calcination and its conversion into a pure earth. Baumé also believes that the retention of a little phlogiston permits the revivification of the metal by the addition of more phlogiston to the calx. (Presumably, the *rationale* for this is the Stahlian doctrine of the attraction of like by like in chemical reactions.[114]) The presence of the saline principles may be responsible for the fusibility of certain calces. Once synthesized, the metallic matter may be transported into the Earth by the sea.

Mountains are formed originally of calcareous earth, but on losing their primitive organization they may be converted into vitrifiable earth—as revealed by the common occurrence of siliceous fossil remains. Argillaceous matter, by the slow loss of phlogiston, may be converted into talc, mica, diamond or asbestos. The heat of the interior of the globe may be explained by the matter of fire evolved in these various transmutations!

9. *T. O. Bergman* (1735–1794)

Let us pass on from these speculations, and as our final example of phlogistic mineralogy consider very briefly some of the ideas of the Swedish chemist, T. O. Bergman, though no detailed commentary is required here, since his work has already been discussed elsewhere in a number of publications.[115] Bergman is one of the most important figures in the history of mineralogy in that he was the first to devise a systematic scheme for the separation, identification and quantitative determination of the earthy components of mineral substances.[116]

[114] See: H. Metzger, *Newton, Stahl, Boerhaave et la Doctrine Chimique*, Paris, 1930, p. 139.

[115] See, for example: E. J. Ferguson, 'Bergman, Klaproth, Vauquelin, Wollaston', *Journal of Chemical Education*, 1940, **17**, 555–562 and 1941, **18**, 3–7; J. A. Schufle, 'Torbern Bergman, Earth Scientist', *Chymia*, 1967, **12**, 58–97; F. Szabadváry, *History of Analytical Chemistry*, Oxford, 1966, pp. 71–81; J. R. Partington, *op. cit.*, (note 4), vol. iii, pp. 179–199; A Fredga and S. Rydén, 'The notes of Juan José de Elhuyar from the lecture notes of Torbern Bergman in 1782', *Lychnos*, 1959, **18**, 161–208.

[116] A *résumé* of Bergman's general method for the analysis of minerals has been given in W. A. Campbell and C. E. Madden, 'The Development of Qualitative Analysis from 1750 to 1850 Part 1', *Proceedings of the University of Durham Philosophical Society*, 1959, **13** (Series A, No. 12), 108–118, and in somewhat more detail in my paper 'Some Eighteenth Century Methods for the Chemical Analysis of Minerals', *Journal of Chemical Education*, 1973, **50**, (5), 337–340.

Five elementary earths (siliceous, calcareous, argillaceous, magnesian, and ponderous[117]) are recognized, and their various chemical properties are utilized in the formation of a crude but viable analytical scheme, utilizing a definite sequence of reactions to produce dissolutions, precipitations and separations. The development of this scheme was surely one of the most important steps in the whole history of chemistry—certainly in the history of analytical chemistry.

We have noticed above in many places that during the course of the eighteenth century there was a gradual change in the meaning of the word ' earth '. We have seen the concept evolving from a hypothetical elementary principle (which we have designated by the term ' EARTH ') to a class of individual earthy substances, separable by chemical procedures, and capable of being placed in individual bottles. This theoretical evolution culminated in the work of Bergman. The analysis of minerals then became a genuine conceptual possibility, and, as we know, this possibility was soon actualized. It is clear, then, that it was not phlogistonism *per se* that stood in the way of the development of a chemical system of mineralogy, for Bergman retained his belief in phlogiston until the end of his life. It was the renunciation of the speculative tendency to reduce all earthy substances to a few ultimate EARTHS, and the willingness to concede any number of empirically distinguishable simple earths, that made the analytical breakthrough possible. This is not to say, however, that Bergman did not display any speculative tendencies. His ideas on heat, solar light, colour and photosynthesis, for example, exhibit many of the fanciful characteristics of the declarations of the earlier phlogistonists. But so far as the question of ' earths ' was concerned, Bergman's thinking was securely linked to his laboratory procedures.

## 10. *Conclusions*

It would be possible to describe many other phlogistic mineralogical schemes. The works of Woltersdorff, Cronstedt, Kirwan, and numerous other writers, provide excellent examples for further discussion, but enough has been written here, I trust, to give the proper flavour of the phlogistic mineralogists. The most characteristic feature of their

[117] T. O. Bergman, *Outlines of Mineralogy, translated from the Original of Sir Torbern Bergman by William Withering*, Birmingham, 1783. In his earlier work, Bergman had included the diamond in his list of elementary earths: ' Disquisitio Chemica de Terra Gemmarum ', *Nova Acta Regiae Societatis Scientiarum Upsaliensis*, 1777, **3**, 137–170. This paper was republished in Bergman's collected *Physical and Chemical Essays*: *Opuscula Physica et Chemica* . . . , 6 vols., Stockholm, Uppsala and Åbo, 1779–1790, vol. ii, pp. 72–117.

thinking seems to have been a *penchant* for sweeping generalizations, or assertions that were not based upon a close attention to the results of experience—this despite their constant claims that their theories were firmly grounded in experience. At the same time, however, we must note that despite the high-flown synthetic schemes promulgated by these phlogistic mineralogists, knowledge of rocks and minerals was developed, and useful classificatory schemes, based on external characteristics and simple chemical tests, were established. Becher's theory of the three EARTHS was never the basis of a satisfactory system of mineral classification. As a result, theory and practice were not properly integrated in the schemes that we have been considering. Nevertheless, it was the phlogistic chemists (notably Marggraf[118]) who were responsible for the discovery of the many reactions that were utilized by Bergman in his all-important system of analysis.

In describing the development of mineral theories from Becher to Bergman, I have drawn attention to the gradual decline in ' organic ' modes of explanation, and the concomitant increase in emphasis on chemical processes and changes. Such ' organic ' features, with roots in the Neo-Platonism and Hermeticism of the Renaissance, gradually declined in the eighteenth century, and where we do find them in later writers (such as Lehmann) they seem to be largely vestigial in character.

We have noted that the phlogistic mineralogists did make occasional use of aspects of the mechanical philosophy, despite the official Stahlian view that mechanical explanations were inappropriate in accounts of chemical phenomena. Thus, it would appear that on occasions the phlogistonists were compelled to have recourse to mechanical similes, in order to make their accounts intelligible to themselves and to their readers, even though Stahl himself ostensibly disapproved of such theorizing.[119]

Finally, I reiterate that the key to the development of a viable chemical system of mineralogy was the recognition of the individual earths, the importance of which has been emphasized by Siegfried and Dobbs.[120] In this paper, then, I have tried to display something of the unsteady and uncertain—but all-important—development of the notion of ' earth ', from the hypothetical triad of Becher to the five white powders, recognized by Bergman as the last terms of his chemical analyses. This seems to me to be a crucially important factor in the history of mineralogy in the eighteenth century.

[118] On Marggraf's contributions, see F. Szabadváry (*op. cit.*, note 115), pp. 55–59.

[119] G. E. Stahl, *Fundamenta Chymiae Dogmaticae & Experimentalis*, . . . , Nuremberg, 1723, preface.

[120] See note 68.

*Acknowledgments*

I am most grateful for the generous help that I have received, during the preparation of this paper, from the following persons: Dr. R. M. Gascoigne, Mr. J. B. Howes, Professor G. Seddon, and Associate Professor T. G. Vallance.

I also wish to acknowledge the permission granted by the Librarian of the Science Schools, Clifton College, to reproduce passages from the Rouelle Lecture notes in his care.

# V

## AN EXAMINATION OF G. E. STAHL'S *PHILOSOPHICAL PRINCIPLES OF UNIVERSAL CHEMISTRY*

THE phlogiston theory of Becher and Stahl and their followers, and its overthrow at the hands of Lavoisier and his colleagues, has, rightly or wrongly, probably occasioned more detailed investigations than any other feature of the history of chemistry; yet surprisingly little has been written about the philosophical and methodological commitments of the Stahlian chemists.[1] The following paper attempts to throw some light on these matters by means of an examination of Stahl's major exposition of his philosophy of chemistry, given in his *Philosophical Principles of Universal Chemistry*.[2] Although, as Professor Guerlac has suggested,[3] it was perhaps Holbach's translations of Stahl's treatises on sulphur and salts[4] which were of particular significance in stimu-

[1] A notable exception to this statement is obviously provided by the classical investigations of Hélène Metzger: "La Philosophie de la matière chez Stahl et ses Disciples", *Isis*, **8**, 427–64, 1926; "La Théorie de la composition des sels et la théorie de la combustion d'après Stahl et ses Disciples", *Isis*, **9**, 294–325, 1927; *Les Doctrines chimiques en France du début du XVIIe à la fin du XVIIIe siècle*, Paris, 1923 (reprinted Paris, 1969). Two recent studies which incorporate material which enhances our understanding of Stahl's work should also be mentioned: R. E. Schofield, *Mechanism and Materialism. British Natural Philosophy in An Age of Reason*, Princeton, 1970, ch. 9; and A. Thackray, *Atoms and Powers: An Essay on Newtonian Matter-Theory and the Development of Chemistry*, Cambridge (Mass.) and London, 1970, 171–5. Detailed expositions of Stahl's chemistry are to be found in: L. J. M. Coleby, *Studies in the Chemical Work of Stahl*, Ph.D. thesis, University of London, 1938; J. R. Partington, *A History of Chemistry*, II, London, 1961, 653–88; and N. Koertge, *A Study of Relations between Scientific Theories: A Test of the General Correspondence Principle*, Ph.D. thesis, University of London, 1969, 131–78, 361–74.

[2] G. E. Stahl, *Fundamenta Chymiae Dogmaticae & Experimentalis* . . . , Nurenburg, 1723. I have used the English translation of Dr. Peter Shaw, one of the chief spokesmen for Stahlian ideas in England in the earlier part of the eighteenth century: *Philosophical Principles of Universal Chemistry: or, The Foundation of a scientifical Manner of Inquiring into and Preparing The Natural and Artificial Bodies for the Uses of Life: Both in the smaller Way of Experiment, and in the larger Way of Business. Design'd as a General Introduction To the Knowledge and Practice of Artificial Philosophy: or, Genuine Chemistry in all its Branches. Drawn from the Collegium Jenense of Dr. George Ernst Stahl. By Peter Shaw M.D.*, London, 1730.

[3] H. Guerlac, "Some French Antecedents of the Chemical Revolution", *Chymia*, **5**, 73–112, 1959.

[4] G. E. Stahl, *Ausführliche Betrachtung und zulänglicher Beweiss von den Saltzen, dass dieselbe aus einer Zarten Erde mit Wasser innig verbunden bestehen*, Halle, 1723, which in French became: *Traité des Sels, Dans lequel on démontre qu'ils sont composés d'une terre*

lating the interest of French chemists in the phlogiston theory,[5] these works do not tell us very much about Stahl's more deep-seated intellectual commitments. In this respect, the *Philosophical Principles* is much more revealing, and its examination can do much to assist our understanding of the conditions of German chemistry in the early years of the eighteenth century.

As is well known, the earlier historians of chemistry did not think kindly of the phlogiston theory. Thus a nineteenth-century writer, Rodwell, informs us that:

> It was not the first chemical theory; it did not give the first explanation of combustion; it was established in the face of facts which carried with them its refutation. When the first stage of its development was passed facts were adapted to the theory, and phenomena were tortured and garbled so as to fit in with it, by means of which the progress of chemical science was somewhat [*sic*] retarded.[6]

Stahl stood condemned for establishing a theory "in the face of facts". By contrast, a recent writer, Fichman,[7] has suggested that the phlogistonists made important contributions to the development of chemical theory and practice. He sees them as "effecting a significant clarification in the methodology of chemistry", and believes that they "played a fundamental role in preparing the synthesis brilliantly completed by Lavoisier". Thus the phlogistic hypothesis is no longer regarded as an extraordinary aberration in the history of science. While this change in attitude is to be welcomed, I believe that a scrutiny of Stahlian chemistry does reveal certain essentially conservative and scholastic elements in the thinking of the leader of the school, when compared with the general trend of thinking in France and England at the same period. In the present paper, the general nature of Stahl's thought will be examined rather than the *minutiae* of the phlogiston theory itself. In particular, the problem of the role of essentialism[8] in Stahlian thought will be considered. The scientific

*subtile, intimément combinée avec de l'eau*, Paris, 1771; G. E. Stahl, *Zufällige Gedancken und nützliche Bedencken über den Streit von dem so genannten Sulphure, und zwar sowohl dem gemeinen verbrennlichen oder flüchtigen als unverbrennlichen oder fixen*, Halle, 1718, translated as: *Traité du Soufre, ou Remarques sur la Dispute qui s'est élevée entre les Chymistes, au sujet du Soufre, tant commun, combustible ou volatil, que fixe, &c . . .*, Paris, 1766. Both these anonymous translations are generally attributed to Baron d'Holbach.

[5] It should be remembered, however, that knowledge of the phlogiston theory had been readily available to French readers since the publication of Macquer's *Elemens de Chymie Theorique* in 1749, well before the appearance of d'Holbach's translations.

[6] G. F. Rodwell, "On the Theory of Phlogiston", *Philosophical Magazine*, **35**, 1–32, 1868.

[7] M. Fichman, "French Stahlism and Chemical Studies of Air 1750–1770", *Ambix*, **18**, 94–122, 1971.

[8] Here, I am using the term "essentialism" in a manner similar to that of K. R. Popper in his work, *The Open Society and its Enemies*, 4th edn., London, 1962, II, chap. 11, which gives a valuable account of the Aristotelian methodology. The Aristotelian essentialist

revolution of the seventeenth century has traditionally been seen as a revolt against the prevailing scholasticism of the earlier period, with its essentialist's metaphysic. Butterfield has referred to "The Postponed Scientific Revolution in Chemistry".[9] Is it proper to construe the eighteenth-century chemical revolution as an overthrow of essentialism in chemistry?

At the time when Stahl wrote, there were broadly three types of explanation which could be invoked to account for chemical phenomena: the corpuscularian explanations of the Cartesians, the Newtonian explanations in terms of inter-corpuscular attractions, and the traditional "chymical" explanations, relying on forms, principles and essences. Stahl rejected the hooks and eyes, needles and wedges of the Cartesians, typified by Lemery's work.[10] Such explanations multiplied entities indefinitely. Each class of phenomena required its own particular characteristic type of particle to serve as an explanatory principle. Moreover, the existence of the particles which Lemery postulated could neither be proved nor disproved by independent experimental investigations. The Newtonian scheme, being explored in some detail in England in the early eighteenth century,[11] eventually failed to fulfil its initial promise, and, in any case, was never particularly influential in German chemical circles. For Stahl, the traditional explanations of the chemists were to be preferred. Mundane objects were made up of "principles" which *conferred* properties upon the materials of which they were constituents. Such a mode of explanation, which can be traced back through the work of Becher to Paracelsus and beyond, was congenial to the chemist with empirical interests, suspicious of too much theory. This essentialist approach would have seemed natural and appropriate to Stahl, with the scholastic education which he would have received at the University of Jena.

Let us first consider the evidence for the influence of scholasticism on Stahl's work. It is, I believe, clearly evident from the manner in which the *Philosophical Principles* was constructed. Stahl attempted to view the whole of chemistry within a unified philosophical scheme. It was as much a philosophy of nature as a text-book of chemistry in our modern sense. Whereas today

hoped to be able to deduce the properties of things from their defining formulae, or essences. (For a more detailed exposition of this point, see M. R. Cohen and E. Nagel, *An Introduction to Logic and Scientific Method*, London, 1934, 234–8.) As will be made evident in the subsequent discussions, I construe the view that a "principle" could confer properties on substances as a manifestation of the traditional essentialist position. (See page 47, below.)

[9] H. Butterfield, *The Origins of Modern Science 1300–1800*, London, 1950, chap. 11.

[10] G. E. Stahl, *Einleitung zur Grund-Mixtion Derer Unterirrdischen mineralischen und metallischen Cörper* . . . , Leipzig, 1720, 51. (Cited by H. Metzger, *op. cit.* (1926), 430.)

[11] On the subject of the influence of Newtonianism on chemistry, see particularly: A. Thackray (1) and R. E. Schofield (1).

very few writers pay explicit attention to the philosophy of chemistry,[12] when Stahl wrote it was considered proper to develop one's ideas within some general philosophical framework, the nature of which was to be made evident to the reader.[13]

The work opens with the following bold statement: "*Universal* CHEMISTRY is the Art of resolving *mixt, compound,* or *aggregate* Bodies into their *Principles*; and of composing *such Bodies* from those *Principles.*"[14] Subsequently, we read: "Its [chemistry's] *Theory,* in general, is acquired by *Information,* axiomatically and demonstratively delivered in the synthetical manner, *à priori*; but analytically, and *à posteriori,* in particular."[15]

These assertions immediately draw our attention to the way in which the objectives and procedures of chemistry could be incorporated into the traditional methodological framework of resolution or composition (or analysis and synthesis). Indeed, this framework was peculiarly appropriate to chemistry, in which the chief operational procedure had for long consisted of attempts to decompose and recompose substances.

In order to assess the significance of Stahl's opening remark, and in particular his use of the terms resolution and composition, it may be helpful to say something about the traditional logical procedures of analysis and synthesis, although a full examination of the matter cannot be attempted here.[16] As is

[12] The literature on this subject is extraordinarily sparse. A pioneering essay by A. D. Ritchie has had few successors: A. D. Ritchie, *Studies in the History and Methods of the Sciences*, Edinburgh, 1958, chap. 7. See also, E. F. Caldin. *The Structure of Chemistry in Relation to the Philosophy of Science*, London, 1961.

[13] In her paper entitled "Structure of Matter and Chemical Theory in the Seventeenth and Eighteenth Centuries" in M. Clagett, ed., *Critical Problems in the History of Science*, Madison, Wisconsin, 1962, paper 16, Marie Boas has pointed out that the text-book writers of the period "began to include a short section on the structure of matter, usually beginning with particles and continuing with elements, presumed particulate. This theory was usually supported . . . with elaborate experimental evidence. Yet many chemists felt that the prefatory sections of a book were the place for theories about the structure of matter, and disapproved of the tendency shown by the more advanced chemists to try to inject into chemical operations explanations in terms of particles." She suggests that this attitude probably explains why eighteenth-century chemists frequently found the writings of Becher and Stahl more satisfying than those of Boyle. It is clear that Stahl followed the custom of separating the theoretical and practical portions of his treatise. He was, however, remarkably successful in uniting the practical and theoretical (or philosophical) aspects of his subject. His "injection" was an amalgam of the notion of chemical essences and principles, together with certain corpuscularian explanations. The scheme proved immensely popular among *chemists*.

[14] Stahl, *Philosophical Principles* (2), 1.

[15] *Ibid.*, 2.

[16] The clearest exposition of the matter with which I am familiar is: C. M. Turbayne, *The Myth of Metaphor*, Yale, 1962, 28–53. For a discussion of the lexicographical vicissitudes of the terms analysis and synthesis see: L. M. Régis, "Analyse et synthèse dans

well known, the Aristotelian methodological enterprise was characterized by attempts to find correct definitions, which would express the essences of things.[17] If the correct definition could be supplied, then one might hope that other attributes of things could be obtained deductively from the given definition. In subsequent writers, attempts to penetrate to the essences of things became known as analysis or resolution.[18] If, subsequently, deductive chains could be set up, moving from the first principles, established by the preliminary analysis, to the phenomena under investigation—the process of synthesis or composition—then a completely scientific account of the matter would have been offered. The double pathway of analysis and synthesis can be traced through the writings of Galen, Chalcidius, Boethius, Scotus Erigena, Albertus Magnus, Aquinas,[19] and the methodologists of the Paduan medical school[20] (notably Zabarella). A prime (or notorious) example of the procedure is to be found in the work of Hobbes, who believed that by a mental analysis one could reduce all phenomena to the two principles of matter and motion. Starting from these, he hoped that one could, by synthesis, account for all physical phenomena, and even for the mental and moral aspects of the world.[21]

Thus we face the problem: Does Stahl's suggestion that chemistry operates within the traditional methodological framework of analysis and synthesis imply that he was an essentialist, choosing to seek first principles by the procedures of chemical analysis, rather than mental resolutions? In answering this question it should be remembered that the dual methodological pathway of analysis and synthesis was invoked by many writers until well into the

l'oeuvre de saint Thomas", in *Studia Mediaevalia in honorem admodum Reverendi Patri Raymondi Josephi Martin, O.P., S.T.C.*, Brugis Flandrorom, 1948, 303–30. Other valuable discussions of the methodological pathways of analysis and synthesis are to be found in: A. C. Crombie, *Robert Grosseteste and the Origins of Experimental Science*, 1953, 28–9, 61–79; R. McKeon, "Philosophy and the Development of Scientific Methods", *Journal of the History of Ideas*, **27**, 3–22, 1966; J. H. Randall (Jr.), "The Development of Scientific Method in the School of Padua", *Journal of the History of Ideas*, **1**, 177-206, 1940; N. W. Gilbert, *Renaissance Concepts of Method*, New York, 1960. See also, W. P. D. Wightman, "Quid sit methodus? Method in sixteenth-century medical teaching and discovery", *Journal of the History of Medicine and Allied Sciences*, **19**, 360–76, 1964.

[17] Thus Aristotle says "there is knowledge of each thing only when we know its essence" (*Metaphysica*, 1031b), and the "definition is the formula of the essence" (*Metaphysica*, 1031a). (W. D. Ross, *The Works of Aristotle translated into English* . . . , viii, 2nd edn., 1928.) See also, note 8, above.

[18] *Cf.* Turbayne (16).

[19] *Cf.* Régis (16).

[20] *Cf.* Randall (16).

[21] T. Hobbes, *The English Works of Thomas Hobbes of Malmesbury; now first collected and edited by Sir William Molesworth*, Bart., 11 vols., London, 1839, i, 69 and *passim*. (Reprinted Scientia Aalen, 1962.)

nineteenth century—men who cannot possibly be described as scholastic essentialists. It is mentioned in the work of Hooke[22] and Newton[23] (although they confused matters by using the words analysis and synthesis in exactly opposite ways[24]). It is to be found in methodological writings of the late eighteenth century,[25] and again in the nineteenth century,[26] though the method has largely fallen into disregard in the present century.[27] Thus it would appear that one cannot claim that Stahl was an essentialist, merely because he attempted to use the methodological pathways of analysis and synthesis. One may say, however, that his utilisation of these pathways reveals his belief in the efficacy of traditional methodological procedures.

As Stahl's exposition of the theory of chemistry unfolds, it becomes clear that he is attempting to ground it firmly on the traditional theory of fourfold causation. As early as the first page of the *Principles*, we are informed that chemistry has a "subject"—"all the *mix'd*, *compound*, and *aggregate Bodies* that are *resolvable* and *combinable*". It has "ends" in view—both theoretical or philosophical, and practical, the latter having "medicinal, mechanical and oeconomical" aspects. It has its own particular "means", which are "either *remote* or *immediate*; that is, either *Instruments* or the *Operations* themselves". And finally, it has its "efficient" cause—the chemist!

A somewhat similar approach is adopted when Stahl discusses the factors

[22] R. Waller, *The Posthumous Works of Robert Hooke, M.D., S. R. S. Geom. Prof. Gresh, etc. Containing his Cutlerian Lectures, and other Discourses, . . . ,* London, 1705, 330. (Reprinted by the Johnson Reprint Corporation, New York and London, 1969.)

[23] I. Newton, *Opticks: or, a Treatise of the Refractions, Inflexions and Colours of Light*, 4th edn., corrected, London, 1730, 404–5. (Reprinted Dover, New York, 1952.)

[24] This problem has been examined by Mary Hesse in her paper, "Hooke's philosophical Algebra", *Isis*, **57**, 67–83, 1966. I have also considered it in my paper, "Robert Hooke's Methodology of Science as Exemplified in his *Discourse of Earthquakes*", *British Journal for the History of Science*, **6**, 109–30, 1972.

[25] See, for example, Nicholson, W., *An Introduction to Natural Philosophy. . . . In Two Volumes*, 4th ed., London, I, 101–2.

[26] See, for example, A. Bain, *Logic Part Second Induction*, 2nd. ed., London, 1873, Appendix F, Analysis and Synthesis, pp. 397–402.

[27] Doubtless most chemical tyros in this century assiduously learn the text-book definitions of chemical analysis and synthesis while being quite unaware of the methodological ancestry of these terms. Greater emphasis was given to the procedures in the elementary texts of the previous century, closer to their methodological roots than are the analogous works of the present day. For example, in Samuel Parkes's well known *Chemical Catechism*, we find: "In order to ascertain the exact nature of bodies, we have often recourse to *synthesis* as well as analysis. Whenever the component parts of any body are reunited in order to form a similar substance and a similar substance is actually produced, the nature of that body is said to be proved by synthesis. When a body admits of being examined in both these ways, the result is very satisfactory." (S. Parkes, *The Chemical Catechism, with Notes, Illustrations, and Experiments*, 5th edn., London, 1812, 25.)

which cause particles to come together to form mixts (or compounds, as we say today). As is well known, Stahl adopts the four "principles" of his master, Becher—water and three "earths"—as the components of material substances.[28] These, then, function as the material causes of mixts. The efficient cause is subdivided into the "principal and less principal".[29] The former is no less than the "constant Action of God" or, in short, "Nature". The latter is, more prosaically, motion. Motion it is that brings the particles of the four "Principles" together to form mixts. This may occur naturally, in Nature's own time, place and manner, or artificially, owing to the activities of the chemist: "In short, *Nature* acts by the means of general motion; but the *Artist* applies particular motions to his Subjects, at a certain time, and in a certain place, and so can operate where and when he pleases; but *Nature* only when and where opportunity spontaneously allows."[30] Evidently, the chemist must play an important role in the scheme of things if he is able to emulate the activities of the divine artificer![31]

It is noteworthy that Stahl does not accept either Cartesian/Gassendist hooks and eyes or the Newtonian notion of intercorpuscular attraction as the efficient cause of mixtion. Imbued as he was with German pietism, the "constant Action of God" doubtless seems more acceptable than these alternative abstractions.

Only the formal cause of mixtion remains to be discussed, for Stahl does not in this instance specify a final cause. The form of mixts, he tells us, corresponds to the "different figure, number and situation of the *material Principles*".[32] Here the twentieth- and eighteenth-century chemists will doubtless find agreement as to a proper mode of explanation, although our moderns do not care to use the term formal cause. But Stahl is not able to tell us how the formal cause of a mixt is to be ascertained by the chemist.

It is clear, then, that Stahl has chosen to present his work within the framework of the traditional theory of causation. But again, this does not prove that Stahl was an essentialist, though once more it may point in that direction. Doubtless he learned to think in terms of the traditional four causes when he

[28] The ontological characteristics of these are discussed below, pp. 43–5.

[29] Stahl, *Philosophical Principles* (2), 8.

[30] *Ibid.*, 9.

[31] Possibly here (and also in his assertion that the function of the chemist is that of an efficient cause) Stahl is betraying the influence of the long alchemical tradition which preceded him. The alchemist saw himself as a *collaborator* in processes which were believed to occur slowly in nature—the growth and maturation of metals. He could, by means of the instruments of his art, speed up such changes. See, M. Eliade, *The Forge and the Crucible*, New York, 1956, in which this interpretation of the role of the alchemist is developed at length.

[32] Stahl, *Philosophical Principles* (2), 9.

was a student at Jena, but he may have continued to use this framework in his later life without necessarily being committed to an essentialist metaphysic.

Next we look at his matter theory, which is treated in the first section of the book, following the preliminary introduction.

> All *natural Bodies* are either *simple* or *compounded*; the *simple* do not consist of *physical parts*; but the *compounded* do. The *simple* are *Principles*, or the first material causes of *Mixts*; and the *compounded*, according to the difference of their mixture, are either *mix'd*, *compound* or *aggregate*: *mix'd*, if composed merely of Principles; *compound*, if form'd of Mixts into any determinable single thing; and *aggregate*, when several such things form any other entire parcel of matter, whatsoever it be.[33]

Evidently, Stahl believes in a hierarchy of matter, from "principles", through mixts and compounds, to aggregates. The notion of such a hierarchy was not, of course, new. It may be found in earlier writers such as Sennert[34] and Boyle.[35] But, if such a theory is to be really satisfactory, one must know something of the "principles". What are they? The traditional methodologist would have hoped to arrive at some first principles, which could be used as starting points for deductive chains of reasoning, by means of some kind of mental resolution of phenomena. However, such a programme, of the kind envisaged by Hobbes, was surely too ambitious. Perhaps the chemist in his laboratory, with his fires and furnaces, distillatories, files, menstrua and the like, would be able to break through to the desired first principles, where mental resolution had failed to penetrate. It would appear that Stahl did think in such terms, though he was only too aware of the difficulties involved. We read:

> A *Principle* is defined, *a priori*, [as] that in mix'd matter, which *first existed*, and *a posteriori*, [as] that into which it is at last resolved.
>
> Both these Definitions are exact, if we allow of a *pure, natural resolution*: but as this is not easily obtainable from the Chemistry of these days, and so can hardly be come at by Art; a difference, at present, prevails between the *physical* and *chemical Principles* of mix'd Bodies.
>
> Those are called *physical Principles* where of a Mixt is really composed; but they are not hitherto settled; for the four *Peripatetical Elements*, according to their vulgar acceptation, do not deserve this title. And those are usually term'd *chemical Principles*, into which all Bodies are found reducible by the chemical operations hitherto known.[36]

[33] *Ibid.*, 3.

[34] *Danielis Sennerti Opera Omnia* . . . , 3 vols., Paris and Venice, 1641, I, 154. (Cited in, A. G. Van Melsen, *From Atomos to Atom: The History of the Concept Atom*, New York, 1960, 86.)

[35] *The Philosophical Works of the Honourable Robert Boyle Esq; Abridged, . . . In Three Volumes. By Peter Shaw, M.D.*, London, 1725, I, 209–10.

[36] Stahl, *Philosophical Principles* (2), 4.

This passage has a number of interesting features. As Stahl points out, "principles" (or elements) may be viewed either as substances acting as starting points for the primordial syntheses, or as the last products of analysis or resolution. The two should be identical, provided that a complete resolution can be achieved successfully. But in actuality such attempts to carry through resolutions to the ultimate components of matter were never successful, and, it would appear, one could never know with certainty whether or not the process had been carried out to the required limits. The essentialist, however, would naturally focus his attention on the ultimate first principles—such as Stahl's "physical Principles". By contrast, a nominalist (and I shall suggest below that Lavoisier falls into this category) would be concerned with the approach from the other end—from the *a posteriori* definition, as Stahl termed it. It will be noted that Stahl's definition of a "chemical Principle" ("that into which . . . Matter is at last resolved") is extraordinarily similar to Lavoisier's pragmatic "last point which analysis is capable of reaching".[37] Thus it is not the operational concept of an element which is present in Lavoisier's work, but lacking in Stahl's. The "chemical Principles" of the latter correspond operationally to the "simple substances" of the former. It is, rather, that the two men have different intellectual stances. Stahl still sees the search for "physical Principles" as a worthy, albeit difficult, aim. Lavoisier has given up the search in disgust. Stahl's "chemical Principles" have certain essentialist features whereas Lavoisier's "simple substances" have, for the most part, lost this characteristic. But about this more will be said below.

Stahl, then, is well aware that the eighteenth-century chemist cannot penetrate to the "physical Principles", and he does not pretend that the "vulgar" earth, water, air, and fire function as such. More surprisingly, he has doubts as to the nature of the "chemical Principles". Salt, Sulphur and Mercury are proposed—but perhaps Salt, Oil and Spirit have better claims. In any case, one can only speak "generically". Thus one obtains sulphureous or mercurial products by fire analysis,[38] rather than the specific substances, sulphur or mercury. He believes that the difficulties in ascertaining the "chemical Principles" correctly arise from failure to distinguish properly between compounds and mixts. But this is perhaps an excuse for the failure to obtain consistent results in the traditional procedures of analysis by fire.

Having thus shown that he was fully aware of the difficulties involved in

[37] A. L. Lavoisier, *Elements of Chemistry in a New Systematic Order, Containing all the Modern Discoveries . . . Translated from the French, by Robert Kerr, . . .*, Edinburgh, 1790, xxiv. (Republished with a new introduction by D. McKie, Dover, New York, 1965.)

[38] For a discussion of the traditional techniques of analysis by fire, and of the theoretical presuppositions of the art, see A. G. Debus, "Fire Analysis and the Elements in the Sixteenth and Seventeenth Centuries", *Annals of Science*, **23**, 127–47, 1967.

arriving at knowledge of the "physical Principles" and that there is even uncertainty as to what one should take as the "chemical Principles", Stahl then blandly adopts the "principles" of his master, Becher, as the "immediate material Principles of Mixts", *viz.*, water and earth, there being three kinds of earth—vitrifiable, inflammable and liquifiable (or "specifically mercurial"). Whether in fact these are to be taken as chemical or physical principles is not clearly stated, but since there is a suggestion that there is a substructure below the level of these four, it would appear that he is thinking of them as "chemical Principles". This interpretation is confirmed by the discussions of the experimental portion of the work, some of which will be considered below. It may be noted that Stahl has decided in favour of the opinions of his master, Becher, giving more weight to these than to all the experimentation which is subsequently described. Or, one may say, the experiments ostensibly lead Stahl to conclusions, which have already been arrived at from other considerations.

Having thus settled to his satisfaction the nature of the principles from which the matter of the globe is composed, Stahl proceeds to specify how they are combined together in the various mundane substances known to the chemist. The following table gives a summary account of my interpretation of Stahl's views.[39]

$W$ = water $\qquad$ $E_2$ = inflammable earth

$E_1$ = vitrifiable earth $\qquad$ $E_3$ = mercurial earth

$$E_1 + E_2 \xrightarrow{\qquad\qquad} \text{Ruby}$$

$$E_1 + E_2 \xrightarrow[\text{cohering}]{\text{loosely}} \text{Bole}$$

$$E_1 + \underbrace{(E_2 + E_3)}_{\text{in excess}} \xrightarrow{\qquad\qquad} \text{Gold}$$

$$\underbrace{(E_1 + E_2)}_{\text{in excess}} + E_3 \xrightarrow{\qquad\qquad} \text{Silver}$$

$$E_1 + E_2 + \underset{\text{(excess)}}{E_3} \xrightarrow{\qquad\qquad} \text{Lead}$$

$$\underset{\substack{\text{(coarser and}\\ \text{more chalky}\\ \text{than usual)}}}{E_1} + E_2 + E_3 \xrightarrow{\qquad\qquad} \text{Tin}$$

$$E_1 + E_2 + E_3 \xrightarrow[\text{sandy earth}]{\text{sulphur and}} \text{Iron}$$

$$E_1 + E_2 + E_3 \xrightarrow[\text{so much sandy earth}]{\text{ditto, but not}} \text{Copper}$$

$$W + E_1 \xrightarrow{\qquad\qquad} \text{Universal acid (U)} \; [\text{sulphuric acid}]$$

[*continued overleaf*

[39] Equations similar to those given here, but relating to the chemical opinions of Becher, were drawn up by Rodwell (6), 22.

TABLE—*continued*

| | | |
|---|---|---|
| U + $E_2$ | ⟶ | Bitumen (B) |
| W + $E_2$ | ⟶ | Oil |
| W + $E_1$ + B | ⟶ | Sulphur (S) |
| U + $E_3$ | ⟶ | Salt (the "foundation of arsenic") |
| U + $E_2$ | ⟶ | Nitre |
| Metal + $E_3$ | ⟶ | Quicksilver (Q) |
| Q + S | ⟶ | Cinnabar (C) |
| C + $E_2$ | ⟶ | Realgar [red sulphide of arsenic] |
| U + Chalky earth | ⟶ | Alum |
| U + $E_3$ | ⟶ | Sea salt (SS) |
| SS + Clay | ⟶ | Sal gem [rock salt] |
| U + Metal | ⟶ | Vitriol |
| Metals + Arsenical earth | ⟶ | Immature metals [Bismuth, zinc, marcasite, regulus of antimony] |

These proposals may appear speculative in character to us, and doubtless it was work of this kind which contributed to the severe censures passed on the Stahlian chemical philosophy by nineteenth-century historians of chemistry. It should be noted, however, that Stahl does try to offer chemical evidence for these assertions in the latter portions of his book. Two examples will suffice.

Alum can be made by the action of the Universal acid (our sulphuric acid) on "a chalky earth, resembling Lime-stone".[40] This gives him a synthesis of alum. An analysis is also possible:

> Oil of Tartar *per deliquium*, or the common alkaline lixivium, being poured to a Solution of *Alum*; the alkaline particles by this means are combined with the acid ones of the *Alum*; which being disposed to unite more firmly and intimately with those, they drive their former associated terrestrial particles of the Alum from their places; and these now coming together in larger cluster, first muddy the liquor, and afterwards fall to the bottom.[41]

We should note, however, that the claim that the Universal acid consists of water and earth remains unsubstantiated.

As our second example we consider nitre. As was well known, this could be distilled (with vitriol or alum) to yield an acid liquor. If fixed alkali were added to this liquor, nitre could be regenerated. This suggested that nitre was composed of some earthy substance and the acidic liquor. The acidic liquor was

[40] Stahl, *Philosophical Principles* (2), 87.

[41] *Ibid.*, 88.

oily, saline and volatile, so that the principles giving rise to these properties were presumed to be present in it. Nitre was formed when urinous matter and earth were left in contact for some time. Hence the requisite qualities from the urinous matter and the earth were supposed to be combined in the nitre. Presumably, however, it was not a wholesale combination of the earth with the urinous matter, for extreme inflammability and volatility were lacking in the nitre. Thus only certain of the component principles of the reagents were assumed to have gone to form the nitre. The system was flexible—too flexible, we might be inclined to say. Also, it was not even consistent. In most cases $E_2$ is found in inflammable substances, but inspection of the Table shows that this rule had a number of exceptions.

These examples show us something of Stahl's methods. He was evidently endeavouring to follow his methodological programme of analysis and synthesis. Moreover, it is clear from the last example which we have considered that he was thinking as an essentialist, in the sense that his principles conferred particular properties on the substances of which they were constituents.

Such an interpretation is, I believe, confirmed by the opening paragraph of the second part of the work, "*The* PRACTICE *of* CHEMISTRY". Here we read: "We have hitherto considered the *Subject, Object* and *Means of Chemistry*, in general; it remains that we next particularly lay open the several *Species of its Subjects*; . . . And this we shall do in the *analytical Method*; that each of these *Species* may be considered according to its essence, and component parts . . . "[42]

Thus, from the overall evidence which has been considered above, it does appear that Stahl reveals certain significant essentialist features in his thinking. By the analytical method, he does hope to be able to penetrate to the essences of things. In saying this, I do not, of course, mean to brand him as a medieval scholastic. It is, rather, that one can discern some evidence for the influence of the old essentialist doctrines on his thinking. It would be fair, perhaps, to view him as an intermediate figure, between the former scholastic thinking and the new chemical world personified by Lavoisier.

Stahl gives considerable attention to the practical details of analytical procedures. His section entitled "*The* OBJECT *of* CHEMISTRY, *consider'd under the* NOTION *of* DESTRUCTION *or* RESOLUTION"[43] tells us how to convert aggregates into compounds or mixts, and compounds into mixts (or "compounds of higher degree" into "compounds", in modern parlance). Here, the obvious ways in which a chemist might divide his materials into smaller parts are listed—limation, trituration, extinction, sublimation, calcination,

[42] *Ibid.*, 76.
[43] *Ibid.*, 21–42.

solution, corrosion, amalgamation, and fusion. These can break up aggregates. Resolution of compounds may be achieved by distillation, rectification, decantation, evaporation, filtration, amalgamation, sublimation, deflagration, difflation, corrosion, fusion, cementation, gradation, edulcoration, precipitation, cupellation, and filtration. Here, as we see, Stahl is largely referring to processes which we customarily associate with the separation of mixtures.

In his section on "The OBJECT *of* CHEMISTRY, *consider'd under the Notion of* GENERATION *or* COMBINATION"[44] Stahl teaches us how to bring matter together again, that is, to produce materials by synthesis. Here one might employ: collective combination, coagulation, fermentation, and cohobation. Evidently his whole practical scheme is geared to the methodological programme.

Something should also be said about Stahl's "element/instrument" theory, which occupies an important part in his whole theoretical scheme. It is discussed in his section entitled "The COMMON INSTRUMENTS *of* CHEMICAL OPERATIONS".[45] The notion did not originate in the work of Stahl, but under his influence it came to occupy a position of particular importance in the chemical theories which were in vogue during the eighteenth century, prior to the time of Lavoisier. As has been mentioned above, Stahl rejected the four peripatetical elements as the principles of mixts, preferring the earth and water of Becher. But these two not only functioned as the material components of mixts. They also acted as instruments. The chemist, functioning as the efficient cause in chemistry, required instruments to achieve his aims. Here the traditional tetrad of elements would serve his needs, acting as physical agents for chemical changes. Thus, earth and water could serve as both chemical and physical agents, whereas air and fire were restricted to the role of physical agents for chemical actions. Fire, water, and air, says Stahl, are instruments capable of imparting motion whereas earth is the instrument of rest. Fire is the principal agent, though air and water can also impart motion. Stolid earth plays a more humble rôle, serving as the lute for furnaces and vessels, as stable furnace bricks, *etc.* It should be noted that although air and fire are not deemed to be able to act as the components of mixts, they are corpuscular in nature. Igneous particles agitate the particles of mixts, setting them in motion, consequently bringing about chemical changes. Just as water can be seen moving earthen particles when a suspension of ashes in water is heated, so air particles can agitate water particles and thereby cause evaporation or ebullition. The following passage shows how Stahl accounted for the process of distillation. The treatment is a kind of forerunner of the kinetic

[44] *Ibid.*, 43–57.
[45] *Ibid.*, 58–75.

theory, and is indicative of some influence of the mechanical philosophy on his thinking.

> The *Fire* or *Heat*, that is a multitude of small particles in actual motion, move and impel the small constituent particles of the solid vessel; which being thus impell'd, agitate the particles of the contain'd liquor; and these again impel the air contain'd in the vessel, and being also rubb'd together by their own intestine motion, they increase the motion of the contain'd *Air*, which being now agitated in its smallest parts, with a fluctuating motion drives off the aqueous parts in great numbers, or carries them about with itself, till at length striking against the upper part of the vessel, which is colder, and has its small parts actually less moved, they here amass and again collect into *aqueous Aggregates* or *Drops*.[46]

One may note, however, that it is not the motion of the water particles which gives rise to their own movement across the air/water interface. They have, so to speak, to be dragged across by the aerial corpuscles, which are themselves set in motion by the particles of fire.

We see, then, how Stahl attempted to reconcile the many conflicting traditions with which he was familiar. The important role of water in the Helmontian chemistry is incorporated in his scheme. So, too, are the teachings of Becher, which themselves derive from the Paracelsian trichotomy of matter, which is perhaps an extension of the earlier sulphur/mercury theory of Jābir and his followers, which itself appears to be traceable to the theory of exhalations of Aristotle's *Meteorologica*. Simultaneously, Stahl has incorporated the Aristotelian theory of the four elements. The traditional chemical techniques of the "chymists" are united with the standard methodological pathways of analysis and synthesis. The influence of scholasticism is evident in the utilisation of the venerable theory of causation. We have commented on the role of essentialism in the scheme. In addition, the influence of the corpuscularianism of the mechanical philosophy has been discerned.

However, the student of Stahlian chemistry, perhaps unaware of the niceties of the eclecticism which we have been considering, might have been dissatisfied with the apparent asymmetry of the "element/instrument" theory, and, as Rhoda Rappaport has demonstrated,[47] the scheme was subsequently tidied up by Rouelle, in the version which he taught to the young Lavoisier. Thus, for Rouelle, earth, air, fire and water were all held to function both as chemical elements and instruments, or physical agents. This was, of course, a significant change, for it enabled one to contemplate the air as a chemical as well as a

[46] *Ibid.*, 61.

[47] R. Rappaport, "Rouelle and Stahl—the Phlogistic Revolution in France", *Chymia*, **7**, 73–102, 1961.

physical agent; this was one of the important steps leading to the chemical revolution.[48]

The practical sections of the *Philosophical Principles of Universal Chemistry* will not be examined here, despite the fact that they form the major portion of the whole. They provide a detailed account of the reactions and techniques known in Stahl's time, but since such descriptions are to be found in most of the chemical manuals of the period, it is not necessary to discuss them here. They are perhaps of less interest to us than Stahl's theoretical discussions, though his influence was in large measure due to the way in which he was able to weave the experimental material into his theoretical framework. His theories were admirably designed to suit the chemists, rather than the physicists. Indeed, as we know, most chemists up until Lavoisier's time were satisfied with the general theoretical scheme which Stahl erected.

Finally, a few remarks about the post-Stahlian era are offered. Did later chemists forego their hopes of developing their subjects in the traditional manner, basing their work on the operations of analysis and synthesis? It would appear not. The opening paragraph of Macquer's *Elements of Chemistry*[49] adopts exactly the same attitude towards the roles of these two traditional procedural moves as that which we have noted in Stahl's *Philosophical Principles*. But Macquer was a Stahlian. What, then, does Lavoisier tell us about these matters?

It is a somewhat difficult problem to ascertain Lavoisier's views on the subject. His general philosophy of science has perhaps received insufficient attention, by comparison with the energy which has been expended in the elucidation of the details of his *curriculum vitae*.[50] If, however, we take the introduction to his *Traité* as our text, since it gives us the clearest exposition of his views on the philosophy of science, we note immediately that all traces of the scholastic causal framework have gone. This, in itself, in my view, marks a clear distinction between the French chemist and his German predecessor.

[48] See Fichman (7).

[49] P. J. Macquer, *Elements of the Theory and Practice of Chymistry*, 2nd ed., London, 1764, I, 1. Here we read: "The Object and chief End of Chymistry is to separate the different substances that enter into the composition of bodies; to examine each of them apart; to discover their properties and relations; to decompose those very substances, if possible; to compare them together, and combine them with others; to reunite them again into one body, so as to reproduce the original compound with all its properties . . . "

[50] The writings of Hélène Metzger offer one of the most fruitful sources of information on Lavoisier's philosophy of chemistry: "La Philosophie de la Matière chez Lavoisier", *Actualités Scientifiques et Industrielles*, No. 218 (*Exposés d'Histoire et Philosophie des Sciences Publiés sous la direction de Abel Rey*), Paris, 1935; "Introduction a l'étude du rôle de Lavoisier dans l'histoire de la chimie", *Archeion*, **14**, 31–56, 1932. Note also, M. Daumas, *Lavoisier Théoricien et Expérimentateur*, Paris, 1955, esp. chap. VII.

On the other hand, one still finds the interest in analysis, and in principles and elements. But, as has been indicated above (page 44), the character of the Lavoisierian element (or "simple substance", as he preferred to call it) differs markedly from its Stahlian forerunner. The search for "physical Principles" has been abandoned entirely. It is true that Lavoisier's simple substances, pragmatically defined as "the last point which analysis is capable of reaching", bear an apparent correspondence to Stahl's "chemical Principles". However, as has been noted, even Stahl's "chemical Principles" were essentialist in character. Generally speaking, Lavoisier's elements do *not* confer properties on the substances of which they are constituents. A notable exception is, of course, oxygen—the acid former. Here the chemical style of the earlier period is retained. But, on the whole, there is a significant difference between the Stahlian "chemical Principle" and the Lavoisierian "simple substance". Can the reason for this change in viewpoint be demonstrated? The matter cannot, perhaps, be settled conclusively, but the following points are worthy of consideration.

Lavoisier, ever interested in the language of chemistry, may be regarded as a nominalist. This is evident from the manner in which he singled out Condillac for special attention in the preface to the *Traité Élémentaire*.[51] However, Lavoisier was not, I think, a radical nominalist of a Buffonian hue, believing that species are wholly figments of the human imagination. Lavoisier did believe in "natural kinds", and he quoted with approval Condillac's dictum that by observation of particulars, a child may come to the abstract or general notion of a genus. Further observation may lead one to recognize the different species. But if Condillac's nominalism was not as extreme as that of the early Buffon, he did nevertheless, like Locke, reject the notion of "real essences". It is not perhaps necessary to argue this in detail here; one quotation from *De L'Art de Raisonner* of 1775 is probably sufficient to sustain the point. Condillac wrote:

> L'or est jaune, ductile, malléable. Mais pourquoi un métal a-t-il des qualités qu'un autre n'a pas? Vous ne sauriez remonter à une qualité première, qui vous en rende raison. Vous ne sauriez donc démontrer avec précision le rapport d'un métal à un métal. Par conséquent, il ne vous reste qu'à faire l'énumération de leurs qualités, et à comparer celles de l'un avec celles de l'autre.[52]

Obviously, this is pure Lockean philosophy.[53] If, then, we accept the clear evidence that Lavoisier was a follower of Condillac, we may feel confident that

[51] Lavoisier (37), xiii, xxvi, and xxxv-vii.

[52] G. Le Roy, ed., *Oeuvres philosophiques de Condillac*, 3 vols., Paris, 1947–51, I, 628.

[53] *Cf.* J. Locke, *An Essay Concerning Human Understanding*, Bk. 3, Ch. 6. (*The Works of John Locke. A New Edition Corrected. In ten volumes*, London, 1823, II, 208 and *passim*. Reprinted Scientia Verlag Aalen, 1963.)

he would have adopted a similar attitude to the question of essences, preferring the nominalist variety of Locke to the real ones of the scholastics. Thus Lavoisier's change in emphasis from a search for underlying essences, which would give one a means of accounting for all the properties of things, to analysis to tangibles such as iron, copper, sulphur and the like, can be seen to be one of the consequences of the epistemological revolution of Locke, or at least a change which was strongly influenced by the new philosophy.[54] Lavoisier's ultimates are things in crucibles, flasks and bottles.[55] They can be corked up for inspection, later to be recombined to form new substances. By contrast, Stahl's "physical Principles" lie way beyond the reach of his senses, or the analytical techniques of his time. His "chemical Principles" are "generic" rather than "specific", and they have the essentialist characteristics which we have been discussing.

In conclusion, then, I am inclined to accept the view, recently expressed by Siegfried and Dobbs,[56] that the major aspect of the "chemical revolution" was the change in matter theory, with its accompanying methodological shift. By comparison, the change in the theory of combustion was perhaps, in the long run, a matter of subsidiary importance.

*Acknowledgements.*

I am indebted to my colleague, Dr. R. M. Gascoigne, for the valuable comments which he made on the first draft of this paper, and also to the most helpful suggestions subsequently made by the Editor's referees.

[54] The influence on biology of the epistemological shift in emphasis from essentialism to nominalism has been emphasised in a valuable article by D. L. Hull, "The Metaphysics of Evolution", *British Journal for the History of Science*, **13**, 309–37, 1967, where it is maintained that this change in philosophical stance was one of the major factors which influenced the development of an evolutionary theory. As can be seen from the arguments which have been presented above, I believe that the movement from essentialism towards nominalism played a rôle of similar significance in chemistry.

[55] We must, of course, except the "oxygen principle" (i.e. the "base" of oxygen gas), and also heat and light. It does not seem surprising that it was the "oxygen principle", with its "essentialist" character, which defied capture.

[56] R. Siegfried, and B. J. Dobbs, "Composition, a Neglected Aspect of the Chemical Revolution", *Annals of Science*, **24**, 275–93, 1968.

# VI

## A NOTE ON THE STATUS OF A. F. CRONSTEDT'S SIMPLE EARTHS AND HIS ANALYTICAL METHODS

It is well known that Axel Fredrik Cronstedt's *Försök til Mineralogie*, first published in 1758, stands as an important landmark in the history of mineralogy.[1] His system of classification was based to a large extent on the results of chemical tests performed with the simple blowpipe. "External" characteristics, such as hardness, color, and crystalline form, though not ignored, were given much less emphasis than in earlier mineralogical texts. Consequently, it has been customary to think of Cronstedt as one of the founders of mineral analysis, and Bartow has suggested that he might well be called "the father of analytical chemistry."[2]

I would, however, like to consider whether the tests which Cronstedt performed with his blowpipe amounted to genuine chemical *analyses*. Did they involve chemical decompositions and separations, or did they in fact merely provide convenient means for the identification of minerals by the characteristic chemical changes which occurred when the tests were performed? These questions are, I think, worth some consideration.

The originality and importance of Cronstedt's work are not in question. Although he was not the first to make use of chemical criteria for the purpose of mineral classification and was probably not the first to make use of the blowpipe in mineralogical investigations, his work with this instrument enabled him to bring about radical changes in mineral taxonomy, giving it an orientation toward chemical criteria that was to last until the latter part of the eighteenth century, when Haüy's crystallographical methods were introduced.

The nature and extent of Cronstedt's revision of mineral taxonomy may be gauged from Table 1, which shows the development of systems of mineral classification in the years immediately preceding the publication of the *Mineralogie*.[3] This table clearly reveals the strong and continuing influence exerted

[1] A. F. Cronstedt, *Försök til Mineralogie, eller Mineral-Rikets upställning* (Stockholm, 1758). There were three English editions of this work (London, 1770, 1772, and 1788), of which I have used the second of 1772. The full title of this edition is given in n. *j* of Table 1.

On Cronstedt's mineralogy see Virginia Bartow, "Axel Fredrik Cronstedt," *Journal of Chemical Education*, 1953, *30:* 247–252; E. M. Gilsenius, "Beginnings of Greatness in Swedish Chemistry (II). Axel Fredrik Cronstedt (1722–1765)," *Transactions of the Kansas Academy of Science*, 1969, *72:* 476–484; C. S. St. Clair, "The Classification of Minerals: Some Representative Mineral Systems from Agricola to Werner" (Ph.D. dissertation, University of Oklahoma, 1966).

[2] Bartow, "Cronstedt," p. 251. (Incidentally, in listing the divisions of the earths proposed by Cronstedt, Bartow mistakenly referred to the garnets as "granites" and the manganese kind as "magma.")

[3] It has been necessary in the compilation of this chart to alter the order of tabulation of some of the authors in order to display the common features of the different schemes to

by Linnaeus, Henckel, and Pott on systems of mineral classification in the middle years of the eighteenth century. Also apparent is the unanimous agreement that the categories of salts, sulfurs, and metals were satisfactory taxa for the classification of the mineral kingdom. It will be recalled that these subdivisions (together with the tribe of earths) originated in the work of Avicenna[4] (though it is not until 1772 that one finds a really clear exposition of the practical procedures that were to be employed in the placing of minerals in their correct pigeonholes, according to the Arab's quadrichotomy of the mineral kingdom[5]).

The earths and stones were evidently much more troublesome than the salts, sulfurs, and metals. We see that there was continuing disagreement as to whether they should be classified together or separately and whether external characteristics or chemical properties should be used as criteria for their discrimination and classification. In general, however, it is apparent that the authors listed in Table 1 tried to utilize the categories determined by Linnaeus' trichotomy of stones,[6] or Pott's quadrichotomy of earths and stones. But when we come to Cronstedt's system we find very significant revision of the earlier schemes, with a complete restructuring of the classification of earths and stones, nine subdivisions of these being established, including six entirely new categories.[7]

On examining Cronstedt's account it is apparent that he was attempting to offer a *chemical* classification of the earths rather than a "natural history" based on external characteristics. He wrote:

> I flatter myself with so much success, that students, who intend to follow this proposed method, will not be so easily mistaken in the subjects of the mineral kingdom, as has happened with me and others in following former systems; and I also hope to obtain some protectors against those who are so possessed with the *figuromania,* and *so addicted to the surface of things,* that they are shocked at the boldness of calling a *Marble* a *Limestone,* and of *placing the Porphyry amongst the Saxa.*[8]

---

the best advantage. In doing this, however, I have not altered the mutual classificatory relationships of any of the authors' taxa.

[4] See *Avicennae De Congelatione et Conglutinatione Lapidum. Being Sections of the Kitâb Al-Shifâ. The Latin and Arabic texts edited with an English Translation of the latter and with critical notes by E. J. Holmyard . . . and D. C. Mandeville* (Paris: Paul Geuthner, 1927), p. 33. The salts, sulfurs, and mercuries of the mid-eighteenth-century mineralogists were also clearly dependent on the *tria prima* of the Paracelsians and the three "earths" of chemical tradition established by J. J. Becher.

[5] Such an exposition is to be found in J. R. Forster, *An Easy Method of Assaying and Classing Mineral Substances. Containing Plain and Easy Instructions for any person to examine the Products of his own Land, . . .* (London: E. and C. Dilly, 1772).

[6] This trichotomy, based on the action of heat on stones, did not originate in the work of Linnaeus. It appeared also in the *Mineralogia* of Magnus von Bromell (Stockholm, 1730), Urban Hiärne's *Een kort Inledning till åtskillige Malm-och Bergarter, . . .* (Stockholm, 1694), and J. J. Becher's *Physicae subterraneae* (Leipzig, 1738; 1st ed. 1669).

[7] The other three (calcareous, siliceous, and argillaceous) were traditional, as shown in Table 1. It should be noted, however, that Cronstedt *did* consider it possible that all earthy substances might eventually be found to be reducible to two simple kinds, the calcareous and the argillaceous. (See Cronstedt, *Mineralogy,* p. xvii.)

[8] *Ibid.,* p. xxi.

**Table 1. The Evolution of Systems of Mineral**

| *Linnaeus*[a] *1735* | | *Henckel*[b] *?* | | *Pott*[c] *1746* | | *Wallerius*[d] *1747* | | *Woltersdorff*[e] *1748* | |
|---|---|---|---|---|---|---|---|---|---|
| | | waters | | | | waters | | | |
| stones | refractory | earths | refractory | earths and stones | alkaline and calcareous | earths | dry | earths | argillaceous |
| | | | fusible | | | | argillaceous | | alkaline |
| | | | | | | | mineral | | |
| | | | | | | | sandy | | |
| | calcareous | stones | calcareous | | | stones | calcareous | stones | alkaline |
| | | | argillaceous | | gypseous | | refractory | | gypseous |
| | | | calcareous-siliceous | | argillaceous | | glassy | | argillaceous |
| | vitrescible | | siliceous | | siliceous | | rocks | | vitrifiable |
| minerals | salts | salts | | | | minerals | salts | salts | |
| | sulfurs | earthy juices and sulfureous substances | | | | | sulfurs | bitumens | |
| | | | | | | | semi-metals | semi-metals | |
| | mercurials | metals | | | | | metals | metals | |
| fossils | earths | | | | | concretions | | petrifactions | |
| | concretions | | | | | | | | |
| | petrifactions | | | | | | | | |

[a]Extracted from C. Linnaeus, *Carolus Linnaeus Systema naturae 1735. Facsimile of the First Edition With an Introduction and a first English translation of the "Observations" by Dr. M. S. J. Engel-Ledeboer and Dr. H. Engel,* Dutch Classics on History of Science, No. 8 (Nieuwkoop: B. De Graaf, 1964).

[b]Extracted from J. F. Henckel, *Introduction a la Mineralogie; ou Connoissance des Eaux, des Sucs Terrestres, des Sels, des Terres, des Pierres, des Mineraux, et des Métaux: Avec une Description abrégée des opérations de Métallurgie. Ouvrage Posthume de M. J. F. Henckel, publié sous le titre de Henckelius in Mineralogiâ redivivus, & traduit de l'Allemand,* 2 vols. (Paris: Chez Guillaume Cavelier, 1756). The first edition, published posthumously at Dresden by J. E. Stephanus in 1747, contained an account of Henckel's mineralogical lectures at Freiberg. Since Henckel died in 1744, the scheme of classification that is tabulated here cannot be dated precisely.

[c]Extracted from J. H. Pott, *Lithogéognosie ou examen chymique des Pierres et des Terres en général, Et du Talc, de la Topaze, & de la Stéatite en particulier, Avec une Dissertation sur le Feu & sur la Lumière,* 2 vols. (Paris: Chez Jean-Thomas Herissant, 1753). The first German edition was published in Potsdam in 1746.

[d]Extracted from J. G. Wallerius, *Mineralogie, Ddscription* [*sic*] *Générale des Substances du Regne Mineral. Par Mr. Jean Gostchalk Wallerius . . . Ouvrage traduit de l'Allemand,* 2 vols. (Paris: Chez Durand et Pissot, 1753). The first edition (in Swedish) was published in Stockholm in 1747. (Wallerius' class of "waters" was discussed in a separate treatise, *Hydrologia eller Wattu-Riket,* published in Stockholm in 1748).

[e]Extracted from J. L. Woltersdorff, *Systema Minerale in Quo Regni Mineralis Producta Omnia Systematice per Classes, Ordines, Genera et Species Proponuntur* (Berlin: Christian Ludewig Kunst, 1748).

## Classification from Linnaeus to Cronstedt

<table>
<tr><th colspan="2">Gellert[f]<br>1755</th><th colspan="2">Cartheuser[g]<br>1755</th><th colspan="2">D'Argenville[h]<br>1755</th><th colspan="2">von Justi[i]<br>1757</th><th>Cronstedt[j]<br>1758</th></tr>
<tr><td rowspan="4">earths</td><td rowspan="2">argillaceous</td><td rowspan="4">earths</td><td rowspan="4">divisions based on external characteristics</td><td rowspan="4">earths</td><td rowspan="4">divisions based on external characteristics</td><td rowspan="9">earths and stones</td><td rowspan="2">precious (noble)</td><td>calcareous</td></tr>
<tr><td>siliceous</td></tr>
<tr><td rowspan="2">alkaline</td><td rowspan="2">semi-precious (semi-noble)</td><td>garnet kind</td></tr>
<tr><td>argillaceous</td></tr>
<tr><td rowspan="5">stones</td><td>calcareous</td><td rowspan="5">stones</td><td rowspan="5">divisions based on external characteristics</td><td rowspan="5">stones</td><td rowspan="5">divisions based on external characteristics</td><td rowspan="2">calcareous</td><td>micaceous</td></tr>
<tr><td>"fluores"</td></tr>
<tr><td>gypseous</td><td rowspan="2">refractory</td><td>"asbestus" kind</td></tr>
<tr><td>argillaceous</td><td rowspan="2">fusible</td><td>zeolites</td></tr>
<tr><td>vitrifiable</td><td>manganese kind</td></tr>
<tr><td colspan="2">salts</td><td colspan="2">salts</td><td colspan="2">salts</td><td colspan="2">salts</td><td>salts</td></tr>
<tr><td colspan="2">sulfurs</td><td colspan="2">inflammables</td><td colspan="2" rowspan="2">sulfurs</td><td colspan="2">combustibles</td><td>inflammables</td></tr>
<tr><td colspan="2">semi-metals</td><td colspan="2">semi-metals</td><td colspan="2">semi-metals</td><td>semi-metals</td></tr>
<tr><td colspan="2">metals</td><td colspan="2">metals</td><td colspan="2">metals</td><td colspan="2">metals</td><td>metals</td></tr>
<tr><td colspan="2"></td><td colspan="2">heteromorphs</td><td colspan="2"></td><td colspan="2">petrifactions</td><td>petrifactions</td></tr>
</table>

[f]Extracted from C. E. Gellert, *Metallurgic Chymistry. Being a System of Mineralogy in General, and of all the Arts arising from this Science. To the great Improvement of Manufactures, And the most capital Branches of Trade and Commerce. Theoretical and Practical. In Two Parts. Translated from the Original German of C. E. Gellert, By I.* [John] *S.* [eiferth] (London: T. Becket, 1776). The first German edition was published in Leipzig in 1755.

[g]Extracted from F. A. Cartheuser, *Elementa mineralogiae systematice disposita A Friderico Augusto Cartheuser Med. Doct.* (Frankfurt: Pauli Sigismundi Gaebler, 1755).

[h]Extracted from Anon. [A. J. Dezallier d'Argenville], *L'Histoire naturelle eclaircie dans une de ces parties principales, l'oryctologie, qui traite des terres, des pierres, des métaux, des minéraux, et autres fossiles. Ouvrage dans lequel on trouve une nouvelle methode Latine & Françoise de les diviser, & une notice critique des principaux ouvrages qui ont paru sur ces matières* (Paris: Chez De Bure l'Aîné, 1755).

[i]Extracted from J. H. G. von Justi, *Grundriss des gesamten Mineralreiches worinnen alle Fossilien in einem, ihren wesentlichen Beschaffenheiten gemässen, Zusammenhange vorgestellet und beschrieben werden* (Göttingen: Im Verlag der Wittwe Vandenhöck, 1757).

[j]Extracted from A. F. Cronstedt, *An Essay towards a System of Mineralogy, by Axel Fredrik Cronstedt. Translated from the Original Swedish, with Notes, By Gustav von Engestrom. To which is added, A Treatise on the Pocket-Laboratory, Containing An Easy Method, used by the Author, for Trying Mineral Bodies, Written by the Translator. The Whole Revised and Corrected with some Additional Notes, By Emanuel Mendes da Costa* (2nd English ed., London: Edward and Charles Dilly, 1772).

And again:

> . . . earths are here arranged according to their constituent parts, as far as hitherto discovered, and are divided into nine orders.[9]

So Cronstedt's intentions seem to have been clear enough. Nine categories of earths were established, according to their several responses to a variety of chemical (and also some physical) tests. These nine categories were to correspond with nine types of "constituent parts"; and thus it appears that he envisaged that there were at least nine (and doubtless possibly more) individual simple earthy substances, even though he also entertained the possibility of the ultimate reduction of the earths to a very few simple kinds.[10] The establishment of Cronstedt's nine orders of individual earths seems to accord well with the account given by Siegfried and Dobbs of the gradual breakdown during the eighteenth century of the original speculative "ideal" earth of the alchemists and early chemists into a *class* of several simple earthy substances capable of being separated from each other by chemical analysis so that they might be stored in separate bottles.[11]

However, before trying to apply the account of Siegfried and Dobbs to the work of Cronstedt, one should look at the actual procedures that he used in order to establish his orders. As already indicated, his favorite weapon in the analytical attack upon the mineral kingdom was the simple blowpipe.[12] Cronstedt did not actually give a formal account of the use of this instrument, but his compatriot Gustav von Engeström appended an illustrated essay on this topic to his English translation of the *Mineralogie.*[13] From this one can, I think, gain a good impression of the methods employed by Cronstedt himself, for von Engeström's account meshes well with the descriptions that one can, with more trouble, tease out from the *Mineralogie* proper.

Now according to von Engeström's account the tests to be applied in the examination of a mineral were these:

1. An examination with the magnifying glass to discover heterogeneous parts
2. A test for hardness
3. The test of the action of heat on a sample at the charcoal block
4. The test of the action of the heat of a blowpipe flame, using soda as flux
5. Ditto, using borax as flux
6. Ditto, using microcosmic salt as flux

Essentially, then, the tests were simplified versions of the standard assayers' techniques, though they enabled the "mineralist" to make simple identifications cheaply and conveniently and also quite reliably. For example, considering

[9] *Ibid.*, p. 9.

[10] See n. 7.

[11] R. Siegfried and B. J. Dobbs, "Composition, a Neglected Aspect of the Chemical Revolution," *Annals of Science*, 1968, *24:* 275–293.

[12] For some account of the history of the use of this instrument, see my paper "Edward Daniel Clarke, 1769–1822, and his Rôle in the History of the Blow-Pipe," *Ann. Sci.*, 1972, *29:* 213–235.

[13] Von Engeström's essay was entitled "Description and Use of a Mineralogical Pocket Laboratory; and especially the Use of the Blow-Pipe in Mineralogy. By Gustav von Engestrom."

the action of heat using a blowpipe at the charcoal block we have the results shown below.

| ***Substance under investigation*** | ***Responses to the action of heat on a charcoal block*** |
|---|---|
| Calcareous substances | These do not melt but become friable. On cooling and treatment with water, considerable heat is evolved. |
| Siliceous substances | These become brittle but do not melt. They sometimes lose their color. |
| Garnets | These melt and give a black slag. |
| Argillaceous substances | These do not melt but become white and hard. If much iron is present, they may darken and fusion may occur. |
| Mica and "Asbestus" | These become hard and brittle and are more or less refractory, though there may be some signs of fusion. |
| "Fluores" | These emit light and melt to give a white, opaque slag. |
| Zeolites | These foam and form a frothy slag. |
| "Manganese" | No information given. |
| Iron ores | These may become magnetic after heating. They do not fuse. |
| Ores containing sulfur or arsenic | Fumes are evolved and also a smell of burning sulfur or garlic. |
| Lead ores | Metallic lead is produced. |
| Some copper ores | Metallic copper is produced. |
| Antimony ores | These volatilize, with the production of a characteristic smell. |
| Zinc ores | White flowers of zinc are produced. If any metal is formed, it burns with a blue flame. |

By a battery of such tests, using different fluxes, Cronstedt, von Engeström, and their colleagues would have been able to characterize their specimens with very reasonable success. I think it must be admitted, however, that these methods did not really result in genuine chemical *analyses.* There was no guarantee that they produced decompositions or resolutions to simpler substances. Rather, the new technique extended the scope of the "natural historian" of mineralogy by providing an increase in the range of characteristics that might be taken into consideration during the identification and classification of minerals.[14] Clearly, substances like mica, "asbestus," zeolites, and the like must have been recognized in the first place as "natural kinds" by their leading external physical characteristics.[15] Then when the whole known range of mineral substances was tested at the blowpipe, certain characteristic behavior patterns must have been observed, and these would have been linked with the other, independently observed, external physical characteristics. But when Cronstedt implied that a particular kind of earth (corresponding, I take it, to a particular kind of "constituent part") was associated with each characteristic set of responses to the action of the heat of the blowpipe, then a considerable theoretical leap was involved, not following strictly from the experimental evidence.[16]

Certainly, the tests employed by Cronstedt involved chemical changes and in that sense provided chemical means for the characterization of substances. But, strictly speaking, they did not amount to chemical analyses. "Blowpipe characterizations" would be a more fitting expression. Apart from the case of the metallic ores, Cronstedt's methods enabled him only to divide and sort substances rather than analyze them in the true chemical sense of the word.[17] By contrast, the earlier work of A. S. Marggraf,[18] which laid the foundations for the analytical scheme of Torbern Bergman,[19] constituted the beginnings of the chemical analysis of minerals. Here genuine chemical separations were performed, yielding substances that really could be placed in separate bottles.

It is true that if Cronstedt's methods were applied to a *rock,* then they might give an indication of the mineral content of that rock, and in that sense one might say that an analysis had been performed. This would be on a par with the identification of the components of a mineral water by the actions of various indicators or precipitants. In this limited sense, then, one might say that Cronstedt was performing analyses. But in other respects I think it better to say that his tests simply enabled him to characterize minerals.

[14] It is a task beyond the scope of this short note to give an exposition of the development of natural histories of the mineral kingdom to the middle of the eighteenth century. Suffice it to say that by that period a number of common mineral "natural kinds" had been recognized. See, e.g., J. Hill, *A General Natural History, or New and Accurate Descriptions of the Animals, Vegetables, and Minerals of the Different Parts of the World,* Vol. I: *A History of Fossils* (London: Thomas Osborne, 1748).

[15] This was probably not the case for the zeolites, for which the characteristic intumescence which occurs on heating—due to a chemical change involving liberation of water—was an obvious pointer to a new group of minerals. The name *zeolite* (from ζέω, "I boil," and λίθος, "a stone") was bestowed by Cronstedt.

[16] Cronstedt does not state *explicitly* that there is a special kind of earth associated with each of his nine orders of earthy minerals, but this hypothesis is implicit in what he says throughout his text.

[17] The word *analysis* has had a multiplicity of meanings in philosophy, mathematics, and scientific methodology. This has also been the case in chemistry, for the word has come to mean either the *separation* or the *identification* of the components of a mixture or compound. In many cases the identifications necessarily involve separations. In the procedures used by Cronstedt, however, identification of a mineral by "analysis" with a blowpipe often did not involve any separations at all, so I question whether the word analysis is appropriate.

[18] See Marggraf's collected papers: *Opuscules chymiques de M. Margraf* [*sic*], *de l'Académie de Berlin,* 2 vols. (Paris: Chez Vincent, 1762). Much of Marggraf's pioneering analytical work was performed in the 1740s.

[19] I have given a schematic description of this in my paper "Some Eighteenth Century Methods for the Chemical Analysis of Minerals," *J. Chem. Educ.,* 1973, *50:* 337–340.

VII

# Mineralogy and the 'Chemical Revolution'

On the opening page of a volume well known to historians of science,[1] Marcellin Berthelot claimed that the 'chemical revolution' brought about a transformation of science, and that the effects of this 'revolution' were not restricted to chemistry, but were felt in all the physical and natural sciences. Among such related sciences it is natural to consider mineralogy, and I think it has been customary to assume that the 'chemical revolution' did indeed exert a profound influence on mineralogy, to the extent perhaps that this science was unable to advance satisfactorily until chemical theory was revised according to the maxims of the new chemistry. Such a viewpoint has been expressed in print fairly recently[2], and may be found stated elsewhere somewhat less explicitly in a number of other publications.[3] It is the purpose of the present paper to examine this question in rather more detail than has been attempted hitherto, so that we may gain a somewhat clearer picture of some of the important side-effects of the 'chemical revolution'. This celebrated *volte face* in chemical theory has for long exercized a fascination for historians of chemistry,[4] probably to the extent that its importance has been over-emphasized. But it was, of course, an episode of very considerable significance in the history of the science, and it is, I think, a worthwhile exercise to explore the influence that it had on sciences related to chemistry.

There is no doubt that the participants in the 'chemical revolution' believed that the change in chemical theory would have a marked effect on studies of the mineral kingdom. For instance, a note of Lavoisier (published posthumously) stated that:

> cette même précision, ce même esprit de recherche que caractérise le siècle où nous vivons, a été porté dans la minéralogie; une nomenclature exacte, fondée sur des caractères non équivoques fournit aux savants les moyens de s'entendre d'une extrémité du monde à l'autre, et l'application de la chimie aux connaissances

minéralogiques développe à nos yeux les mystères de la nature et quels sont les mixtes et les principes dont sont composés la plupart des minéraux.[5]

This extract does not state explicitly that it was the *new* chemistry that was having such a profound effect on mineralogy, but since the note from which it is drawn reads as a kind of panegyric to the new chemical paradigm there is no possibility of mistaking Lavoisier's meaning.

Fourcroy, likewise, stressed the importance of the 'chemical revolution' for mineralogy. In his *Philosophie Chimique*, he drew up a list of twelve subject headings that would serve as a scheme for the teaching of the new chemistry,[6] and a number of topics dealt with in the book clearly impinged on mineralogical theory. He mentioned, for example, as applications of the new theory: the alterations of the colours of minerals by light, the reciprocal combinations of earths by means of fire, the natural history of stones, the compound nature of earths and stones, the history of soils, volcanoes, thermal waters and bitumens, the formation of ores, the influence of acids on mineralization, and so on. It is noteworthy, however, that Fourcroy did *not* envisage that the new theory of combustion would, in itself, have a particularly profound effect on the development of mineralogy.

It is evident, then, that the participants in the 'chemical revolution' recognized its likely significance as an aid to the understanding of the mineral kingdom and the development of mineralogy. It is, however, equally evident that the 'revolution' must be viewed in a much wider context than that of the displacement of the ill-fated phlogiston theory of combustion by the oxygen theory of Lavoisier. Indeed, Siegfried and Dobbs have maintained (rightly, in my view) that the establishment of the correct order of simplicity of substances was "the real centre of the chemical revolution".[7] Certainly, the revision of the chemical theory of combustion was intimately linked with the revision of the order of simplicity of substances. But, for mineralogy, a more important change was the pragmatic acceptance of the various earths (such as those listed in the table of Lavoisier's *Traité Élémentaire*[8]) as "simple substances", which were capable of being separated one from another by chemical analysis. If, for example, we compare the views of Macquer and Lavoisier on the nature of 'earths', we can certainly see the change in attitude, which their opinions reveal, as an important aspect of the 'chemical revolution'.

Macquer, it will be recalled, was still, in the 1770s, an exponent of a descendant of the Aristotelian theory of the four elements, suggesting that the various earths encountered in every-day or laboratory experience consisted of a "terrene principle", intimately combined with various impurities.[9] Lavoisier, by contrast, thought in terms of a number of distinct simple earths, which could be separated from each other by practical procedures, and were not reducible to any fundamentally simple earthy 'principle'.

So we need have no difficulty in understanding why Lavoisier and Fourcroy, and their colleagues, should have felt that their work would have a significant effect on mineralogy. The question, however, is to ascertain who was primarily responsible for the change in outlook that we discern when we compare the views of Macquer and Lavoisier on the nature of earths.

One of the leading features of the paradigmatic Stahlian chemistry of the period before the 'chemical revolution' was that the existence of certain 'principles' was assumed, which were supposedly capable of conferring or imparting certain properties on substances of which they were constituents. Consider, for example, Becher's celebrated triad of 'earths'. The first vitrescible earth was supposedly responsible for fusibility or solidity, diaphaneity or opacity; the second 'fatty' earth imparted colour, taste, odour and inflammability; the third mercurial earth conferred metallic characteristics, such as malleability and ductility.[10] Although quartz, sulphur and mercury were not held to be one and the same as these three earthy 'principles', it is clear that Becher thought that they were the closest available approximations, or contained the hypothetical 'principles' in high degree. Mercury itself, for example, allegedly possessed metallic characteristics to an unusual extent. But the chemists of the pre-revolutionary period were never really successful in isolating their earthy 'principles', though some claims of this kind were made.[11]

By contrast, the earths of Lavoisier's table were capable of experimental isolation from minerals, and they were not, in general, responsible for conferring any very specific properties to the minerals of which they were constituents. As already indicated, such a change in theoretical stance clearly formed an important aspect of the 'chemical revolution', but it should be noted that it was neither Lavoisier himself, nor any of his French colleagues, who brought about this very significant change. It occurred in the analytical work of mineralogists and chemists such as Marggraf, Scheele and Bergman, all of whom adhered to the theories of the phlogiston theory of combustion.

As is well known, it was the Swedish chemist, T. O. Bergman, who was the first to put forward a general scheme for the analysis of minerals, in a paper published in 1777.[12] In this, he showed how to extract the earths lime, baryta, magnesia, silica and alumina from gems, getting the pulverized materials into solution by means of fusion with alkali, and then carrying out a definite sequence of reactions in order to bring about a separation of the several earths. To devise such a scheme, a wide knowledge of precipitation reactions, and the actions of various acidic and alkaline solvents, was required. It is noteworthy, therefore, that only two years previously Bergman had systematized his knowledge of such reactions for the publication of his influential *De attractionibus electivis disquisitio*.[13] And in fact, the reactions described in Bergman's system of mineral analysis can, on inspection, be seen to be consonant with the reactions summarized in his earlier affinity tables.

None of the five simple earths,[14] listed above, were discovered by Bergman, though in 1775 he helpfully clarified the difference between lime and magnesia,[15] a clarification that was still necessary despite the fact that since the beginning of the eighteenth century magnesia had been recognized as being in some way different from lime.[16] It was, however, Bergman's special contribution firstly, to perceive these substances as a class, the members of which could be separated from minerals and from each other by "humid" methods of analysis, and secondly, to devise a definite practical procedure for carrying out such separations. Moreover, Bergman maintained[17] that these 'earths' did pre-exist in the minerals of which they were constituents and were not somehow brought into being by the "humid" analytical procedures. He argued, for example, that chalk, after having been dissolved in acid, could be formed again without loss by the precipitating action of fixed alkali. Similarly, a metal, dissolved in acid, could be recovered (again without loss) by precipitation with another metal. Such reactions implied that the earth or metal, dissolved in an acid, existed in a recoverable form in the solution. Hence, it might be argued by analogy that if earths or metals were obtained by precipitation in the analysis of a mineral they pre-existed as such in that mineral. This important theoretical assumption underlay the whole of Bergman's analytical work.

Bergman believed that his five earths, when combined in twos, threes, fours, or all five together, could provide sufficient variety to account for the known earthy species of the mineral kingdom. Thus, designating the five earths by 'p', 'c', 'm', 'a' and 's',[18] his double species were 'pa', 'ps', 'cm', 'ca', *etc.* But 'ps' and 'sp' (for example) were held to represent different substances, 'ps' indicating an excess of ponderous earth, and 'sp' an excess of siliceous earth. Working in this way, Bergman predicted the existence of 325 different earthy combinations.

Besides gaining a much clearer idea of the chemical composition of minerals as a result of the application of his methods of "humid" analysis, Bergman could also put forward an associated scheme of mineral classification and nomenclature. He utilized the four broad classes of 'salts', 'earths', 'metals' and 'phlogistica', derived from the categories established in the eleventh century by Avicenna,[19] and added to these the classes of "*Petrae*" and "*Petrefacta*". The class of 'earths' was divided into five genera according to the preponderance of the five simple earths, mentioned above. The "*Petrae*" (or "fossils mixed mechanically") and the "*Petrefacta*" (or "organic fossils") were themselves subdivided according to whether they contained saline, earthy, metallic or phlogistic matter.

Bergman offered systematic names for his salt-like minerals, according to their acidic and earthy constituents (*e.g.*, *Barytes vitriolatus*, *Calcareum fluoratum*, *Calcareum aeratum*), but was unable to carry through the reforms in nomenclature to the whole range of earthy minerals, and fell back on the old well-worn trivial names:

> thus, for example, under the genus magnesia, a species occur[s], . . . [of] the formula, *smca*, composed of siliceous, calcareous, and argillaceous earth, with some admixture of iron, which in systematic authors is denominated asbestos, and treated as a peculiar genus. To this, indeed, the trivial name of asbestos may be properly applied, as it seems to be so well understood, that the youngest mineralogist is in no danger of being misled by it.[20]

But, as we shall see below, the French chemists, followers of Lavoisier, had no more success than Bergman in fitting their ideas on nomenclature to the results of the "humid" analyses of minerals.

Bergman argued for his scheme on pragmatic grounds. A knowledge of the chemical composition of the bodies of the mineral kingdom gave an indication of the practical uses to which they might be put. Moreover, classificatory systems based on external characteristics were unsatisfactory, for minerals with the same chemical composition could appear in several different guises, and, conversely, substances that were similar in appearance might have quite different chemical compositions. Bergman, therefore, believed that classes, genera and species should be based on chemical composition, and only varieties should be defined by external characteristics.

All this sounds very well, though when we look at Bergman's actual system of mineral classification we find that in a way it was rather absurd. For example, in the argillaceous genus, we find a certain species defined by the head: argillaceous earth loosely bound with siliceous earth, making half the weight or sometimes more, and a little lime.[21] This was none other than some member of the zeolite family. Faced with such definitions of species, it is perhaps in no way surprising that many *mineralogists* chose to reject the chemical system as the fundament of mineral classification. They met zeolites in the field, and wished to classify them. The starting point for their work was an actual specimen of zeolite. Certainly, the chemical analysis might be a guide to precise classification, but the macro-specimens must be given prior consideration. To draw up the heads of a system of classification according to the uncertain results of chemical analysis seemed quite unwarranted. Moreover, as was pointed out by Dolomieu,[22] if one did not have a prior concept (based on external characteristics) of a mineral's species in mind when one performed an analysis, one was in effect analyzing 'mineral X', and the analysis would be almost valueless. (Today the situation is rather different. With very accurate methods of mineral analysis available, the first intimations of a new mineral species may arise as a result of chemical analysis.) Nevertheless, despite such objections, chemical analysis did come to form the basis of most systems of mineral classification (other than those of the Wernerians) in the last two decades of the eighteenth century.

The points I have made above concerning Bergman's system of chemical mineralogy are, no doubt, not unfamiliar to historians of science. I have mentioned them here at some length simply because I wish to emphasize that this chemical system was devised by one wholly committed to the phlogiston theory.[23] Moreover, similar schemes were utilized by several opponents of the new chemistry of Lavoisier and his colleagues, without any particular inconvenience. Monnet, for example, who as late as 1788[24] believed that water was converted into earth in the formation of minerals, in 1779[25] adopted four of the simple earths utilized by Bergman (neglecting the ponderous earth discovered by Scheele in 1774[26]) and classified minerals accordingly. Kirwan, in his *Elements of Mineralogy* of 1784[27], adopted similar principles of mineral classification, and offered a new general system of mineral analysis.[28] He also was a phlogistonist at that stage. By the time he published the second edition of this work,[29] Kirwan had come to accept the new chemistry, but this did not in itself have any very marked effect on either his system of analysis or his scheme of classification, though both of these showed considerable elaboration in the second edition, owing to recent discoveries of new metals and earths, and improved techniques of analysis.

My point, then, is that these analytical schemes and these chemical systems of mineral classification were developed by men who adhered to the old phlogistic theories. Moreover, we find that Lavoisier himself drew from Bergman and Kirwan in his own work. For example, the earths of Lavoisier's table of "simple substances" were very likely derived from Bergman.[30] Indeed, in discussing the metals, Lavoisier[31] followed Bergman's suggestion[32] that heavy earth was in fact a metallic calx (or oxide, for Lavoisier) which could not be reduced to the metal by charcoal. Thus, the important conceptual shift that occurred when the old idea of earthy property-conferring 'principles' was given up and replaced by the practical view of a range of earths, appearing as the last terms of chemical analysis, was a change that took place among and within the school of 'phlogistic' mineralogists, and before Lavoisier's system of chemistry became generally accepted. Consequently, the phlogistonists themselves overthrew one of the main features of the old chemistry of 'principles'. Lavoisier followed here; he did not lead.

To confirm this opinion it is instructive to examine Lavoisier's memoir of 1783,[33] in which he described the results of his investigations of the effects of high temperatures on a considerable range of rocks, minerals and metallic substances, the high temperatures having been achieved by the action of a blow-pipe flame enhanced by an oxygen blast. In this memoir, Lavoisier published his own scheme of mineral classification. If we are to believe Lavoisier's own statement, he was led to his classificatory scheme by the results obtained in these high-temperature investigations:

> J'avais d'abord suivi, pour l'arrangement des matières, le système adopté par M. Daubenton au Cabinet du Roi; mais le résultat même de mes expériences m'a forcé d'y apporter quelques modifications, et je me suis trouvé insensiblement rapproché de l'ordre qui depuis a été adopté par M. Bergman dans sa Sciagraphie, et par M. Kirwan dans ses Éléments de minéralogie.[34]

I regret to say, however, that I am not entirely convinced that this statement represents the situation in its true light. Daubenton's scheme[35] was largely based on traditional external characteristics, and was commonly used by French students of mineralogy at the end of the eighteenth century.[36] The schemes of Bergman and Kirwan were, as we have seen, chemical systems, centred round the individuality of the several simple earths. Lavoisier's scheme seems to me to have been an amalgam of the 'natural history' of Daubenton and the chemical systems of Bergman and Kirwan, and I do not think that the results of the high-temperature investigations had very much influence on the final shape of the scheme that Lavoisier proposed. Nor indeed, from what we know today of the physical and chemical properties of minerals, could these high-temperature reactions have served as a basis of a really useful system of mineral classification. It appears, therefore, that we find Lavoisier drawing from the phlogistic mineralogists, rather than achieving a notable breakthrough in mineralogy as a result of his experiments or his new theoretical ideas.

Let us now consider the question of the influence of the new system of chemical nomenclature on mineralogy. As before, let us try to find what the participants in the 'chemical revolution' thought about the matter. Guyton de Morveau, for example, wrote with some optimism:

> La nomenclature souvent fausse, toujours arbitraire et incohérente, des anciens minéralogistes, était l'un des plus grands obstacles aux progrès de cette étude, à laquelle il fallait consacrer une partie de sa vie, comme dans la botanique, pour en apprendre seulement la synonymie. Ces dégoûts o[n]t été écarté par l'appropriation des dénominations de la chimie française aux substances fossiles congénères, et en adoptant pour les autres les rectifications dont le citoyen *Haüy* a si bien senti la nécessité dans ses mémoires sur la structure des cristaux.[37]

This sounds encouraging, but it should be noted that in the last sentence cited here, Guyton seems to be speaking of the application of the new terminology in a restricted, rather than a general, sense. Evidently, Guyton found that it was not possible to apply the new chemical terms to all minerals, so that he had to have recourse to the names adopted by the crystallographer, Haüy, who, in naming the species of his class of earthy minerals, had used trivial names, rather than ones based on the new chemistry. And, as has been pointed out by W. A. Smeaton,[38] even Guyton, despite his zeal for terminological reform, eventually had to admit (in his lectures delivered at the *École Polytechnique* in 1805–6) that the complete revision of mineral nomenclature according to the maxims of the new chemistry was impracticable. It is easy to see why this was so.

If mineralogical systems were to be based on the principles established by Bergman, according to the earthy and metallic components revealed by chemical analysis, a separate systematic name would be required for each quantitative variation in composition. As we have seen, Bergman did make some attempt to do this, distinguishing, for example, 'ps' from 'sp', according to the earthy component that was in excess. Clearly, however, this system could not cope with situations in which there were several recognized minerals, each containing the same earthy components, but in different proportions.[39] Again, where minerals displayed a range of composition (as did, for example, the olivine family) the new nomenclature was placed in a powerless position.

Despite such difficulties, however, the new nomenclature was introduced into mineralogy quite rapidly, wherever possible and appropriate, which was in fact in the segments of the schemes of mineral classification that dealt with salts. It is interesting to note, therefore, that towards the end of the eighteenth century we begin to see some separation of chemical mineralogy and what we now call inorganic chemistry. For example, writing in 1800, Fourcroy stated:

> Aujourd'hui, l'ordre nouveau qu'on adopte en minéralogie[,] excluant du nombre des pierres toutes les terres ou bases en général contenant des acides, et présentant celles-ci en particulier sous le nom de substances acidifères, les minéralogistes français ne regardent comme vraies pierres et ne comprennent sous la dénomination de substances terreuses que de purs assemblages ou combinaisons entre elles.[40]

He followed Haüy's system of nomenclature for the 'earthy' minerals, and of the forty-five types listed only four (*silex*, *petro-silex*, *zircon*, and *argile*) could in any way be said to give an indication of their chemical composition by means of their names. And none of them gave any indication whatsoever of their properties.

Nevertheless, as I have said, the new nomenclature was utilized in many mineralogical treatises, where this was possible. According to Guyton de Morveau,[41] the first person to do this was Ignaz von Born, in his catalogue of the mineral collection of Éléonore de Raab, published in Vienna in 1790,[42] and I have not seen an earlier example than this. Daubenton introduced the new nomenclature into the later editions of his influential *Tableau Méthodique*, giving the old and new names as synonyms. However the real success (or failure) of the programme for the reform of the mineralogical nomenclature by means of the new chemistry may most easily be gauged by examining an interesting, and not very well known, volume by L. J. Sevrin, published in 1807.[43]

Sevrin's monograph very helpfully provides us with a full list of the synonyms of the old and new systems of nomenclature, both for chemistry and for mineralogy. Like Guyton de Morveau in his earlier work, Sevrin displayed some optimism about the prospects of the application of the new nomenclature in mineralogy:

> Maintenant que nous possédons en minéralogie, et principalement en chimie, des noms dont la signification est bien déterminée, relativement aux principes qui constituent un corps composé, il faut donc nous acoutumer de bonne heure avec ces sortes de noms, et s'en occuper souvent, si nous voulons nous les rendre familiers; par-là, nous pourrons mieux augmenter l'activité de l'imagination, l'étendue de la mémoire, et faciliter l'exercise de la réflexion, sur-tout en s'occupant des objets qui, exerçant davantage l'attention, lient ensemble un plus grand nombre d'idées.[44]

Our expectations heightened by this asseveration, we may find ourselves somewhat disappointed when, with the advantage of hindsight, we examine Sevrin's comparative table of old and new mineral names. In fact, for the 471 minerals that were listed, the new nomenclature was only utilized in 110 cases (23%), and these were nearly all salts, metals or alloys. The minerals formed from a combination of earths (according to Bergman's theory) proved intractable, so far as the new terminology was concerned. This twentieth-century view of the situation is confirmed by statements published at the beginning of the nineteenth century. Alexandre Brongniart, for example, expressed the opinion that:

> la nomenclature chimique qui peut être très bonne pour la chimie, dans laquelle la synthèse est une preuve irrécusable de l'analyse, est souvent plus gênante qu'utile dans la Minéralogie; on ne pourroit pas même espérer qu'elle y devînt jamais générale, à cause de la nature très-composée de certains minéraux. D'ailleurs, les genres des minéraux ne sont pas seulement fondés sur les principes constituans; ils sont encore établis sur la différence des propriétés importantes des composés. Ainsi l'alumine dans l'alun, et l'alumine dans le corindon, devra toujours former la base de deux genres distincts qu'il faudra nécéssairement désigner par les noms différens.[45]

It is true that some writers were perhaps more optimistic about the applications of the new nomenclature. Dolomieu, for example, with a characteristically Gallic caste of mind, hoped to be able to set mineralogical principles into some kind of logical order, so that by starting at the beginning of the logical thread, so to speak, the whole science might be unravelled. In this undertaking, the importance of proper nomenclature was considered to be paramount. If the correct systematic names could be given, then one might hope:

> faire une espèce de série méthodique pour tous les mots, en les mettant sous une sorte de dépendance les uns des autres, afin de former une chaîne d'idées qui remonte des moindres particularités aux premières généralités, et qui réunisse les notions les plus complexes aux considérations les plus simples. Le but de pareilles séries semblables à celle qui ont été formées pour *l'Encyclopédie*, étant d'arriver à la désignation d'un petit nombre d'expressions, qui sont en quelque manière les clefs de la science.[46]

This programme sounds almost grotesque,[47] and it certainly was not carried out by Dolomieu himself, though this is not, perhaps, an entirely fair comment, for he died in 1801, not long after these words were written. We cannot, however, imagine a small number of expressions which would, if known, provide keys to the whole science of mineralogy. This is reminiscent of Aristotelian essentialism, and I am not aware of any such mineralogical programme having been seriously, let alone successfully, attempted.

Perhaps the most troublesome problem that faced mineralogists and crystallographers in the first decade of the nineteenth century was the attempted reconciliation of the chemical system of mineralogy and the crystallographical system of Haüy. As is well known,[48] the problem centered round the anomaly of calcite and aragonite. These were shown by Klaproth,[49] Fourcroy and Vauquelin,[50] and others, to have the same chemical composition, but Haüy showed that they had different "primitive forms"[51] (a rhomboid for calcite and an irregular octahedron for aragonite). According to the results of chemical analysis, the two should have been given the same name, but according to Haüy's crystallographical system the two belonged to different species, and had to have different names. Haüy was a firm advocate of the new chemistry, and chemistry formed the broad basis of his system of mineral classification.[52] However, in the case of the calcite/aragonite pair he was forced to give priority to crystallography in the determination of mineral species. The problem was eventually resolved to Haüy's satisfaction[53] by the suggestion of Laplace[54] that the same chemical constituent molecules might be joined together by different faces, thus giving rise to a variety of crystal structures. We must note, however, that the problem could not have been resolved by the application of the new chemistry. Moreover, the new nomenclature could not offer guidance as to whether chemistry or crystallography should take precedence in the specification of mineral species. Again, we find that the 'chemical revolution' was not of much help in mineralogy. The calcite/aragonite polymorphism would have provided an enigma for proponents of either of the two chemical systems.

On a few occasions, however, we do find the principles of the new nomenclature being applied in an apparently helpful way and these are worth discussing for they throw some light on the nature of Lavoisier's system of chemistry. We have already noted that the main developments that made possible the establishment of chemical systems of mineralogy were the recognition of the various individual simple earths, and the development of viable systems of "humid" analysis. A number of other simple earths were recognized in the closing years of the eighteenth century, to add to Bergman's list of five. Zirconia was described by Klaproth in 1789;[55] strontia was recognized by Crawford in 1789;[56] yttria was discovered by Gadolin in 1794;[57] and glucina (or "*glucine*") was described for the first time by Vauquelin in 1798.[58] Of these, only glucina was named by French chemists, according to the canons of the new chemistry, and it is interesting to consider why the name that they selected was preferred.[59] The newly discovered earth was named after the Greek word γλυκαινω meaning "I render sweet", because it formed sweet-tasting compounds in combination with other substances. In a sense, then, this earth of the French chemists was named according to its property-conferring powers, so that in this respect "*glucine*" was rather like the old 'principles' of the pre-Lavoisier chemistry. But the cases were not exactly parallel. "*Glucine*" did not itself display the property of sweetness to an abnormally high degree, as sulphur displayed inflammability, or mercury displayed metallicity. The proper *combinations* had to be made for the sweetness to become manifest.[60] The element chromium, also discovered by Vauquelin,[61] provides us with a somewhat similar example. It was named (on the advice of Fourcroy and Haüy) because of the coloured compounds that it formed, not because it was itself coloured.

These two examples show that simple substances were named by the adherents of the new chemistry, according to the natures of the compounds which they were capable of forming. This is suggestive of a degree of 'property-conferring' character on the part of the simple substances of the new chemistry, but this does not, I think, make them the same as the 'principles' of the Stahlian chemists, which were not isolable as pure substances, and which supposedly exhibited to an exceptional degree the properties that they conferred. And chemists who adopted the new chemistry could *not* expect to predict in a general way the properties of minerals on the basis of a knowledge of their constituent simple substances.

It seems, then, that we do not find much evidence for the profoundly beneficial effect of the new chemistry on mineralogical investigations. Nevertheless, some significant changes can be discerned. The 'pre-revolutionary' crystallographical writings of Linnaeus[62] and Romé de l'Isle[63] reveal a belief in certain saline 'principles' which were allegedly responsible for the regularity of structure of crystalline substances. Opinions of this kind died with the coming of the new chemistry. And the writings of the 'post-revolutionary' period exhibit what the modern reader will probably take to be a considerably improved clarity of understanding and expression. Consider, for example, the following passage by an exponent of the new chemistry, J. A. Chaptal:

> We shall immediately perceive that these earths [*i.e.*, lime, magnesia, barytes, alumine and silex] are, in some instances, combined with acids, which form saline stones; that in other instances they are mixed with each other, and form stones properly so called; and that in other instances, again, these stones, so formed by the mixture of primitive earths, are united together, or fixed in a gluten or cement, which forms rocks, pebbles or compound stones.[64]

This may appear attractively clear to the modern reader, but one's judgement should not, Butterfield informs us, be a reflection of 'whiggish' predilections.[65] Certainly, the mineralogists of the 'post-revolutionary' period were closer to our ways of thinking than were their predecessors. But this is not to say that viable systems of mineral analysis and nomenclature had to wait on the establishment of Lavoisier's theories. In fact, as we have seen, such systems were devised by the adherents of the older chemistry. Or, putting the matter another way, if we see the change in matter theory associated with the shift in the concept of 'earth' as a significant aspect of the chemical revolution, then we must look for this in the work of Lavoisier's predecessors, and note that to a large extent Lavoisier derived his ideas on these matters from the work of his phlogistic predecessor, Torbern Bergman. Mineralogists who adopted Rouelle's theory of salts,[66] Cronstedt's methods of blow-pipe 'analysis',[67] and Bergman's theory of earths (and his methods for finding them in minerals), possessed the basic equipment for the establishment of chemical systems of mineralogy—equipment that was found adequate at least until the end of the eighteenth century.

## NOTES

1. P. E. M. Berthelot, *La Révolution chimique Lavoisier* . . . (2nd edn., Paris, 1902), p. 1.
2. V. A. Eyles, in C. J. Schneer (ed.), *Toward a History of Geology* (Cambridge, Massachussets, and London, 1969), p. 174. Dr. Eyles's remark was, I think, only intended as a generalized statement, and should be considered in its context. It has, however, provided a useful focus of attention for the present paper.
3. See, for example: R. Hooykaas, "The species concept in 18th century mineralogy", *Archives Internationales d'Histoire des Sciences,* 5th year (1952) pp. 45–55. Here it is stated (p. 48): "Romé's chemistry is very defective; he is a phlogistonist and his analytical data are faulty." There is here, I think, the implication that adherence to phlogistonism was inimical to the progress of mineralogy.
4. See, for example: P. E. M. Berthelot, *op. cit.* (note 1); A. N. Meldrum, *The Eighteenth Century Revolution—The First Phase* (Calcutta, 1930); D. McKie, *Antoine Lavoisier, Scientist, Economist, Social Reformer* (London, 1952); H. Guerlac, *Lavoisier—The Crucial Year: The Background and Origin of His First Experiments on Combustion in 1772* (Ithaca, New York, 1961).
5. From: "Note pour l'article chimie", *Œeuvres de Lavoisier* . . . (6 vols., Paris, 1864–1893), vol. v, pp. 298–301.
6. A. F. Fourcroy, *Philosophie Chimique, ou Vérités Fondamentales de la Chimie Moderne,* . . . (3rd edn., Paris, 1806).
7. R. Siegfried and B. J. Dobbs, "Composition, A Neglected Aspect of the Chemical Revolution", *Annals of Science, 24* (1968), 275–293 (p. 281).
8. A. L. Lavoisier, *Traité Élémentaire de Chimie, présenté dans un Ordre nouveau et d'après les Découvertes modernes* (2 vols., Paris, 1789), vol. i, p. 192.
9. P. J. Macquer, *Dictionnaire de Chymie, contenant la théorie et la pratique de cette science,* . . . (2nd edn., 4 vols., Paris, 1778), vol. iv, p. 66. ("On peut juger par ce que nous avons dit des principales especes de terres qui forment la surface du globe, qu'a l'exception de la terre vitrifiable la plus pure, il n'y en a aucune qui ne soit altérée & alliée de quelque matiere hétérogene. Ce sont ces restes de matieres étrangeres encore unies à la terre primitive, qui spécifient & caractérisent des différentes especes de terre: . . .") It should be noted, however, that even Macquer was quite prepared to consider the possibility of there being a number of individual simple earths: P. J. Macquer, *Elements of the Theory and Practice of Chymistry* . . . (2 vols., London, 1758), vol. i, p. 6. The first French edition of this work was published in Paris in 1749. It should be noted that Macquer's Aristotelian earth was apparently envisaged from an empiricist's standpoint. It differed significantly from the hypothetical triad of earths of Becher.
10. J. J. Becher, *Joh. Joachimi Beccheri,* . . . *Physica Subterraneae* . . . *Editio Novissima* . . . (Leipzig, 1703), p. 117 and *passim.*

11. See, for example: J. G. Wallerius, *Mineralogie ou Description Générale des Substances du Regne Mineral* (2 vols., Paris, 1753), vol. i, pp. 524 and 576.
12. T. O. Bergman, "Disquisitio Chemica de Terra Gemmarum", *Nova Acta Regiae Societatis Scientiarum Upsaliensis, 3* (1777), pp. 137–170 (republished in: T. O. Bergman, *Opuscula Physica et Chemica* ... (6 vols., Uppsala and Åbo, 1779–1790), vol. ii, pp. 72–117).
13. T. O. Bergman, "De Attractionibus Electivis Disquisitio", *Nova Acta Regiae Societatis Scientiarum Upsaliensis, 2* (1775), pp. 159–248 (republished in: T. O. Bergman, *Opuscula Physica et Chemica* ... (6 vols., Uppsala and Åbo, 1779–1790), vol. iii, pp. 291–470).
14. *Viz*, calcareous, magnesian, aluminous, ponderous, and siliceous earths.
15. T. O. Bergman, *op. cit.* (note 12), 1779–1790, vol. i, pp. 365–404.
16. The physician, F. Hoffmann of Halle, is usually credited with being the first to distinguish between magnesia and lime. (See: J. R. Partington, *A History of Chemistry* (4 vols., London, 1961–1970), vol. ii, p. 696).
17. See T. O. Bergman, "On the Investigation of Truth" in: *Physical and Chemical Essays; translated from the original Latin of Sir Torbern Bergman ... By Edmund Cullen, M.D.* ... (2 vols., London, 1784–1788), vol. i, pp. xix–xl.
18. T. O. Bergman, *Meditationes de Systemate Fossilium Naturali* ... (Florence, 1784) and *Physical and Chemical Essays: translated from the original Latin of Sir Torbern Bergman* ... (vol. iii, Edinburgh, 1791), pp. 205–316 ("Thoughts on a natural system of fossils").
19. See: *Avicennae de Congelatione et Conglutinatione Lapidum being sections of the Kitâb al-Shifâ. The Latin and Arabic texts edited with an English Translation of the latter and with critical notes by E. J. Holmyard ... and D. C. Mandeville* ... (Paris, 1927), pp. 33–34.
20. T. O. Bergman, *op. cit.* (note 18, 1791), pp. 314–315.
21. T. O. Bergman, *Manuel du Minéralogiste, ou Sciagraphie du Règne Minéral, distribué d'après l'analyse chimique ... traduite ... Par M. Mongez le jeune*, ... (Paris, 1784), p. 138.
22. D. Dolomieu, "Lettre sur la nécessité d'unir les connaissances chimiques à celles du minéralogiste; avec des observations sur la différente acception que les auteurs allemands et français donnent au mot chrysolithe, Addressé au citoyen Haüy, membre de l'institut national, par le citoyen D. Dolomieu", *Journal des Mines*, No. 29 (*an* 5), pp. 365–376.
23. The extent of Bergman's commitment to this theory, even in his closing years, is well displayed in his lecture notes of the period. See: A. Fredga and S. Rydén, "The notes of Juan José de Elhuyar from the lectures of Torbern Bergman in 1782", *Lychnos* (1959), pp. 161–208.
24. A. G. Monnet, "Memoire Sur La Formation des Minéraux", *Memoires de l'Académie Royale des Sciences. Années MDCCLXXVI–LXXXVII*, Turin (1788), pp. 337–356.
25. A. G. Monnet, *Nouveau Systême de Minéralogie, ou Essai d'une nouvelle exposition du Regne minéral, auquel on a joint un Supplément au Traité de la dissolution des Metaux, avec des observations relatives au Dictionnaire de Chymie*, (Paris, 1779).

26. See: *The Chemical Essays of Charles-William Scheele translated from the Transactions of the Academy of Sciences at Stockholm* ... (London, 1901), pp. 75–76.
27. R. Kirwan, *Elements of Mineralogy. By Richard Kirwan, Esq*; *F.R.S.* (London, 1784).
28. *Ibid.*, pp. 160 ff. ("Of the general Examination and Analysis of Earths and Stones").
29. 2 vols., London, 1794.
30. The actual theoretical notion of "simple substances" may, however, have been acquired from Guyton de Morveau, who spoke of "*corps simples*" as undecomposable substances (*Élémens de Chymie, Théorique et Pratique*, ... (3 vols., Dijon, 1777), vol. i, p. 10).
31. A. L. Lavoisier, *Elements of Chemistry ... translated from the French, By Robert Kerr* ... Edinburgh, 1790), p. 159.
32. T. O. Bergman, *op. cit.* (note 18, 1791), pp. 246–247. Bergman believed that the heavy earth was in fact a metallic calx because solutions of it in acid formed precipitates with "phlogisticated alkali".
33. A. L. Lavoisier, "Del'action du feu animé par l'air vital sur les substances minérales les plus réfractaires", *Œuvres de Lavoisier* ... (6 vols., Paris, 1864–1893), vol. ii, pp. 451–502.
34. *Ibid.*, p. 452.
35. L. J. M. Daubenton, *Tableau Méthodique des Minéraux, suivant leurs Différentes Natures, Et avec des caractères distinctifs, apparens ou faciles à reconnoître*, (Paris, 1784).
36. L. J. M. Daubenton (1716–1800) was appointed professor of mineralogy at the *Muséum d'Histoire Naturelle* in 1793, and began teaching natural history at the *École Normale* in 1794.
37. L. B. Guyton de Morveau, "Substances Minérales", *Journal Polytechnique, ou Bulletin du Travail fait à l'École Centrale des Travaux Publics*, ..., *1* (1) (*an* 4), p. 139.
38. W. A. Smeaton, "The Contributions of P.-J. Macquer, T. O. Bergman and L. B. Guyton de Morveau to the Reform of Chemical Nomenclature", *Annals of Science, 10* (1954), p. 106.
39. This point was made by Richard Kirwan: "Of Chymical and Mineralogical Nomenclature. By Richard Kirwan, Esq; L.L.D. F.R.S. and P.R.I.A.", *Transactions of the Royal Irish Academy, 8* (1800), pp. 53–76.
40. A. F. Fourcroy, *Systême des Connaissances Chimiques, et de leurs Applications aux Phénomènes de la Nature et de l'Art* (11 vols., Paris, 1800–1802), vol. ii, pp. 285–286.
41. L. B. Guyton de Morveau, *op. cit.* (note 37), p. 139.
42. I. E. von Born, *Catalogue Méthodique et Raisonné de la Collection des Fossiles de Mlle Éléonore de Raab* (2 vols., Vienna, 1790).
43. L. J. Sevrin, *Dictionnaire des Nomenclatures Chimique et Minéralogique Anciennes, Comparées Aux Nomenclatures Chimique et Minéralogique Modernes, D'après les Ouvrages des Chimistes et le Traité de Minéralogie de Mr. Haüy*; ... (Paris, 1807).
44. *Ibid.*, p. iv.
45. A. L. Brongniart, *Traité Élémentaire de Minéralogie, Avec des Applications aux Arts*; *Ouvrage Destiné a l'Enseignement dans les Lycées Nationaux* (Paris, 1807), p. 66.
46. D. G. S. T. G. de Dolomieu, *Sur la Philosophie Minéralogique, et sur l'Espèce Minéralogique* (Paris, *an* 9), p. 19.
47. It was, however, quite in keeping with the traditions of French philosophy of science in the eighteenth century. Condillac, for example, remarked that: "Une première idée

fondamentale est liée à deux ou trois autres; chacune de celles-ci à un égal nombre, ou même à un plus grand, et ainsi de suite." (G. le Roy (ed.), *Œuvres Philosophiques de Condillac* ... (3 vols., Paris, 1947–1951) vol. i, p. 726) Possibly, however, Dolomieu did not have anything as subtle as this in his mind. He might, for example, have been thinking of something like the large systematic tabular representations of the doctrines of the new chemistry that were prepared by Fourcroy for teaching purposes. (See: A. F. Fourcroy, *Synoptic Tables of Chemistry, Intended to Serve as a Summary of the Lectures delivered on that science, in the Public Schools at Paris* [translated by William Nicholson] (London, 1801)). Similar efforts had been made earlier in the century to display a sequence of chemical reactions in a kind of 'logical' order. (See: A. L. Brongniart, *Tableau Analytique Des combinaisons & des décompositions de différentes substances, ou Procédés de Chymie, pour servir à l'intelligence de cette science* (Paris, 1778).

48. For an account of the history of this problem, see: J. G. Burke, *Origins of the Science of Crystals* (Berkeley and Los Angeles, 1966), pp. 126–132.
49. See: J. R. Partington, *op. cit.* (note 16), vol. iii, p. 657.
50. A. F. Fourcroy and L. N. Vauquelin, "Expériences Comparées Sur l'Arragonite d'Auvergne, et le Carbonate de chaux d'Islande", *Annales du Muséum d'Histoire Naturelle, 4* (1804), pp. 405–411.
51. R. J. Haüy, "Sur l'Arragonite", *Annales du Muséum d'Histoire Naturelle, 11* (1808), pp. 241–266.
52. This can be extracted from Haüy's *Traité de Minéralogie* ... (5 vols., Paris, 1801). The system of classification is not very different in the later *Tableau Comparatif des Résultats de la Cristallographie et de l'analyse chimique, relativement à la classification des minéraux* (Paris, 1809), despite the fact that by 1809 Haüy had reached the definite conclusion that mineral species defined in chemical terms, and species defined in terms of his crystal theory, did not necessarily correspond with each other in a one-to-one relationship; in some instances the crystallographical species had to have precedence.
53. R. J. Haüy, *op. cit.* (note 52, 1809), p. xxj.
54. P. S. Laplace, *Traité de mécanique céleste, par P. S. Laplace* ... (5 vols., Paris, 1798–1825), vol. iv (1805), p. 70: "Les molécules intégrantes peuvent encore s'unir par diverses faces, et produire ainsi des cristaux différens par la forme, la dureté, la pesanteur spécifique et leur action sur la lumière".
55. For Klaproth's account, see: M. H. Klaproth, "Chemical Examination of Circon, or Jargon of Ceylon", *Analytical Essays Towards Promoting the Chemical Knowledge of Mineral Substances* (2 vols., London, 1801–1804), vol. i, pp. 175–194.
56. A Crawford, "On the Medicinal Properties of the Muriated Barytes. By Adair Crawford, M.D. F.R.S. Physician to St. Thomas's Hospital. Read November 10, 1789", *Medical Communications, 2* (1790), pp. 301–359.
57. J. Gadolin, "Undersökning af en svart tung Stenart ifrän Ytterby", *Kongl. Vetenskaps Academiens nya Handlingar, 15* (1794), pp. 137–147.
58. N. Vauquelin, "Analyse De l'Aigue marine, ou Béril; et découverte d'une terre nouvelle dans cette pierre; Lue à l'Institut, le 26 Pluviôse, an 6 [14 February, 1798], Par le cit. Vauquelin", *Annales de Chimie*, ..., *26* (*an* 6), pp. 155–169.
59. *Ibid.* p. 169 (footnote). The editors of the journal were Guyton, Monge, Berthollet, Fourcroy, Adet, Hassenfratz, Séguin, Vauquelin, Prieur, Chaptal and Van Mons.

60. I am indebted to my colleague, Dr. W. R. Albury, for his suggestions on this point.
61. L. N. Vauquelin, "Mémoire Sur une nouvelle substance métallique contenue dans le plomb rouge de Sibérie, et qu'on propose d'appeler Chrôme, à cause de la propriété qu'il a de colorer les combinaisons où il entre. Par le cit. Vauquelin: Lu à la première classe de l'Institut national, le 11 Brumaire, an 6 [1 November, 1797], *Annales de Chimie, ..., 25* (*an* 6), pp. 21–31; and "Second Mémoire Sur le métal contenu dans le plombe rouge de Sibérie; Par le cit. Vauquelin", *Ibid.*, pp. 194–204.
62. C. Linnaeus, *Caroli a Linné ... Systema Naturae per Regna Tria Naturae, Secundum Classes, Ordines, Genera, Species, cum Characteribus & Differentiis* (12th edn., 3 vols., Stockholm, 1766–1768), vol. iii, pp. 5,8 and *passim.*
63. J. B. L. Romé de l'Isle, *Essai de Cristallographie, ou Description des Figures Géométriques, Propres à différens Corps du Regne Minéral, connus vulgairement sous le nom de Cristaux, Avec Figures et Développemens* (Paris, 1772), p. 7, 35, 110 and *passim.*
64. J. A. Chaptal, *Elements of Chemistry* ... (2nd edn., 3 vols., London, 1795), vol. ii, p. 8.
65. See: H. Butterfield, *The Whig Interpretation of History* (London, 1931).
66. See: G.-F. Rouelle, "Memoire sur les Sels Neutres, Dans lequel on propose une division méthodique de ces Sels, qui facilite les moyens pour parvenir à la théorie de leur crystallisation", *Mémoires de l'Académie Royale des Sciences* (1744), pp. 353–364; "Mémoires sur les Sels Neutres, Dans lequel on fait connoître deux nouvelles classes de sels neutres, & l'on développe le phénomène singulier de l'excès d'acide dans ces sels", *Mémoires de l'Académie Royale des Sciences* (1754), pp. 572–588. (These references apply to the Amsterdam edition.)
67. See: A. F. Cronstedt, *An Essay towards a System of Mineralogy: ... Translated from the Original Swedish, with Notes, By Gustav von Engestrom. To which is added, a Treatise on the Pocket-Laboratory, containing An Easy Method, used by the Author, for Trying Mineral Bodies, Written by the Translator. ..* (2nd English edn., London, 1772), pp. 175–318. (Pages 189–288 are omitted from the numbering of the pages of this edition, owing to a printer's error.)

## ACKNOWLEDGEMENTS

I am most grateful to Drs. V. A. Eyles, R. M. Gascoigne and W. R. Albury for their comments on earlier versions of this paper. It was read (in a rather different form) at a meeting of the Melbourne branch of the Australasian Association for the History and Philosophy of Science in August, 1973.

# VIII

## SOME EIGHTEENTH CENTURY METHODS FOR THE CHEMICAL ANALYSIS OF MINERALS

This paper examines briefly the development of mineral analysis in the eighteenth century, and describes some of the early inorganic analytical procedures, a number of which are still in use today. It also attempts to portray some of the difficulties that faced the early analysts, and shows how, by the end of the century, many of the theoretical difficulties at least had been overcome.

Even at the beginning of the century, it was possible to do something towards a chemical analysis of mineral substances. The traditional methods of the assayers (1) could give reasonably accurate information about the proportions of metals that could be extracted from ores, though this was not necessarily the same as the proportions that were present in those ores. In addition, it was customary to submit the materials being investigated to as high a temperature as possible in a furnace, without embarking on a full assay. These processes of "analysis by fire" were very uncertain, however, and numerous disputes arose as to whether the substances obtained when a material was submitted to the action of heat were actually "in" the material prior to the heating, or were produced in some way by the heating. Thus at this period most systems of classification were based upon external characteristics. By contrast, by the end of the eighteenth century, chemical methods of analysis were widely used and provided the basis for various systems of mineral classification. These advances became possible through a widening empirical knowledge of precipitation reactions, and of ways of bringing substances into solution. Also, most importantly, the recognition of the chemical individuality of the various "earths," serving as the last terms of chemical analysis, enabled chemists to develop a viable scheme of mineral analysis.

Where did knowledge of these individual "earths" come from'? Knowledge of calcareous earth as an individual substance dates back for centuries, and we cannot tell who was the first to recognize its individuality. The first recognition of magnesia as an earth distinct from the calcareous kind appears to be due to the German medical chemist, Friedrich Hoffmann (1660-1742). Hoffmann showed that the white precipitate obtained when he added "salt of tartar" to the mother-liquor from evaporated brine, dissolved in "vitriolic acid" to give a bitter-tasting solution (Epsom salt), whereas quicklime gave an almost tasteless, sparingly soluble salt (gypsum) when similarly treated with acid. He viewed Epsom salt as a combination of "sulphurous acid" and a "mild and alkaline

Reprinted with permission from the *Journal of Chemical Education,* Vol. 50, No. 5 (1973), pp. 337–340. 

calcareous earth" (2). Evidently, Hoffmann realized that there was something that made this earth different from ordinary calcareous earth, though the significance of the observation was not perhaps realized at this stage.

According to the Swedish chemist, T.O. Bergman (1735-84) (3), "terra ponderosa" was first recognized as an individual substance by his countrymen, J.G. Gahn (1745-1818) and C.W. Scheele (1742-86), whereas the German chemist, A.S. Marggraf (1709-82), had previously considered it to be a calcareous earth. The individuality of the heavy earth was subsequently confirmed by Bergman's own experiments. As might be expected, the main distinction was that "terra ponderosa" gave a sparingly soluble "ponderous spar" with "vitriolic acid," whereas the corresponding gypsum was appreciably soluble.

It was Marggraf who first recognized "argillaceous earth" as an individual entity. He informs us (4) that in his day alum was generally thought to be a salt composed of an "earth" and "vitriolic acid." His predecessors had regarded the "earth" as a calcareous substance, but Marggraf detected significant differences. Thus a solution of "earth of alum" in "nitrous acid" gave no precipitate on treatment with "vitriolic acid," whereas a corresponding solution of chalk did give of a precipitate. Thus he recognized that he was dealing with a new kind of "earth." Marggraf made numerous other observations of fundamental importance, investigating many of the commoner inorganic precipitation reactions and the solvent actions of acids and alkalis upon his precipitates. In this work, he laid the foundations for what later became the tables of qualitative analysis that we studied in our youth.

The first recognition of silica as an individual "earth" seams to have been due to the German chemist, J.H. Pott (1692-1777) (5). He recognized four "primitive earths": calcareous, argillaceous, gypseous, and "vitrifiable," the last corresponding to our silica. His "vitrifiable earth" was largely characterized by its behavior on heating, though he noted its ready fusion on heating with alkali, and its resistance to attack by acids.

By the middle years of the eighteenth century, then, a mass of qualitative data had been assembled, which could be put to good use in the formulation of a general scheme of mineral analysis, and in the work of Bergman we find the first published exposition of such a scheme. The secret, of course, was to bring the whole of the mineral being investigated into solution, which could be achieved by fusion with alkali. This reaction was not new. It had been known for centuries from the work of glass manufacturers. Nevertheless, it was Bergman who first made use of the reaction as part of a general analytical procedure. His method is described in Volume 2 of the English translation of his "Physical and Chemical Essays," under the head: "Method by which the Proximate Principles of Gems are Most Easily Determined." From this, his general analytical scheme may be extracted and exhibited as in Table 1 (6).

**Table 1. Bergman's Analytical Method**

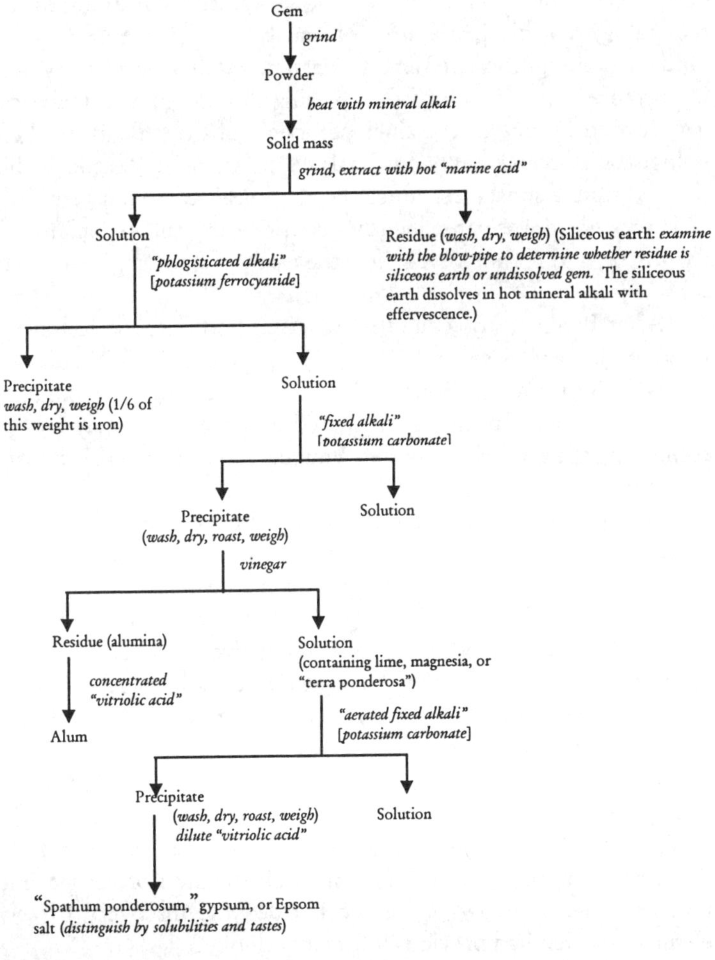

Sadly, Bergman's published results were not at all accurate, the errors, in part, being due to the fact that he had not taken some of the less common "earths" into consideration in his analyses. Also, some of the analytical results of his pupils were included in his own works. An additional source of error seems to have been due to the fact that his materials became contaminated with silica as a result of the grinding of his gems in his mortars. Nevertheless, Bergman's authority as an analytical chemist was so considerable that in many cases his work remained uncorrected for several years. However, the contribution offered to mineralogy by his procedures of "analysis in the humid way" was of fundamental significance, and the actual results obtained were of lesser importance. It may be noted in passing that Bergman's analytical scheme was devised by one of the chief exponents of the phlogiston theory. The phlogistonist could quote his analyses in terms of "earths," thinking of them as simple substances, since they appeared as the last terms of analysis. This procedure was subsequently taken over by the exponents of the new chemistry, and retained even after the earths were recognized as compound substances.

After Bergman took the first vital step, others soon followed his lead, notably Richard Kirwan (1733-1812) in Ireland, and Nicolas Vauquelin (1763-1829) in France. Both of these men used "humid" techniques. Some idea of the rapid progress in analytical procedures may be gauged by comparing the general method of Vauquelin, given in Table 2 (7), with that of Bergman.

Vauquelin's work was undoubtedly impressive, though his methods must have been exceedingly tedious, demanding very considerable skill on the part of the analyst. However, it is in the work of the great German chemist, M.H. Klaproth (1743-1817), that one finds the culmination of eighteenth century methods of mineral analysis. When one considers the apparatus and materials that were available to him, Klaproth's skill and accuracy appear almost uncanny. Unlike Bergman, he quoted the actual weights of materials obtained in his analyses, so that errors were not concealed by arbitrarily making sure that the figures added up to 100 per cent. He took scrupulous care to purify his reagents, was the first to dry or ignite to constant weight, and (unlike Bergman) allowed for the silica acquired from his mortars during the pulverization of his specimens. The precision and accuracy of his work aided the discovery of several new elements, which had previously been overlooked.

## Table 2. Vauquelin's Analytical Method

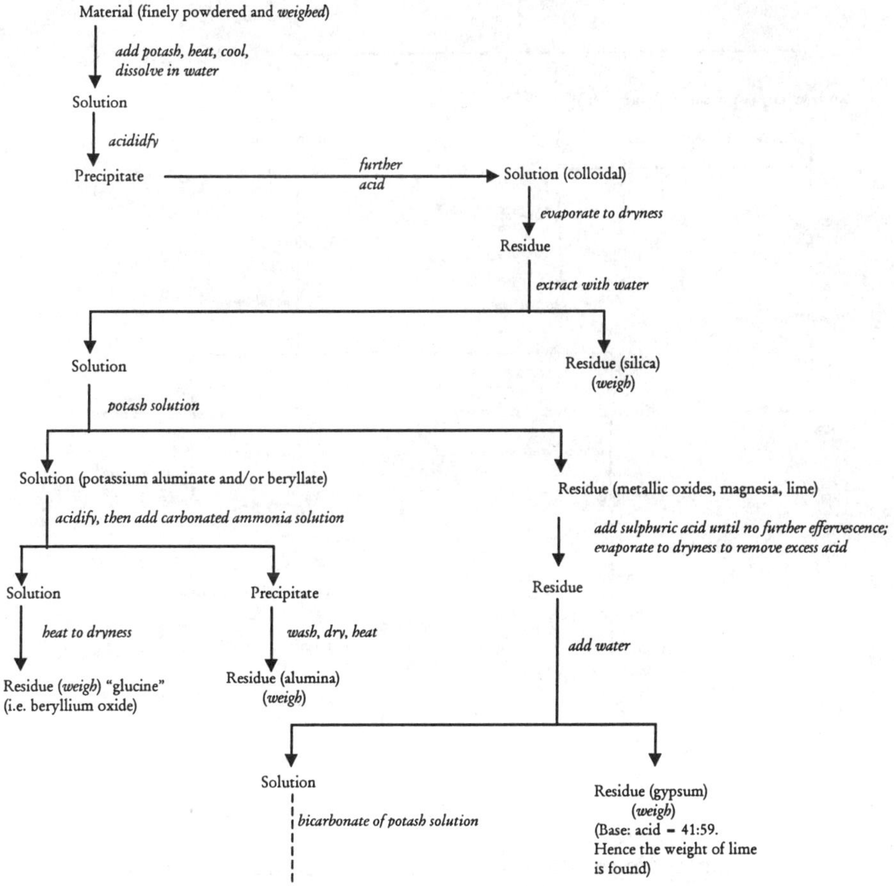

**Table 2. Vauquelin's Analytical Method (cont'd)**

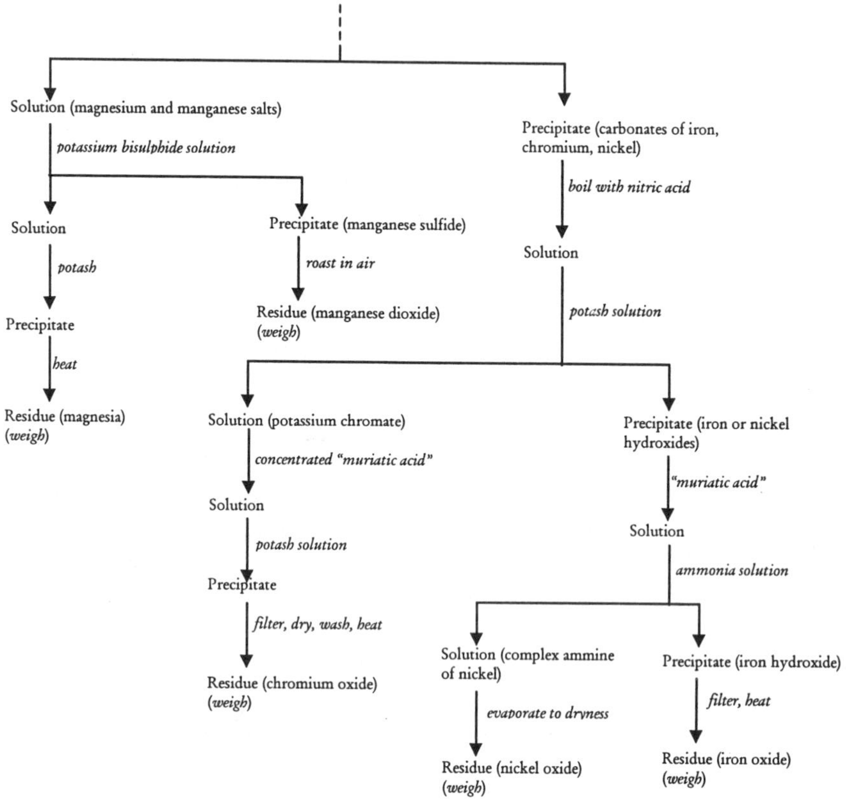


It is impossible to summarize Klaproth's methods briefly, for, unlike Kirwan and Vauquelin, he did not publish a general analytical scheme. Each of his analyses had its own individuality, each being subtly adapted to the nature of the substance under investigation. We take his analysis of the sapphire as our example (Table 3) (8), though it should be emphasized that this was one of the simplest investigations described in his work. Even here, the complexities of the problem are very apparent. A check on the operations had to be maintained at all times with the aid of a balance, so that mistakes of the kind made by Bergman (when he contaminated his gems with silica from his pestle and mortar) would he avoided.

## Table 3. Klaproth's Analytical Method

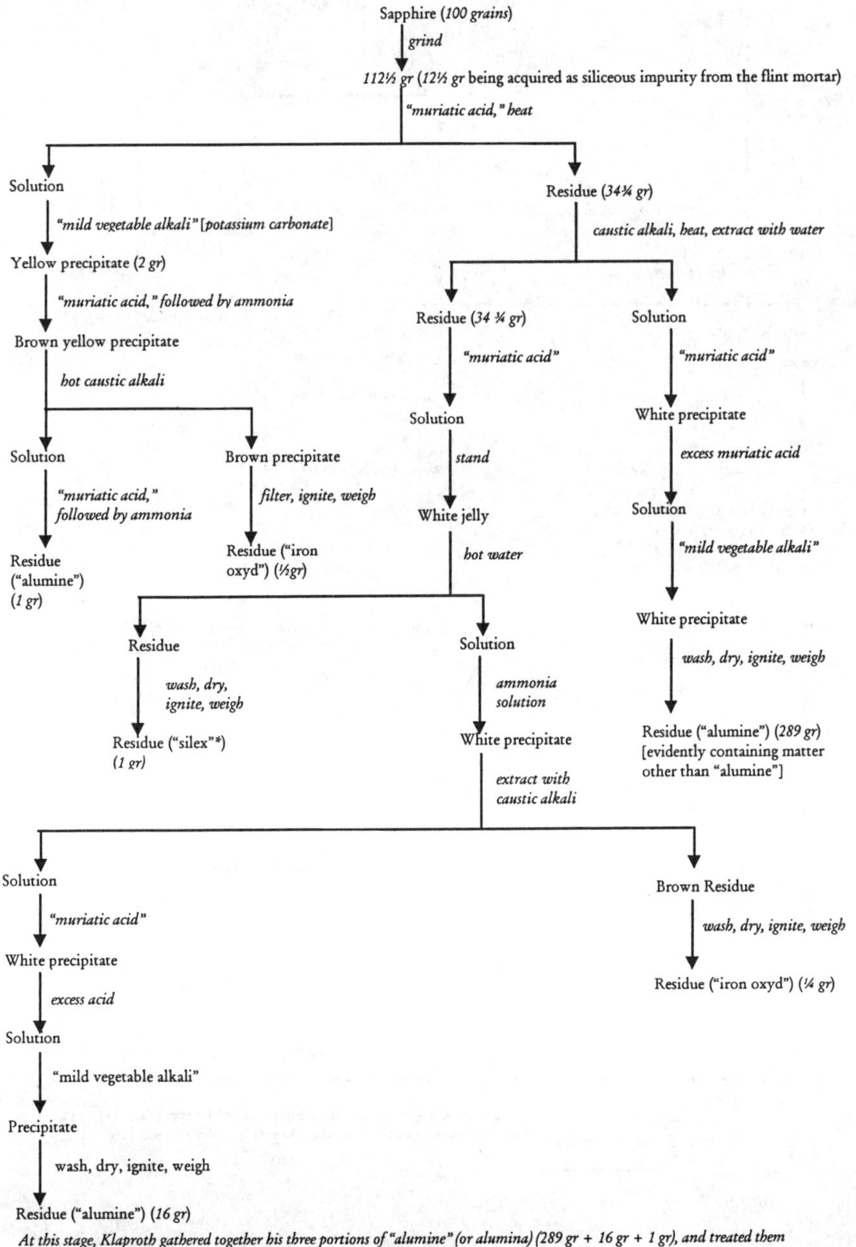


*At this stage, Klaproth gathered together his three portions of "alumine" (or alumina) (289 gr + 16 gr + 1 gr), and treated them with dilute "vitriolic acid." 8 gr remained undissolved, and was trated as silica:*

## Table 3. Klaproth's Analytical Method (cont'd)

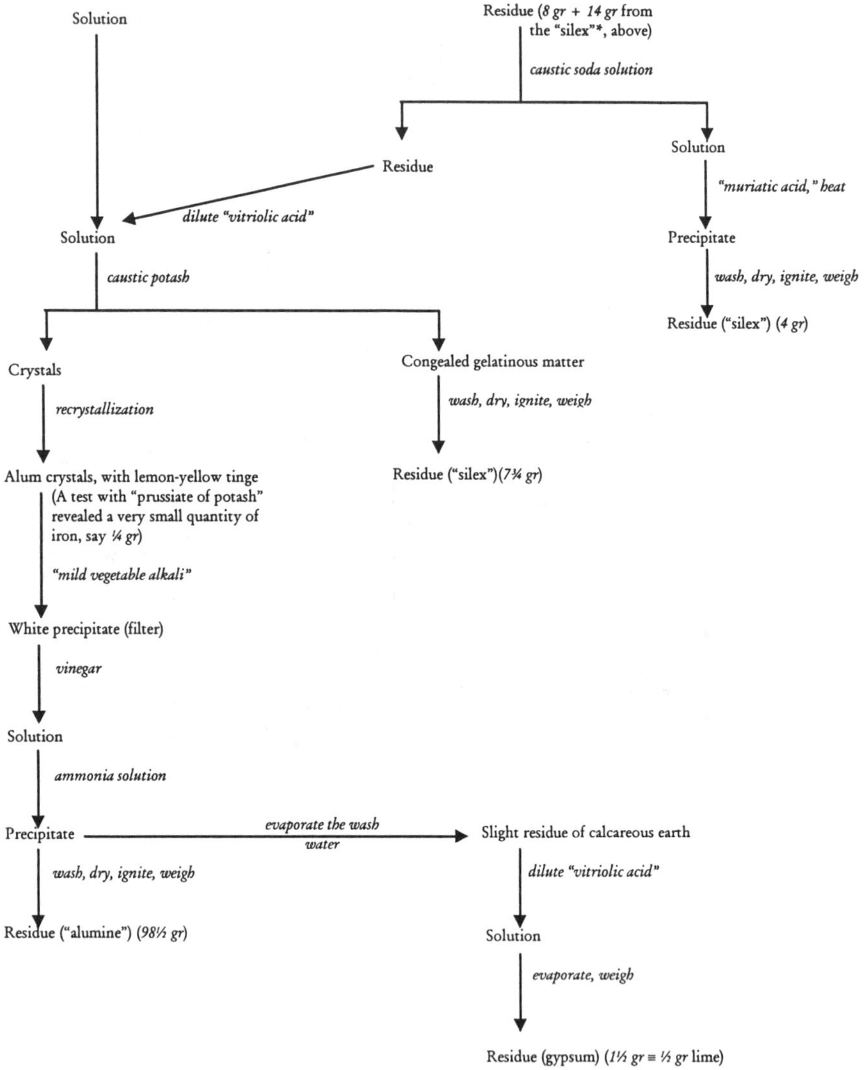


*Summary of the Analysis*

| | |
|---|---|
| Alumine | 98½ |
| Oxyd of iron | ½ + ¼ + ¼ |
| Lime | ½ |
| | 100 |

It may be noted that of the 12½ gr of "silex," known to be present at the beginning of the analysis, 11¾ gr (7¾ + 4) were recovered at the end.

It can be seen, then, that once a technique for bringing minerals into solution had been established, and the general notion of separating earthy substances by precipitation reactions had been put forward, it soon became possible to offer a genuine analysis. No longer, as at the beginning of the eighteenth century, need one fear that one's "analysis by fire" would alter the "ingredients" of a substance during the course of an investigation. Minerals could he viewed as combinations of the several known earths and metals, and the relative proportions of each could be determined with considerable accuracy. With patience and skill, which few can emulate today, it had become possible to carry out satisfactory mineral analyses, thereby laying the basis for the chemical classification of the mineral kingdom.

**Literature Cited**

(1) See: Greenaway, F., Proc. 10th Int. Congr. Hist. Sci., 819 (1962).
(2) Cited in Partington, J.R., "A History of Chemistry," London (1961), Vol. 2, p. 696.
(3) Bergman, T.O., "A Dissertation on Elective Attractions," London (1785), p. 180.
(4) Marggraf, A.S., "Opuscules Chymiques," Paris (1762), Vol. 2, p. 103.
(5) Pott, J.H., "Lithogéognosie," Paris (1753), p. 7 and passim.
(6) Bergman, T.O., "Physical and Chemical Essays," London (1784), Vol. 2, pp. 95-103.
(7) Vauquelin, N., Ann. de Chim., 30, 66 (1799).
(8) Klaproth, M.H., "Analytical Essays Towards Promoting the Chemical Knowledge of Mineral Substances," London, 1801, pp. 72 ff. (first German ed., Berlin, 1795).

# IX

## EDWARD DANIEL CLARKE, 1769–1822, AND HIS RÔLE IN THE HISTORY OF THE BLOW-PIPE

THE common blow-pipe is not, today, an instrument calculated to excite the interests of aspiring chemists and mineralogists, and, although it may still be used from time to time by mineral prospectors, it is generally to be found, if anywhere, dusty and neglected at the backs of the drawers of school chemistry laboratories. By contrast, it was an all important tool for chemists and mineralogists in the eighteenth and nineteenth centuries, and numerous publications were formerly devoted to expositions of the virtues of this deceptively simple piece of apparatus.[1] The present paper examines, albeit briefly, the early history of the instrument, and considers in more detail the various later elaborations that were introduced in order to permit the utilization of the mutual combustion of hydrogen and oxygen in mineralogical analyses. The chief exponent of this art in England was Edward Daniel Clarke,[2] the first professor of mineralogy at Cambridge, whose investigations provide the principal focus for our present considerations.

Clarke is an interesting figure, and one who has not received a great deal of attention from historians of science, possibly because, by spreading his energies over a very wide field in a characteristically eighteenth-century manner, he did not achieve the highest distinction in any particular area of investigation. Nevertheless, we consider that his personality and work, so characteristic of their time, have inherent interest, and warrant

[1] Of these, the most important early expositions of the art were: an Appendix by Gustav von Engeström to his English translation of Cronstedt's *Försök till Mineralogie* (see note 30 below); T. O. Bergman, *Commentatio de Tubo Ferruminatorio, eiusdemque usu, in explorandis corporibus praesertim mineralibus*, Vienna, 1779; and J. J. Berzelius, *Afhandling om Blasrörets användande i kemien och mineralogien*, Stockholm, 1820. I have used the English translation of this work: J. J. Berzelius, *The Use of the Blowpipe in Chemical Analysis, and in the Examination of Minerals, translated from the French of M. Fresnel by J. G. Children . . . with a sketch of Berzelius' system of mineralogy . . .*, London, 1822. Another useful early work in English, with a brief historical account of the development of the blow-pipe is: J. Griffin, *A Practical Treatise on the Use of the Blowpipe*, Glasgow, 1827.

[2] See note 3. For a summary account of Clarke's life and work see: *Annual Register*, London, 1822, pp. 322–324.

Plate I. Edward Daniel Clarke—the frontispiece from William Otter's biography. (See note 3.)

our attention. Clarke was the subject of a sympathetic biography[3] from the pen of his life-long friend and admirer, William Otter, Bishop of Chichester, from whom I have drawn freely for details of Clarke's career.

Clarke was born in 1769 at the village of Willingdon in Sussex, where his father, Edward Clarke (1730–1786) was vicar. The family had, for several generations, been linked with the church, although not with high ecclesiastical offices. Several Clarkes achieved minor reputations as scholarly divines, with literary and antiquarian interests. Edward Daniel's paternal grandmother was the daughter of the notable Dr. William Wotton,[4] from whom, so Otter suggests, Clarke may have inherited an unusually retentive memory. His father was something of a traveller, and spent time in Spain and Minorca, his published descriptions of which[5] achieved passing recognition.

Edward Daniel was at first educated privately, and subsequently at Tonbridge grammar school. His education was completed at Jesus College, Cambridge, where he was a chapel clerk. It appears that he gave little attention to his formal studies during his youth, although he probably read widely in a desultory fashion throughout this period. At Cambridge, Clarke studied mathematics and a little classics, and, more particularly, ancient and modern history, English poetry and natural science. In the final examinations, he ranked no higher than third Junior Optimes. Nevertheless, this qualification was later deemed sufficient to secure his election to a Fellowship of his College, though it would appear that, in the event, the authorities were far from mistaken in their choice, their disregard of the more formal academic qualifications being fully warranted by Clarke's subsequent career.

Since he lost his father shortly after going up to Cambridge, Clarke found it necessary to support himself by various small scholarships and bursaries, which were not, however, sufficient to prevent his getting into some debt while at College. Happily, he succeeded in freeing himself from these encumbrances when, after taking his degree, he obtained a series of employments as private tutor. One might perhaps have expected him to enter the church, in which his family was well connected, but, instead, he chose to take up a succession of positions as tutor to the sons of various members of the aristocracy, employment which enabled him to exercise his taste, or passion one might say, for foreign travel. It is not

[3] W. Otter, *The Life and Remains of Edward Daniel Clarke Professor of Mineralogy in, the University of Cambridge*, 2 vols., 2nd edition, London, 1825.

[4] William Wotton (1666–1727), celebrated scholar and divine, a remarkable child prodigy and noted for his prowess of memory.

[5] E. Clarke, *Defence of the Conduct of the Lieutenant-Governor of the Island of Minorca,....* London, 1767; E. Clarke, *Letters, concerning the Spanish Nation; written at Madrid, during the years 1760 and 1761*, London, 1763.

possible to enter here into all the details of Clarke's several foreign journeys, for which the reader should consult Clarke's own most interesting accounts,[6] but it is upon these volumes that his literary fame chiefly rests. Letters describing this heroic period of his life (1791–1802) are to be found in Otter's biography. Clarke's travels took him over much of the British Isles (his accounts of the Scottish Highlands being particularly interesting), and, overseas, he ranged from Egypt, in the south, to Lapland in the north, and to Asiatic Russia in the east. He had the good fortune to meet many eminent mineralogists and chemists in Sweden, a country of which he formed a high opinion, by contrast with Russia, which was severely censured. Throughout, Clarke was an indefatigable observer and collector, his interests ranging from plants, minerals, medals, coins, statuary and various *objets d'art*, to some quite important classical manuscripts which he acquired in the Middle East. It should be noted that the basis of Clarke's mineral collection (which was purchased for £1500 by the University of Cambridge after his death) was laid during these extensive travels.

In 1793, Clarke was in Italy, accompanying Lord Berwick, and, in company with a number of aristocratic companions and Italian guides, he observed eruptions of Vesuvius, climbing the mountain while it was in an active condition, and examining the crater and the several lava flows. Also, at this time, he had the benefit of conversing with the diplomat, Sir William Hamilton, who had previously made numerous investigations of Italian volcanoes and had adopted Vulcanist and 'uniformitarian' opinions[7]. These observations and conversations appear to have exerted a considerable impression on Clarke's mind, for, later, partly as a result of his consideration of the volcanic theory of Davy, and partly as a result of his own work on the 'gas blow-pipe', he ascribed the activity of volcanoes to the subterranean interaction of hydrogen and oxygen. It is noteworthy,

[6] E. D. Clarke, *Travels in Various Countries of Europe, Asia and Africa.* (*Part the First, Russia, Tartary and Turkey*; *Part the Second, Greece, Egypt and the Holy Land*; *Part the Third, Scandinavia*), 6 volumes, London, 1810–1823. Several subsequent editions were published, the last appearing in 1848.

[7] Sir William Hamilton, *Observations on Mount Vesuvius, Mount Etna, and Other Volcanos*: *in A Series of Letters, Addressed to the Royal Society*, . . . , New Edition, London, 1774. Hamilton wrote (p. 92): 'Nature, though varied, is certainly uniform in her operations, and I cannot conceive that two such considerable Volcanos as Etna and Vesuvius should have been formed otherwise than every considerable Volcano of the known world.' It is interesting to note that he envisaged volcanic activity as a means of maintaining the fertility of the globe by a constant, slow, 'ploughing' of the earth. The treatment adopted (p. 160, 'Such wonderful operations of Nature are certainly intended by all-wise Providence for some great purpose.') is reminiscent of Hutton's similar use of the 'argument from design'.

however, that, in spite of his close acquaintance with volcanic phenomena, Clarke, in the earlier part of his career, adopted certain Neptunist opinions.

Thus, in the first volume of his *Travels*, he stated:

> 'From all these facts it seems evident, that the basaltic pillars of Ineada [in Turkey] were the result of an aqueous deposition; and that their prismatic configuration, . . . is entirely owing to a process of CRYSTALLIZATION . . .'[8]

The problem of the origin of basalt was referred to again on several occasions in subsequent volumes, generally in somewhat ambiguous terms, but, on balance, giving qualified support to the Neptunist theory. On the other hand, in some respects Clarke seems to have moved away from the Neptunist position, at least so far as the geology of Scotland was concerned. Thus, in 1797, Clarke travelled extensively in Scotland, in company with the Honourable Berkeley Paget, but, although he kept a full journal of his activities, doubtless with a view to publication, this intention was never realized. Otter published some account of the Scottish travels, but, in deference to Clarke's wishes, most passages of geological interest (which one must presume, were of a Neptunist hue) were excised.

> '. . . the leaning of his judgement in the course of his tour,' Otter remarked, 'seems to have been in a different direction from that which it afterward took, when in a maturer state of his own knowledge, the learned and accurate labours of Dr. Macculloch had been submitted to him.'[9]

Macculloch, it will be recalled, adopted the Vulcanist point of view in his accounts of the geology of Scotland,[10] from which we may infer that Clarke moved towards this position in the latter part of his career. It appears, however, that he remained in a state of some uncertainty on the matter, and we find that in his later mineralogical lectures at Cambridge he refrained from speaking out boldly on the matter. (See note 15 below.) It is, of course, possible to view many geological phenomena in a Neptunist or Vulcanist manner with almost equal facility,[11] and it is well known that many of the leading geologists of Clarke's day were unable to determine the matter to their satisfaction.[12] It appears, then, that Clarke reflected, rather than illuminated, the uncertainties in the geological theories of his time.

[8] E. D. Clarke, *op. cit.* (footnote 5), vol. i, p. 667.

[9] W. Otter, *op. cit.*, vol. i, p. 279.

[10] J. Macculloch, *A Description of the Western Islands of Scotland, including the Isle of Man*, 2 volumes, London, 1819.

[11] I have illustrated this point by means of a simple example in a recent paper: 'The Vulcanist Neptunist Dispute Reconsidered', *J. Geol. Educ.*, 1971, **19**, 124–129.

[12] See, for example: M. J. S. Rudwick, 'Hutton and Werner Compared: George Greenhough's Geological Tour of Scotland in 1805', *Brit. J. Hist. Sci.*, 1962, **1**, 117–135.

In 1798 Clarke took up residence at Jesus, having been elected to a Fellowship before he went to Scotland. If we are to judge the matter by Otter's account,[13] this move was taken principally to avoid enlistment in the militia at Uckfield, where Clarke was then residing. After only a year's residence in College, Clarke, together with his pupil and patron John Martin Cripps, Thomas Malthus (the demographer), and Otter, set off for an extended tour of Europe and the Middle East. Malthus and Otter soon dropped off, but Clarke and Cripps continued abroad for three years. On his return to Cambridge in 1802, after many hardships, Clarke, by then with a secure reputation as a natural historian and antiquarian, was awarded the honorary degree of LL.D.

In the ensuing years, Clarke was largely occupied with the preparation of his memoirs, and the proper cataloguing and sale of the antiquities that he had collected during his travels. In 1806 he married, and the union ultimately led to a family of seven children. Otter records that the initial financial prospects for matrimony were considered to be unpromising, but that they improved significantly when Clarke took up the College living at Harlton, which had become vacant at about that time. Clarke had been ordained in 1805. It appears that he was zealous in the preparation of a number of attractive sermons for the parish; he also preached on occasions at Great St. Mary's.

Not long after his marriage Clarke commenced a series of lectures on mineralogy under the auspices of the University. Something of the style of these lectures may be gauged from their *Syllabus*, which Clarke published in 1807,[14] and, more particularly, from an annotated copy of this publication in the hand of one of Clarke's students, W. S. Macleay of Trinity, which is now in the possession of the library of the Linnean Society of New South Wales.[15] The *Syllabus* provided a skeleton of the lectures, giving the appropriate topic headings, to which students could add their own notes as Clarke spoke. The treatment was largely descriptive, rather than theoretical, and was, to a considerable extent, devoted to a consideration of the specimens that Clarke had himself collected during the course of his

[13] W. Otter, *op. cit.*, vol. i, p. 436.

[14] E. D. Clarke, *Syllabus of Lectures in Mineralogy*, Cambridge, 1807.

[15] I am much indebted to Associate Professor T. G. Vallance, of the University of Sydney, for drawing my attention to this copy of Clarke's *Syllabus*, and also for allowing me to peruse a transcription which he has prepared of certain portions of Macleay's annotations. Macleay (1792–1865) attended Clarke's lectures in 1816. After going down from Cambridge he served as a diplomat in Paris, where he became acquainted with Cuvier. Subsequently he worked in Havana, and moved to Sydney in 1839, where he soon became recognized as one of the leading naturalists of New South Wales. His principal interests were biological rather than mineralogical.

travels. The subject was, however, treated in a most systematic manner, following a scheme of mineral classification (based on chemical composition, rather than external characteristics) that Clarke had himself devised.[16]

The lectures were apparently highly successful and led to the creation of a chair in mineralogy in 1808, a step that ultimately resulted in the development of the present Department of Mineralogy and Petrology. The series of lectures was continued annually until 1821, and Clarke was still working hard at the beginning of 1822, preparing his lectures for that year. These were never presented, however, for he died on March 9.

Clarke's mineralogical work in Cambridge is largely associated with his development of the so-called 'gas blow-pipe', in which, by means of the combustion of hydrogen and oxygen, he was able to investigate the behaviour of minerals at very high temperatures. It is to a discussion of this work, and its precursors, to which we now turn our attention.

According to Otter's account,[17] Clarke's interest in the high temperatures that could be achieved using hydrogen and oxygen was stimulated by reading a French translation of a German work on the subject by F. Ehrmann. Consequently, we examine something of the history of the blow-pipe,[18] leading to Ehrmann's studies.

[16] It is not possible to enter here into a detailed discussion of Clarke's lectures, though they are of considerable interest, and offer a valuable insight into the Professor's scientific views. Clarke did not lay much emphasis on theory, and many of the items in the printed *Syllabus* which promise some theoretical discussion are not represented in Macleay's notes. Why such theoretical aspects were omitted we cannot now determine, but it seems not unreasonable to suggest that the omission reflects Clarke's interest in the collection and classification of minerals rather than a theoretical treatment of their properties, and this seems to be in keeping with his somewhat 'Baconian' approach to science—a point which will be touched on again below. This is not to say, however, that he ignored the chemical basis of minerology. His chemical system of classification may be gauged from his *Methodical Distribution of the Mineral Kingdom into Classes, Orders, Genera, Species, and Varieties* (Lewes, 1806), a pamphlet of six pages, dedicated to the 'Students of the University of Cambridge'.

[17] W. Otter, *op. cit.*, vol. ii, p. 344. The work by Ehrmann referred to is: *Essai d'un Art de Fusion a l'aide de l'air du feu, ou air vital par Mr. Ehrmann . . . . Traduit de l'Allemand par M. de Fontallard & revu par l'auteur. Suivi des mémoires de Mr. Lavoisier . . . sur le même Sujet*, Strasbourg, 1787.

[18] Some account of the history of the blow-pipe is given in: F. Szabadváry, *History of Analytical Chemistry*, Oxford, 1966, pp. 50–55. A very detailed history, indicative of the importance which was attached to the blow-pipe by the chemists and mineralogists of the period, was written at the end of the eighteenth century: Weigel, C. E., 'Versuch einer Geschichte des Blaserohrs und seine Andwendungen; vom Hrn Profeß. Weigel in Greisswalde', *Beyträge zu den chemischen Annalen* (Crell's *Beyträge*), 1790, **4**, 262–283, 393–419; 1794, **6**, 6–24, 198–216.

It is not, perhaps, absolutely necessary to pursue this matter into the depths of antiquity, but we may remark that the blow-pipe was probably first used by metal workers for providing an air-blast for their primitive furnaces, the technique being depicted in Egyptian paintings from the tombs of Beni Hasan (12th Dynasty, *ca.* 2000–1800B.C.).[19] In our era, the instrument seems to have been first developed by glass blowers,[20] certainly as early as the beginning of the seventeenth century, and possibly earlier. It was undoubtedly used by the craftsmen of the *Accademia del Cimento*[21] for the construction of their beautiful and intricate thermometers. The German alchemist, Johann Kunckel, in his well-known monograph *Ars Vitraria Experimentalis*, which was the standard authority on glass working in the seventeenth century, made the important suggestion that the blow-pipe could be utilized in mineral analysis.[22] For a time, however, this idea was not taken up, and the blow-pipe was still used primarily as an aid to the fusion of metals or glass—as a welding tool. Boyle, for example, described the instrument as 'a small bent tube of glass', and he stated that

> '. . . when the flame of a lamp or candle is directed by it, so as to beat upon a body at a convenient distance, it may be made to melt silver, or even copper itself; . . .'[23]

Stahl, too, mentioned the blow-pipe *en passant*.[24] There is a more detailed description in Cramer's influential manual of assaying, together with an accompanying illustration, but it appears from the description given that the instrument was still used primarily as a means of melting small

[19] P. E. Newberry, *Beni Hasan* (Published under the auspices of the Egypt Exploration Fund), Part I, London, 1893, Plate XI; and Part II, London, 1894, Plates IV and VII.

[20] Weigel (*loc. cit.*, pp. 264–281) gave a very detailed account of the early application of the blow-pipe in the glass industry, particularly with reference to the construction of thermometers, but was unable to determine the origin of the application with complete precision. As he pointed out, when an instrument of this kind is developed by artisans, it may be in use for a number of years before this fact becomes recorded in the literature.

[21] *Saggi di Naturali Esperienze Fatte nell Accademia del Cimento sotto la Protezione del Serenissimo Principe Leopoldo di Toscana e Descritte dal Segretario di essa Accademia*, 2nd edition, Florence, 1691, pp. II–IV. (The first edition, which I have not seen, was published in 1660.) The Italians termed the blow-pipe a *gonfia*.

[22] J. Kunckel, *Ars Vitraria Experimentalis oder, vollkommene Glasmacher Kunst...*, Frankfurt and Leipzig, 1679, pp. 399–400. (Cited in: F. Szabadváry, *op. cit.*, p. 81).

[23] R. Boyle, *The Philosophical Works of the Honourable Robert Boyle Esq; Abridged, methodised, and disposed . . . . By Peter Swan*, M.D., London, 1725, vol. i, p. 477.

[24] G. E. Stahl, *Philosophical Principles of Universal Chemistry : . . . . Drawn from the Collegium Jenense of Dr. George Ernest Stahl. By Peter Shaw, M.D.*, London, 1730, p. 32.

quantities of metal at the charcoal block.[25] Mineral assays were still made by the time-honoured methods of roasting and cupellation on the furnace,[26] so that the blow-pipe had not, at this period, brought about any striking changes in chemistry, mineralogy, metallurgy, or the art of assaying.

It was in Sweden that the analytical potentialities of the instrument were first fully appreciated, where Anton Swab[27] is generally credited with being the first to make real use of the blow-pipe for mineralogical investigations. Swab did not himself publish a specific account of his methods, but the idea evidently spread in Swedish chemical circles, and, in the hands of A. F. Cronstedt,[28] it assumed a position of very considerable importance. Cronstedt, of course, is best known for his discovery of nickel,[29] but his development of a chemical system of mineral classification, rather than one based on external characteristics, was probably of more far-reaching importance, and was the forerunner of many such schemes, exerting a considerable influence upon mineralogical thinking in the latter part of the eighteenth century. Like Swab, Cronstedt never gave a published description of the manner in which he used the blow-pipe, but fortunately this deficiency was made good by his colleague, Gustaf von Engeström, Warden of the Stockholm Mint, who gave a detailed illustrated account of the new methods in an appendix to his English translation of Cronstedt's *Försök till Mineralogie.*[30]

[25] J. A. Cramer, *Elements of the Art of Assaying Metals . . . , Translated from the Latin . . . .* , London, 1741, p. 99. The Latin edition (which I have not seen) was published in 1739. Cramer's illustration of the blow-pipe is of particular interest since it shows a small bulb in the tube, to collect moisture. This refinement probably indicates that the instrument had been in use for some time, although having received little attention in the literature.

[26] For a discussion of early assay methods see: F. Greenaway, ' The early development of analytical chemistry ', *Endeavour*, 1962, **21**, 91–97.

[27] Here, I am relying on Partington's authority (*History of Chemistry*, London, 1962, vol. iii., p. 174). The issue was confused by Bergman (*op. cit.*, p. 3), who stated that *Andreas* Swab was the first to adopt the blow-pipe in Sweden for mineralogical analyses. Weigel (*loc. cit.*, p. 283) maintained, on the authority of Von Engeström, that Cronstedt should have priority.

[28] For a discussion of the contributions of Cronstedt, see V. Bartow, ' Axel Fredrik Cronstedt ', *Jour. Chem. Educ.*, 1953, **30,** pp. 247–252. A most attractive portrayal of Cronstedt at work with his blow-pipe is to be found in: F. Greenaway, *Chemical Laboratories and Apparatus to* 1850, Science Museum, London, 1966, item 8.

[29] See M. E. Weeks, *Discovery of the Elements*, 5th edition, Easton, Pa., 1945, pp. 70–73,

[30] A. F. Cronstedt, *An Essay Towards a System of Mineralogy, translated from the original Swedish with notes, by Gustav von Engeström. To which is added, a treatise on the Pocket-Laboratory, containing an easy method, used by the author, for trying mineral bodies, written by the translator. The whole revised and corrected, with some additional notes, by Emanuel Mendes da Costa.* 2nd. edition; *with an appendix, containing additions and notes by Professor M. T. Brunnick*, London, 1772.

Essentially, Cronstedt, von Engeström, and the other Swedish chemists, were merely using scaled down versions of the standard assayers' techniques, applied as qualitative tests. Powdered samples of the minerals under investigation were examined by heating alone on the charcoal block, or in conjunction with soda, microcosmic salt, or, more particularly, borax. The several colours obtained with the borax or microcosmic beads could be utilized for the characterization of compounds of the known metals, and, when new minerals were discovered and named, their behaviour under the blow-pipe could rapidly be determined and incorporated into the body of analytical knowledge. In most cases, there cannot have been a very clear understanding of the reactions involved, although the phlogiston theory gave an ostensibly intelligible account of the various reductions. Perhaps more importantly, it soon became possible to establish fairly standard procedures for mineral recognition by means of blow-pipe techniques, so that the mineralogical tyro could have some hope of recognizing new specimens that he might discover, without having to rely on the laborious Wernerian descriptions of the external characteristics of minerals, or a knowledge of the complexities of the newly developing science of crystallography. In the late eighteenth century, then, we find small portable 'travelling laboratories' being designed for mineralogical prospectors, only a blow-pipe, candle, charcoal block, and a few simple chemicals being essential. Von Engeström himself gives us the first record of such travelling kits, and others soon followed his lead.[31] Several of the leading chemists of the late eighteenth and early nineteenth centuries, notably Bergman and Berzelius,[32] paid particular attention to the new techniques, and slowly the rather difficult art of breathing in and out simultaneously (!)[33] (which was required in order to maintain a hot and steady flame on the sample under investigation) spread from Sweden to other parts of Europe, and to Germany, France and England in particular. The blow-pipe received numerous modifications, and, naturally, was adapted for use with a pair of bellows in the laboratory, but the details of these developments will not be entered into here, though the variety of instruments used may be gauged from Plate II in which a number of the early forms of the instrument are illustrated.

[31] For a description of these early portable mineralogical 'laboratories', see W. A. Smeaton, 'The portable chemical laboratories of Guyton de Morveau, Cronstedt and Gottling', *Ambix*, 1966, **13**, 84–91.

[32] See note 1.

[33] J. J. Berzelius, (*op. cit.*, p. 20) wrote: 'It is a difficulty like that which a man experiences when he endeavours at the same time to turn his right arm and his right leg in opposite directions'. The trick required was to expel air from the mouth by means of the cheeks, while, at the same time, breathing into the lungs through the nose.

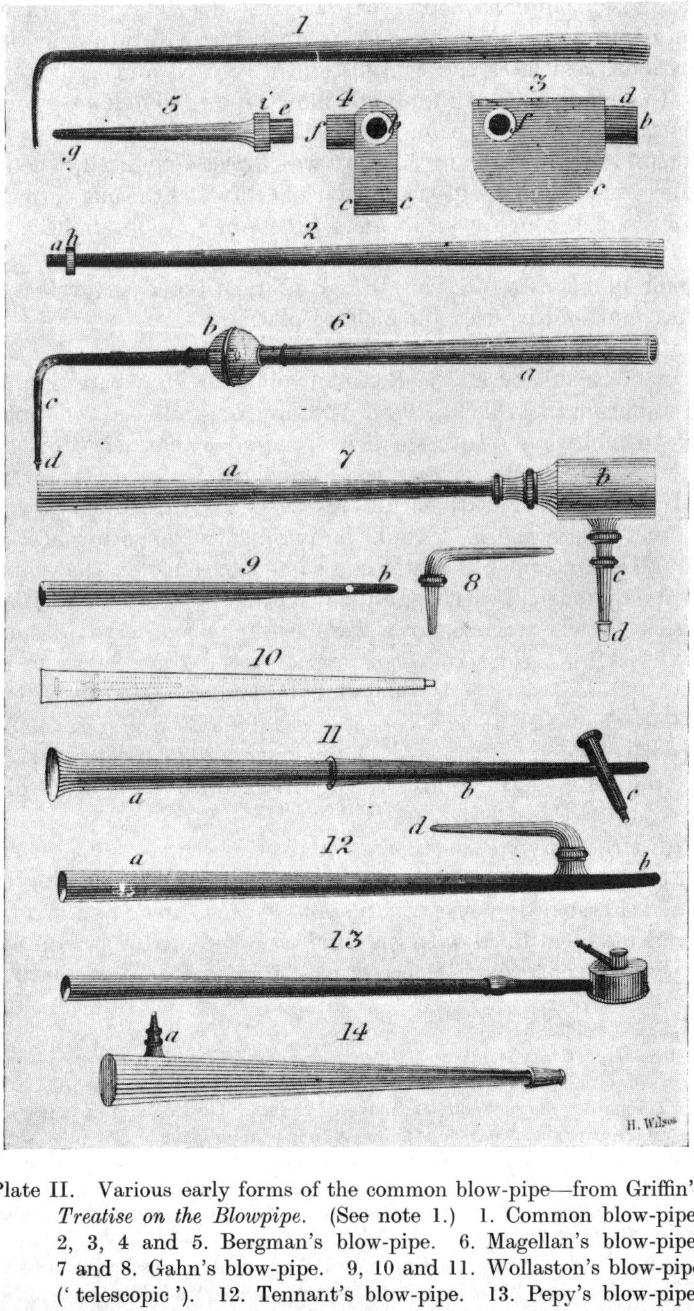


Plate II. Various early forms of the common blow-pipe—from Griffin's *Treatise on the Blowpipe*. (See note 1.) 1. Common blow-pipe. 2, 3, 4 and 5. Bergman's blow-pipe. 6. Magellan's blow-pipe. 7 and 8. Gahn's blow-pipe. 9, 10 and 11. Wollaston's blow-pipe (' telescopic '). 12. Tennant's blow-pipe. 13. Pepy's blow-pipe. 14. Black's blow-pipe.

Our present interest, however, is more closely connected with the attempts that were made to raise temperatures by utilizing oxygen in combustions, and later the combustion of oxygen and hydrogen, rather than with elaborations of the simple blow-pipe to save lung-power (which Berzelius condemned as being 'perfectly useless'). It appears that the first person to use oxygen in this way was Priestley himself, when he filled a bladder with his dephlogisticated air and directed the gas onto a burning piece of wood through a small glass tube, thereby obtaining a very hot flame.[34] He pointed out that it would be easy to supply the gas from a large pair of bellows, and his friend John Mitchell suggested that the method could well be used for melting platinum.

A more complex piece of apparatus was devised by F. C. Achard, professor of chemistry at the Berlin Academy. In a paper published in Rozier's journal in 1782[35] he described how he had used a double pair of bellows to pump air into a small iron cylinder that served as a furnace. The inlet valve of the bellows was connected to an earthenware vessel containing nitre, so that when this vessel was heated, ordinary air, enriched with 'dephlogisticated air', could be pumped into the furnace. Achard placed a Hessian crucible containing some iron nails on the grate of clay within the cylindrical furnace, and surrounded the crucible with burning coals, so that when the bellows were set in motion a vivid combustion occurred. After eight seconds of pumping, the grate began to melt, the crucible vitrified, the nails melted completely, and the cylinder was virtually destroyed. It was a spectacular demonstration of the powers of the recently discovered gas, although, as would be expected for a German chemist publishing in 1782, his interpretations were based on the phlogiston theory.

Lavoisier showed a particular interest in these developments, and carried out numerous experiments at high temperatures, using his 'pure air' and charcoal for his combustions.[36] At first, like Priestley, he employed bladders filled with his 'air', and directed a stream of the gas on to lighted charcoal. Success was immediate, and platinum was

[34] J. Priestley, *Experiments and Observations on Different Kinds of Air, and other branches of Natural Philosophy, connected with the subject*, 3 vols., London, 1790, vol. ii, p. 168.

[35] F. C. Achard, 'Sur un nouveau moyen de produire, avec une très-petite quantité de charbons ou d'autres substances inflammables, une chaleur égale à celle qu'on peut produire par des verres & des miroirs ardens d'une grandeur considerable; avec la description d'un fourneau, qui, en servant a chausser un appartement, purifie l'air qu'il renferme, en le privant de son phlogistique', *Observations sur la Physique, sur l'Histoire Naturelle et les Arts*, 1782, **20**, 371–376.

[36] A. L. Lavoisier, 'Mémoire sur un moyen d'augmenter considérablement l'action de feu et de la chaleur dans les opérations chimiques', *Mem. Acad. Roy. des Sci.*, 1782 (1785), 457–465; reprinted in *Oeuvres de Lavoisier*, Paris, 1862, vol. ii, pp. 423–431.

melted at the first attempt. His further experiments employed the well-known,[37] Heath Robinson like, system of magazines, operating on the principle of the modern gas-holder, in which oxygen was stored over water and delivered automatically at a constant pressure to the blow-pipe table. Having thus provided himself with this most elaborate apparatus, Lavoisier systematically examined the effects of higher temperatures on a considerable range of rocks, minerals and metallic substances. Here he was extending the observations of the German chemist, J. H. Pott, who, earlier in the century, had examined the effects of furnace temperatures on an enormous range of rocks and minerals, and various mixtures of these with different fluxes, in a desperate attempt to discover the secret of the Meissen porcelain.[38] Like Pott, Lavoisier carried out a similar 'Baconian' series of experiments, though operating at considerably higher temperatures. The results achieved suggested that *all* substances were fusible, and it appeared that, apart from platinum, all metals could be volatilized—a point of some interest—but as an *analytical* tool, the apparatus compared unfavourably in some ways with the common mouth blow-pipe, for the distinction between fusibles and infusibles, which could be a useful diagnostic feature, was largely eliminated. Lavoisier achieved no theoretical insights of any particular significance as a result of his extensive series of investigations, and no important new system of mineral classification was developed. To be sure, it was perfectly natural to attempt to extend the range of working temperatures by means of the newly available technique. Who could tell, *a priori*, what might happen at these very high temperatures? Had not the common blow-pipe proved invaluable? But, in the event, the ant-like collection of data, not directed towards the testing of any particular hypothesis, did not lead to any new theoretical insight. Nevertheless, pehaps owing to the excitement and possible dangers of these new high temperature reactions, numerous other investigators took up Lavoisier's lead, their work being described in detail in Ehrmann's monograph. As early as 1785, an anonymous work describing the various devices that had been invented to employ dephlogisticated air in conjunction with the blow-pipe was published in Tubingen.[39]

[37] A. L. Lavoisier, *op. cit.*, ('*Oeuvres*'), plates IV and V, accompanying the text.

[38] J. H. Pott, *Chemische Untersuchungen welche fürnehmlich von der Lithogeognosie oder Erkennung und Bearbeitung der gemeinen einfacheren Steine und Erden, ingleichen von Feuer und Luft handeln*, Potsdam, 1746. I have used the French edition: J. H. Pott, *Lithogéognosie, ou Examen Chymique des Pierres et des Terres en Général, et du Talc, de la Topaze, & la Stéatite en particular, avec une Dissertation sur le Feu & sur la Lumiere*, Paris, 1753.

[39] This work is cited by Ehrmann (see note 17, above, p. 219) as: *Description de quelques machines établies pour l'usage de l'air déphlogistiqué avec le chalumeau & le feu de fonte, & d'une méthode de se procurer beaucoup de cet air*, Tubingen, 1785. I have not seen it.

In his memoir of 1782, Lavoisier mentioned that he had attempted to use two blow-pipes, one delivering vital air and the other inflammable air, but the results were not as satisfactory as those obtained merely with vital air and charcoal. He toyed with the idea of burning inflammable air in an atmosphere of vital air, but did not put it into practice. Perhaps the first to do this with success was Robert Hare of Philadelphia, who, in 1801, invented the so-called hydrostatic blow-pipe, which enabled him to burn oxygen and hydrogen together with comparative safety.[40]

Essentially, Hare's apparatus consisted of a small 'brewery' barrel with a horizontal partition through its mid-point, and a vertical partition dividing the upper portion of the lower chamber into two compartments. The apparatus was filled with water to a depth of about 2/3 of the whole. By means of a special pump, hydrogen and oxygen could be introduced separately into the two lower compartments, the water being displaced by suitable pipes into the upper chamber. The gases could then be delivered to a blow-pipe table, through appropriate tubes, by the pressure of the water in the apparatus. The jets consisted of two holes drilled in the same frustum of silver, so that the two gases could be brought together for combustion. The whole functioned rather like a Kipp's apparatus, except that the gases were introduced from outside, rather than being generated internally. By means of this apparatus, which was subsequently modified over a number of years,[41] Hare could carry out fusions with some facility, and, again, it appeared that all substances were fusible if a sufficiently high temperature were reached. The experiments were shown to Priestley, by then residing in America, who thought that the hydrostatic method was quite original, besides being highly effective. Several American chemists also interested themselves in Hare's work, including Benjamin Silliman, who published the results of his own high temperature experiments carried out with Hare's apparatus.[42]

Following Hare's work, Thomas Thomson also attempted to carry out experiments with the oxy-hydrogen blow-pipe,[43] but he did not pursue

[40] R. Hare, 'Memoir on the Supply and Application of the Blow-pipe', *Philosophical Magazine*, 1802, **14**, 238–245 and 298–306. This was also published in France under the title 'Memoire sur l'usage du chalumeau' (*Annales de Chimie*, an XI, **45**, 113–138).

[41] Numerous forms of his instrument are described by Hare in his: *A Compendium of the Course of Chemical Instruction in the Medical Department of the University of Pennsylvania . . . . For the Use of His Pupils*, Philadelphia, 1827.

[42] Benjamin Silliman, 'Experiments on the Fusion of Various Refractory Bodies by the Compound Blow-Pipe of Mr. Hare', *Phil. Mag.*, 1817, **50**, 106–114.

[43] This is according to the statement of Clarke in his principal exposition of the blow-pipe techniques: E. D. Clarke, *The Gas Blow-Pipe, or Art of Fusion by Burning the Gaseous Constituents of Water: Giving the History of the Philosophical Apparatus so denominated; the Proofs of Analogy in its Operations to the Nature of Volcanoes; Together with an Appendix containing an account of Experiments with this Blow-Pipe*, London, 1819, p. 6.

the matter very far, owing to the dangers involved in such work. It was, then, the adventurous Edward Daniel Clarke who took up this work in England and who eventually established a considerable reputation for himself on the basis of his daring investigations. To an account of Clarke's work in this field we now return.

It was, of course, known that the major problem involved in the mutual combustion of the components of water was that the flame might burn back into the gas reservoir at a rate exceeding the gas flow, and thus cause a destructive explosion. This problem was examined by Tennant and Wollaston, in connection with the problem of gas illumination,[44] and, in 1814, Tennant announced in his public lectures in Cambridge that the flame could be burned in safety if separated from the gas reservoir by a narrow capillary tube. It was this principle that Clarke attempted to utilize in his apparatus, in ignorance of the various rates of propagation of flame fronts in different gaseous mixtures.

In April, 1816, H. I. Brooke wrote to Thomson suggesting an invention which would 'relieve the great inconvenience... felt in using the common blow-pipe with the mouth.'[45] This was simply a metal vessel into which air could be pumped by means of a syringe. The compressed air could then be allowed to escape from the container through a jet, and impinge upon a candle flame. It was a typical example of the type of invention to which Berzelius, who preferred pulmonary control, took exception. Such an apparatus was constructed for Brooke by John Newman, the London instrument maker, who also published a description of the new apparatus,[46] which could deliver a steady stream of gas for about twenty minutes. The whole could be packed conveniently into a small box for itinerant mineralogists, and was thus more versatile than Hare's clumsy barrel. When working in the laboratory, Newman and Brooke planned to use compressed air, oxygen, or possibly hydrogen, in the reservoir.

The new apparatus came to the attention of Clarke,[47] to whom Newman suggested that a mixture of hydrogen and oxygen could probably be used, provided that one took the precaution of passing the mixture through a fine capillary tube. Clarke asked Davy's opinion, and it transpired that

[44] *Ibid.*, p. 3.

[45] H. I. Brooke, 'Description of a new Blow-Pipe', *Annals of Philosophy*, 1816, **7**, 367. See also, *Ibid.*, 1817, **9**, 7–8.

[46] J. Newman, 'Account of a new Blow-pipe in a letter from Mr. John Newman to the Editor', *Quarterly Journal of Literature, Science and the Arts. Edited at the Royal Institution of Great Britain*, 1817, **1**, 65–66.

[47] E. D. Clarke, 'Account of some Experiments made with Newman's Blow-pipe, by inflaming a highly condensed Mixture of the Gaseous Constituents of Water; in a Letter to the Editor, from Edward Daniel Clarke, LL.D. Professor of Mineralogy in the University of Cambridge', *Quarterly Journal of Literature, Science and the Arts*, 1817 ..., **2**, 104–123.

Davy had already tried such an experiment without ill effects, believing that a capillary tube about 3 inches long and with an internal diameter of 1/80 inch would provide an adequate safeguard.[48] Thus encouraged, and in spite of Wollaston's words of caution, Clarke set to work in Cambridge before a number of interested persons, using one of Newman's pieces of apparatus. A two to one ratio of hydrogen to oxygen was employed. He was immediately successful, and it is likely that the obtained somewhat higher temperatures than Hare or other earlier workers. Platinum was fused with ease. Diamonds burned. Gold volatilized. Magnesia, barytes, silica, and 'alumine' all succumbed to the heat of the flame. Some fusion of lime was apparent, and all known minerals, with the possible exception of zircon, were found to be fusible.

> 'I sacrificed the month of August [1816] to chemistry,' Clarke wrote with enthusiasm to Otter. 'Oh, how I did work! It was delightful play to me; and I stuck to it day and night. At last, having blown off both my eyebrows and eyelashes, and nearly blown out both my eyes, I ended with a bang that shook all the houses round my Lecture-room. The Cambridge paper has told you the result of all this alchemy for I have actually decomposed the earths, and obtained them in metallic form.'[49]

Clarke's claim to have isolated the alkaline earth metals (notably barium) will be considered in more detail below. For the present, we continue our account of the development of his experiments.

The violent explosion referred to above could well have proved fatal to Clarke or one of his collaborators, for, according to his report, bits of apparatus were flying round the laboratory like cannon shot. The dangers were overcome to some extent, however, by a safety apparatus devised by one of Clarke's colleagues, the Reverend J. Cumming, occupant of the chair of chemistry at Cambridge. The gases were made to bubble through a chamber of water before ignition, there being a wire gauze over the bottom to this chamber, the intention being to confine any explosion to this chamber, rather than the main gas reservoir. Nevertheless, explosions still occurred occasionally, particularly if the handle of the syringe used to compress the gases was drawn out inadvertently while the intervening stop-cock was open. Consequently, Clarke provided himself with a wooden safety screen and listened to ensure that the bubbling

[48] Davy was allowing Clarke very little safety margin, for, subsequently, he wrote: 'The circumstance [of the combustion of gaseous mixtures at elevated temperatures] is shown in a very elegant manner, in burning concentrated mixtures of oxygen and hydrogen at the end of a long-tube of one-sixtieth or one-seventieth of an inch in diameter, when the experiment begins (the tube being cool) there is no danger; gradually, however, as it become heated, the combustion steals as it were down the tube, and at last reaches the reservoir of the gases.' (H. Davy, *The Collected Works of Sir Humphrey Davy, Bart. LL.D., F.R.S. . . .*, vol. vi, London, 1840, p. 17.

[49] W. Otter, *op. cit.*, vol. ii, pp. 346–347.

of gas through the water could be heard before he ignited the gas mixture. Later, oil was substituted for the water. The final form of the apparatus, as illustrated in Clarke's monograph, *The Gas Blow-Pipe*..., is shown in Plate III.

Since the readers of Thomson's *Annals of Philosophy* were being kept fully informed of all developments with a 'blow by blow' series of reports on Clarke's experiments, other correspondents had willingly proffered their advice. A double-acting pump was suggested by R. W. Barchard[50] so that the pressure in the reservoir would never be reduced by lifting the pump handle of the syringe. Wollaston[51] suggested a faggot of brass capillaries in the tube leading to the blow-pipe jet itself, and J. T. Beale[52] offered the idea of a piece of porous cane that would act as a system of capillary tubes. Even simpler, was the suggestion of T. S. Booth,[53] who proposed that the compression chamber should be dispensed with altogether, gether, being replaced by a bladder of the mixed gases which could be slowly compressed by a weight placed on top of it. One had almost gone full circle, back to the simple bladder of Priestley. Nevertheless Booth's idea was taken up, and eventually a virtually danger-free apparatus was constructed by Goldsworthy Gurney, who drove his gases from a bladder through a Cumming safety apparatus to the blow-pipe jet.[54]

The final stage in the development of apparatus for the safe handling and mutual combustion of oxygen and hydrogen that we shall notice here occurred with the invention of the oxy-hydrogen torch by J. F. Daniell, professor of chemistry at King's College, London.[55] As in the modern instrument, two concentric jets were used, oxygen being passed through the inner one, and hydrogen through the outer. Thus, all the advantages of the hot oxy-hydrogen flame could be achieved, but, by separating the gas reservoirs, all the attendant dangers of explosion were avoided. It should be noted, however, that one of the figures given in Hare's paper of 1802 reveals that he had, at this earlier date, devised a simple version of this form of oxy-hydrogen jet.

We now return to a more detailed consideration of Clarke's actual experiments. Perhaps his most striking claim was that he had succeeded

[50] R. W. Barchard, *Ann. Phil.*, 1817, **10**, 66–67.

[51] E. D. Clarke, *Ann. Phil.*, 1817, **9**, 326–327.

[52] J. T. Beale, *Ann. Phil.*, 1817, **9**, 481–482.

[53] T. S. Booth, *Ann. Phil.*, 1817, **10**, 67–68.

[54] Gurney's apparatus is illustrated in the frontispiece of Griffin's *Practical Treatise on the Use of the Blowpipe.* (See note 1.) It is also depicted in F. Greenaway, *op. cit.*, (note 28), item 12.

[55] J. F. Daniell, 'On a New Oxy-Hydrogen Jet', *Phil. Mag.*, 1833, **2**, 57–60.

Plate III Clarke's representation of his 'Gas Blow-Pipe'—the frontispiece to his work describing this apparatus. (See note 43.) (4 = 'A large Bladder, or silk balloon, varnished with a solution of *caoutchouc*, for containing the *mixed gases* of *hydrogen* and *oxygen*'. 5 = Cap of the Cumming's safety cylinder. 8 = Syringe for transferring the gaseous mixture from the bladder to the pressure chamber. The early form of the apparatus, before the safety screen was adopted, was fitted with a syringe in a vertical position.)

in isolating some of the alkaline earth metals, which had hitherto only been obtained electrolytically. Clarke's first description of his method was as follows :

> 'Having obtained a portion of this *earth* [' barytes '] in a state of purity, I mixed some of it with *lamp-oil*, and rubbed both together in a *porcelain* mortar, into a paste. This *paste* being placed upon *charcoal*, was brought to the *ignited gas*, and kept exposed to its most intense heat, for some minutes. By this means it was fused, and assumed the form of a black shining *slag*, like that of *iron* from a foundry. A small portion of this slag was then held, by means of a little borax, upon the end of the tube of a tobacco-pipe; and again exposed to the *ignited gas*. The *slag* being now firmly fixed upon the pipe-clay admitted the action of the file, and exhibited a shining *metallic* surface, resembling that of silver.'[56]

Clarke designated his metal *Plutonium*, on the grounds that, with a specific gravity of 4, the name Barium, signifying heavy, was unwarranted. The name, Plutonium, was intended to indicate the relation of the metal to the 'dominion of fire'. The new designation was not accepted, however, for the name barium was well entrenched. Indeed, a number of chemists, notably Wollaston, were unwilling to accept Clarke's claim to have isolated the metal by the new method. Wollaston visited Cambridge in April, 1817, to view the experiments, but was not convinced, though the specimen of 'barytes' which he brought with him for analysis had deliquesced, and Clarke maintained that the failure was due to the hydration of the sample. Clarke also visited London, where attempts had been made to repeat his analysis, in order to persuade the chemists at the Royal Institution of the efficacy of his method, but since they did not appear for his demonstration he returned to Cambridge without having convinced them of his views.[57] It would appear, however, that Clarke's claim was well founded. His samples showed metallic lustre tarnishing on exposure to air, were malleable and ductile, and evolved hydrogen when placed in water. His estimated specific gravity of 4·0 agrees with the modern value. Claims to the isolation of strontium were also made, but in less forthright terms, and an early claim to have isolated 'silicium' (silicon) by means of the gas blow-pipe, although supported by Joshua Mantell of Lewes, who repeated the experiment,[58] was rejected by Thomson, and was not maintained. Later, Clarke pointed out that it was easy for substances to acquire a pseudo-metallic lustre when subjected to his high temperatures, and that this was

[56] E. D. Clarke, *loc. cit.*, (note 47), p. 119.

[57] Here I have reported Clarke's version of the affair, from: ' Account of some Experiments made with the Gas Blow-pipe; being a Continuation of former Observations upon the same Subject ', *Ann. Phil.*, 1817, **10**, 133–138.

[58] J. Mantell, *Ann. Phil.*, 1818, **11**, 310–311.

not in itself sufficient to indicate the isolation of a metal. It is interesting to note in this connection that Thomson, a Wernerian,[59] claimed that Clarke's work could be used as evidence against the Huttonian theory, the argument being that tin ores acquired this pseudo-metallic lustre when subjected to fusion. Nevertheless, certain ores of tin, which were found in association with granite, lacked this lustre. Hence, it was concluded, the granite had never been in a state of fusion. The argument appears more reasonable today, in the light of certain modern theories of granitization, than would have been the case, say, fifty years ago.

We have argued above that Lavoisier did not achieve any new theoretical insights of any particular significance as a result of his high temperature investigations. What can be said of Clarke's work in this respect? It is true that he was able to offer a new technique for the preparation of pure nickel, a useful technological advance.[60] But the gas blow-pipe was such an overwhelmingly powerful instrument that it tended to obliterate all mineralogical distinctions. Consequently, no new system of mineralogical classification developed from his work. Indeed, it may be mentioned that the closest Clarke came to discovering a new element (lithium) occurred when he was analyzing the mineral petalite by the customary procedures of wet analysis.[61] Thus we are forced to the conclusion that Clarke, like Lavoisier and Hare before him, did not produce any significant theoretical breakthrough a a result of his investigations. It was largely a technological achievement.

Let us now consider Clarke's claims to priority in these experiments. They were, in fact, somewhat bitterly attacked by Hare in an appendix that he wrote to a chemical text-book for his medical students at the University of Pennsylvania,[62] under the title: *Strictures on a Publication entitled Clark's Gas Blowpipe*. In this, Hare was able to show that many of the results achieved by Clarke were anticipated by himself or by Professor Silliman, and that perhaps Clarke's only claim to originality was that he had been the first to advocate the use of a 2 : 1 ratio of hydrogen and oxygen. But, as Hare properly pointed out: Who would ever think of using any other ratio? Probably, then, all that can be claimed for Clarke in this respect is that he had used an apparatus that was more

[59] Thomson was one of the founder members of Jameson's Wernerian Club in Edinburgh. See the list of members given in: J. M. Sweet, 'The Wernerian Natural History Society in Edinburgh', in *Abraham Gottlob Werner, Gedenkschrift aus Anlaß der Wiederkehr seines Todestages nach* 150 *Jahren am* 30. *Juni* 1967, Freiberger Forschungshefte C223 Mineralogie—Lagerstättenlehre, Leipzig, 1967, pp. 205–218.

[60] E. D. Clarke, *Ann. Phil.*, 1819, **14**, 142–143.

[61] E. D. Clarke, *Ann. Phil.*, 1818, **11**, 194–198 and *Ibid.*, pp. 365–368.

[62] R. Hare, *op. cit.* (note 40).

effectual in producing high temperatures than any achieved hitherto, and that he had succeeded in convincing chemical 'authority' (in the person of Thomas Thomson) that all the earths were fusible. Hare's earlier experiments had not so persuaded Thomson.

Thus, we must ask ourselves, did anything really significant emerge from all this interest in high temperature work? It would appear not, in spite of the great interest that it evoked at the time. What, then, was the motivation for such investigations? In our view, leaving aside the inherent excitement of the dangers associated with the investigations, Clarke's research programme was entirely in keeping with the spirit of his times. Consider, for example, what the 3rd edition of the *Encyclopaedia Britannica* (1797) had to say about the nature of scientific research. In a detailed article under the head 'View of Bacon's Philosophy', it asserted that the object of Natural History was:

> '1. To observe with care, and describe with accuracy, the various objects of the universe.
>
> '2. To determine and enumerate all the great classes of objects; to distribute and arrange them into all their subordinate classes, . . . and to mark with precision the principles of this distribution and arrangement, and the characteristics of the various assemblages.
>
> '3. To determine with certainty the particular group to which any proposed INDIVIDUAL belongs.'[63]

Such a programme, an evident amalgam of Bacon and Linnaeus, appears to have been followed by Clarke, although, as we have seen, his gas blow-pipe did not really facilitate classification.

A similar programme, continued the encyclopaedist, characterizes the endeavours of the *philosopher* (as opposed to the natural historian), the principal difference being that the philosopher deals with *events*, rather than objects. Thus, by grouping together similar events, the laws of nature were supposed to emerge. The groupings were held to be due to the operation of hidden causes, and these it was the task of the natural philosopher to determine. This, of course, was the difficult part, as the writer freely admitted, though here the experimental method could usefully be invoked, whereby one could isolate the phenomenon under investigation from a multitude of confusing factors. The causes, needless to say, were not observed directly, but were to be inferred from the phenomena. How this was to be achieved was not clear, though one thing was certain—hypotheses were *not* to be invoked, unless, merely as temporary conjectures, they might serve to direct new experiments. Thus, the 'official' philosophy of science at the period we are considering, if we may be permitted so to call it, was a legacy of Aristotle (the search for causes), Bacon (the

[63] *Encyclopaedia Britannica*, 3rd edition (1797), vol. xiv, p. 582.

collection of data to form a natural history), Linnaeus, and Newton.[64] It is true that the encyclopaedist saw a few Humean clouds on the horizon, but these were rapidly dismissed. It will be seen that no methodology had been developed which would instruct one how to attain, mechanically, the Baconian 'axioms', the causes, or high level laws. Needless to say, the problem is still with us today, but in the eighteenth century, Newton's admonitions against hypotheses may well have had some deleterious effect. To be sure, the Daltons, Youngs and Huttons of Clarke's day developed bold hypotheses with great freedom and facility, but lesser men, perhaps to some extent inhibited by Newton's maxims, often were restricted to the Baconian collection of data. Clarke, it would appear, his wealth of knowledge and experience notwithstanding, was such a lesser man. Sadly, significant high level laws, or axioms, or causes—call them what you will—did not emerge from the series of investigations with the blow-pipe, which he conducted so assiduously and with such enthusiasm. His contribution was but to the *Sylva Sylvarum* of mineralogy.

Finally, we ask ourselves how we should regard Clarke as a person. He was evidently a bold man, both as an experimenter and as an explorer, and it is perhaps these intrepid aspects of his work that remain among his chief claims to lasting recognition. He does not appear to have been an original thinker, although, like so many men of his time, his industry and his literary style command admiration. Most of the improvements to his celebrated apparatus, and indeed the very apparatus itself, were suggested by others. The general nature of his experiments was far from new, and his most interesting theoretical suggestion—that volcanic activity might be due to the combustion of hydrogen and oxygen in the interior of the earth—was never taken up, though it was certainly no more bizarre than Davy's conjecture that the activity might be due to the interaction of water with sodium or potassium. Clarke was largely content to follow, rather than lead, the scientific theories and opinions of his day.

[64] The exact status of Baconianism in the early nineteenth century certainly warrants further research. While it is generally held that Bacon's maxims were widely accepted at that time—which is the viewpoint which I have myself adopted above—it is possible to find contrary opinions strongly expressed. Brewster, for example, wrote in 1824: 'The opinion so prevalent during the last thirty years, that Lord Bacon introduced the art of experimental enquiry on physical subjects, . . . , appears to me to be without foundation, and perfectly inconsistent with the history of science . . . . The method given by Bacon is, . . . quite useless, and in point of fact has never been used in any successful inquiry. A collection of facts, however, skilfully they may be conjured with, can never yield general laws . . . . ' (M. M. Gordon, *The Home Life of Sir David Brewster*, 2nd edition, Edinburgh, 1870, pp. 128 and 130). I am grateful to my colleague, Mr. J. R. Panter, for drawing my attention to this passage.

From Otter's account, Clarke appears to have been a most agreeable person, devoted to his wife and family, though one may be inclined to suspect that he cultivated persons of influence with some assiduity, and we have noted an apparent sudden enthusiasm for the church at the time of his marriage. Perhaps we may be excused for drawing attention here to a passage that reveals the relationship between his scientific discoveries and theological beliefs in an interesting way :

> '. . . all the constituents of created nature are *combustible*; and although it be not permitted to the *chemist*, " who finds his endeavours to comprehend the works of *creation* checked at every turn ", to pry into "His ways, which are past finding out," yet, as a philosopher, he will probably call to mind the remarkable prediction which ordains, for the final dissolution of the material world, that " THE EARTH, AND THE WORKS THAT ARE THEREIN, SHALL BE BURNT UP." '[65]

Thus, even if Clarke's blow-pipe did not bring about any really significant advances in chemistry or mineralogy, it should please us that, in the heyday of Paleyism at Cambridge, it could play some rôle, albeit a very minor one, in the physico-theological argument!

Clarke was a man of his time, with characteristic energy and optimism. His contribution was largely technological, rather than theoretical, and can be construed as being Baconian in spirit. We believe that his investigations, typifying so much of the work of his period, have proper claims to our attention.

*Acknowledgments*

Plate I is reproduced from the copy of the second edition of Otter's *Life and Remains of Edward Daniel Clarke . . . .*, held at the Fisher Library, University of Sydney.

Plate II is reproduced from the copy of Griffin's *Practical Treatise on the Blowpipe . . . .*, held in the Menzies Library, University of New South Wales.

Plate III is reproduced from the copy of Clarke's *The Gas Blow-Pipe, . . . .* in the possession of Professor Vallance. I would like to record that it was through conversations with Professor Vallance that I first became interested in the work of E. D. Clarke and his rôle in the history of the blow-pipe.

[65] E. D. Clarke, letter to Thomson, dated October 5, 1816, published in *Ann. Phil.*, 1816, **8**, 357–364. Clarke was citing Bishop Watson in the quotations within this passage.

# X

# ROBERT HOOKE'S METHODOLOGY OF SCIENCE AS EXEMPLIFIED IN HIS 'DISCOURSE OF EARTHQUAKES'

A NUMBER of authors have drawn attention to the contributions to geology of Robert Hooke,[1] and it has been pointed out that in several ways his ideas were more advanced than those of Steno,[2] who is sometimes taken to be the founder of geology as a scientific discipline. Moreover, it has been argued that in a number of instances Hooke should receive the credit for ideas which are usually believed to have originated in the work of James Hutton.[3] This recognition of the significance of Hooke's work is regarded by the present writer as being well founded. But, by contrast, the relationship between Hooke's geological ideas and his views on the proper methods for conducting scientific enquiries has been largely overlooked, and his views on the methodology of science, as revealed in his *Posthumous works*,[4] have received little discussion. Moreover, on

[1] The notices of Hooke's work given by Lyell and Geikie are well known: C. Lyell, *Principles of geology, being an attempt to explain the former changes of the earth's surface, by reference to causes now in operation* (1st edn., 3 vols., London, 1830–3), i. 31–5; and A. Geikie, *The founders of geology* (2nd edn., London, 1905), pp. 68–73. The standard biography of Hooke (M. 'Espinasse, *Robert Hooke* [London, 1956]) deals with his geological contributions, although not in great detail. A number of papers have also commended the enlightened character of Hooke's geological thought, including: A. P. Rossiter, 'The first English geologist Robert Hooke (1635–1703)', *Durham University journal*, xxix (1935), 172–81; W. N. Edwards, 'Robert Hooke as geologist and evolutionist', *Nature*, cxxxvii (1936), 96–7; E. N. da C. Andrade, 'Wilkins Lecture. Robert Hooke', *Proceedings of the Royal Society of London*, Section A, cci (1950), 439–73, and Ibid., Section B, cxxxvii (1950), 153–87; G. L. Davies, 'Robert Hooke and his conception of earth-history', *Proceedings of the Geologists' Association, London*, lxxv (1964), 493–8. Davies's views have been restated in his *The earth in decay: a history of British geomorphology 1578–1878* (London, n.d.). Hooke's evolutionary views attracted attention in: A. P. Pavlov, 'Robert Hooke, un évolutioniste oublié du xviie siècle', *Palaeobiologica*, i (1928), 203–10. His influence on the work of Raspe has recently been emphasized by A. V. Carozzi: 'Robert Hooke, Rudolf Erich Raspe, and the concept of earthquakes', *Isis*, lxi (1970), 85–91. Hooke's work is also noticed in several places in: C. J. Schneer (ed.), *Toward a history of geology* (Cambridge, Mass., 1969), and in R. S. Westfall's introduction to the recent reprint of Hooke's *Posthumous works* (see note 4, below). Hooke's methodology of science has been discussed in two valuable papers by M. B. Hesse: 'Hooke's development of Bacon's science', *Actes du Dixième Congrès International d'Histoire des Sciences; Ithaca 1962* (Paris, 1964), i. 265–8; and Hooke's philosophical algebra', *Isis*, lvii (1966), 67–83, but without particular reference to his geological theories.

[2] For example: Davies 1964, op. cit. (1), 397.

[3] Ibid.

[4] Richard Waller, *The posthumous works of Robert Hooke, M.D. S.R.S. Geom. Prof. Gresh., etc., containing his Cutlerian Lectures, and other discourses, read at the meetings of the illustrious Royal Society* (London, 1705). A facsimile of this volume, with an introduction by R. S. Westfall, has been published by the Johnson Reprint Corporation (New York and London, 1969). The six sections are:

I The present deficiency of natural philosophy is discoursed of, with the methods of rendering it more certain and beneficial.

II The nature, motion and effects of light are treated of, particularly that of the *sun* and *comets*.

III An hypothetical explication of memory; how the organs made use of by the mind in its operation may be mechanically understood.

one of the few occasions on which they were discussed, by William Whewell, they were belittled. Whewell regarded Hooke's methodological contribution as merely 'an attempt to adapt the *Novum organon* to the age which succeeded its publication',[5] and he implied that the 'same imperfections' were to be found in the writings of both Hooke and Bacon. The influence of Bacon is, to be sure, most obvious in Hooke's general philosophy. Nevertheless, it will be argued that Hooke had a much clearer idea than did Bacon of the importance of so-called 'hypothetico-deductive' methods in science, and that in many ways his methodological views represent a significant advance on those of Bacon.

Hooke's major geological contribution, 'A discourse of earthquakes',[6] represents the contents of a series of lectures that were probably presented to the Royal Society in 1667 and 1668. It is dated 15 September 1668 and therefore preceded the publication of Steno's *Prodromus* of 1669. The lectures were not published in Hooke's lifetime, but were included in the *Posthumous works* of 1705. Here we find, besides the 'Discourse' proper, a number of other lectures on geological topics which his biographer, Richard Waller, found among Hooke's papers after his death.[7] Not all of these lectures are dated, so that Waller had to use internal evidence

---

IV An hypothesis and explication of the cause of gravity, or gravitation, magnetism, *etc.*
V Discourses of earthquakes, their *causes* and *effects*, and histories of several; to which are annext, *physical explications* of several of the fables in *Ovid's Metamorphoses*, very different from other mythologick interpreters.
VI Lectures for improving navigation and astronomy, with the descriptions of several new and useful *instruments* and *contrivances*; the whole full of curious disquisitions and experiments.

The titles for the six sections, as they appear in the text, differ substantially from the above, which are cited from the title page. In the text, section V is entitled: 'Lectures and discourses of earthquakes, and subterraneous eruptions. Explicating the causes of the rugged and uneven face of the earth; and what reasons may be given for the frequent finding of shells and other sea and land petrified substances, scattered over the whole terrestrial superficies'.

[5] W. Whewell, *The philosophy of the inductive sciences, founded upon their history* (2nd edn., 2 vols., London, 1847), ii. 267.

[6] See note 4. The lectures usually referred to as the 'Discourse of earthquakes' occupy pp. 279–328 of the *Posthumous works.*

[7] These additional lectures were published by Waller in the *Posthumous works* as follows:
(a) Undated, pp. 329–45.
(b) Undated, pp. 346–50.
(c) 2 February 1686/7, pp. 350–4.
(d) 9 February 1686/7, pp. 355–60.
(e) 16 February 1686/7, pp. 360–2.
(f) 9 March 1686/7, pp. 363–4.
(g) 'Of the standing of the mercury in the tube to the height of 75 inches, read May the 28th, 1684', pp. 365–70.
(h) Undated, pp. 371–84.
(i) 'A copy of Dr. Thomas Gale's paper concerning giants', pp. 384–403.
(j) 15 February 1687/8, pp. 403–10.
(k) 29 February 1688, pp. 410–16.
(l) 23 July 1690, 'A discourse of earthquakes in the Leeward Islands', pp. 416–24.
(m) 30 July 1699, 'A discourse of the causes of earthquakes', pp. 424–8.
(n) 18 July 1688, pp. 428–33.
(o) 29 May 1689, pp. 433–6.
(p) Undated, pp. 436–7.
(q) 14 November 1699, 'A description of the Ridge of Mary Burrow in the Queens [*sic*] County in Ireland', pp. 437–8.
(r) 26 May 1697, pp. 438–50. Read to the Royal Society, 25 July 1694.

when attempting to place them in chronological order. All were judged to be later than 1668, but the exact dating of this material is not of vital importance to my argument. It is evident that Hooke returned many times to the geological problem that was most central to his interests—the problem of the presence of marine shells in strata far removed from the sea—and there is a good deal of repetition of his ideas in these additional lectures. Although their discussions of ancient mythology are tedious to the modern reader, some of the later lectures[8] contain important statements and illustrations of Hooke's methodological views, and they provide the focus for the discussions of the present paper. Clearly Hooke found the problem of fossil shells to be one of great interest and scientific significance, and he considered the question repeatedly in his later years. So far as we can judge from the evidence available, he never abandoned his opinion that fossils[9] represented the remains of former living organisms, although the *Posthumous works* does reveal some changes, during his lifetime, in his views on the causes of the entombment of these organisms. It should be noted that although Hooke was deeply religious, he did not exhibit geological inhibitions arising from theological considerations.

Hooke's 'Discourse' begins[10] with a brief critical discussion of the somewhat naïve Baconian methods that characterized the work of many of the early members of the Royal Society, with their undue emphasis on the mechanical collection of data. This introductory section is followed by a description of some of the organic remains that he had found, with emphasis on the ammonites and echinoids to be found in southern England. Evidently figures were prepared and these passed into the hands of Hans Sloane, from whom they were recovered by Waller, so that Hooke's diagrams are now to be found illustrating the 'Discourse'.[11]

Having collected this material, Hooke was immediately faced with the problem: how can it be that these shells, which are evidently the remains of once living marine organisms, are frequently found buried considerable distances above sea level and, on occasions, many miles from the sea? Hooke did not evade this difficult question. The fossils were not to be dismissed as mere 'sports of nature'; they did not arise from some mysterious 'plastick vertue'. For Hooke, the evidence was incontrovertible: fossils could only be the remains of marine organisms which had once been alive. He was thus led to set down eleven propositions, arising from the simple observation of the marine shells in strata far removed from

[8] See, especially, ibid., (a) to (e).

[9] Throughout this paper the word 'fossil' will, for simplicity, be taken to refer only to the petrified remains of a former living organism. Readers are asked to disregard the anachronism involved.

[10] Waller, op. cit. (4), p. 280.

[11] The figures, which reveal Hooke's high standard of draughtsmanship, represent a variety of ammonites ('snake-stones' or *cornua ammonis*), nautilids, belemnites, echinoids ('helmet-stones'), gasteropods, pelycepods, sharks' teeth, crinoid stems, etc. The first of the seven sets of engravings illustrating the *Posthumous works* was reproduced in Andrade, op. cit. (1), 467. The last two of the seven sets were prepared by Waller from specimens which he himself collected.

the present seas. These propositions[12] may be summarized as follows:

1. The 'curiously figured Bodies' are indeed the remains of animals or vegetables, turned to stone by 'some petrifying liquid Substance'.
2. 'Some extraordinary cause' must be responsible for this petrifaction.
3. The cause could be 'either some kind of fiery Exhalation arising from subterraneous Eruptions or Earthquakes; or secondly, a Saline Substance, whither [*sic*] working by Dissolution and Congelation, or Crystallization, or else by Precipitation and Coagulation; or thirdly, some glutinous or bituminous Matter, which upon growing dry or setling [*sic*] grows hard, and unites sandy Bodies together into a pretty hard Stone; or fourthly, a very long continuation of these Bodies under a great degree of Cold and Compression.'
4. Water may be transmuted to stone by some unknown means.
5. Other fluids, at rest, may congeal into hard substances.
6. Since the Creation, there have been several changes in the relative levels of land and sea.
7. Britain has been affected in the past by such changes.
8. Changes of this kind could be due to alterations in the centre of gravity of the globe, to the 'Eruption of some kind of subterraneous Fires', or to elevations and depressions of the earth's crust, brought about by earthquakes.
9. Some of the highest peaks in the world have at one time been under the sea.
10. The inequalities of the earth's surface could arise from movements produced by earthquakes.
11. '. . . there have been many other Species of Creatures in former Ages, of which we can find none at present; and . . . 'tis not unlikely also but that there may be divers new kinds now, which have not been from the beginning.'

These, then, were the initial proposals that Hooke outlined for the consideration of his listeners. The remainder of the 'Discourse' is largely devoted to an elaboration of arguments in favour of the postulates summarized above, with a variety of evidence, ranging from historical records to astronomical observations, invoked in support. As we should expect, Hooke emerges from these discussions as a man fully prepared to accept the evidence of his senses: an object dug from the ground which resembles a shell *is* a shell, and no 'plastick Faculty' is to be admitted in an account of its origin. There is, nevertheless, a difficulty, as he admits,[13] for it is not generally possible to reproduce the processes of

[12] Waller, op. cit. (4), pp. 290–1.

[13] There is some inconsistency in Hooke's thought here, for on p. 293 of the *Posthumous works* we read: 'we find not many Experiments of producing of [the effects of petrifaction and coagulation] . . . when and where we will'. Conversely, on p. 296 we find: 'I could . . . add divers Experiments, by which several of these Concretes may be in a short time made artificially by several Chymical Operations'.

petrifaction artificially. Some of Hooke's remarks reveal the misconceptions that flowed from the generally held chemical views of the time. He accepts, for example, the transmutation of water into 'a stony Substance' as a possible cause of petrifaction.

For Hooke, the most plausible explanation that could be offered to account for the presence of marine shells in strata now distant from the sea was that of earthquakes raising and lowering the level of the land relative to that of the sea. Hence he was led to consider the role of earthquakes in geological processes, although, as will be demonstrated below, he did not stop there and went on to consider possible causes of the earthquakes themselves. In a Baconian manner, reminiscent of the 'tables' of heat of the *Novum organum*, Hooke listed and attempted to classify the various types of earthquakes. His ideas may be paraphrased in tabular form as follows:[14]

| LAND RISING | LAND FALLING |
|---|---|
| 1. The raising of land originally at about sea level | 1. The sinking of part of the land to form a lake |
| 2. The raising of land from beneath the sea | 2. The sinking of the surface of the earth below the level of the sea |
| 3. The raising of mountains out of level country | 3. The sinking of the bottom of the sea to form 'vast Vorages and Abysses' |
| 4. The raising of land by 'the throwing on of a great Access of new Earth' | 4. The uncovering of land by material being thrown off in an earthquake, or the washing away of material by waters liberated during an earthquake |

Hooke was able to offer a variety of historical examples to support this classification of earthquakes, but, as Bacon found with his 'tables' of heat, the mere setting out of the evidence in this systematic way did not lead to an elucidation of the causes of the phenomena under investigation. Perhaps Hooke realized this more clearly than Bacon, for, as will be demonstrated below, he soon turned from attempts to utilize 'pure' Baconian induction to what may be interpreted as a version of the 'hypothetico-deductive' method.[15]

In the subsequent discussion of the effects of earthquakes it becomes

[14] Waller, op. cit. (4), pp. 298–9.

[15] The relationship between the methodologies of Hooke and Bacon will be considered in more detail below. It has already been noted that Hooke repudiated the mere collection of mountains of data as being, in itself, of any value for the progress of science. On the other hand, however, we find the constant influence of Baconian thought operating on his work, as instanced in his classification of the types of earthquakes. These difficulties in interpretation probably arise to some extent from ambiguities in Bacon's own writings. I believe, however, that Hooke's methodology represented a considerable advance on that of his mentor.

evident that Hooke had a clear idea of the processes of fluviatile and marine erosion and of the deposition and consolidation of sediments; and G. L. Davies has argued that the concept of unconformity is implicit in what Hooke wrote and that he had an idea of the cyclic nature of geological processes.[16] In addition, we may mention that Hooke considered the question of aeolian erosion[17] and the problem of the sinking of heavy stones into the ground over long periods of time.[18] It may also be noted that Hooke's observations led him to propose a simple evolutionary concept,[19] a point previously remarked upon by Pavlov, Edwards, Rossiter, and others,[20] but the idea of Hooke as an evolutionist should perhaps be treated with some caution, for, as will be demonstrated below (p. 127), Hooke had an alternative scheme to account for the presence of large shells of apparently extinct animals in English strata.

Towards the end of the first set of lectures (the 'Discourse' proper) Hooke threw out a number of interesting suggestions in an attempt to account for earthquakes and the concomitant alterations in the level of land and sea.[21] Could they be associated with changes in the position of the centre of gravity of the earth? It was known that there had been alterations in the positions of the magnetic poles in historical times. Why should there not be analogous alterations in the positions of the earth's axis of rotation, and also changes in the position of the centre of gravity?

It was these questions that Hooke took up in his subsequent lectures, which contain some most illuminating remarks on his views concerning the proper methods to be adopted by the scientific investigator. Indeed, these papers are in many ways of greater interest than the 'Discourse' itself, owing to the way in which they exhibit a close integration of methodological and geological ideas.

The paper that Waller placed immediately after the 'Discourse'[22] bears no date, though it is obvious that it should be dated later than the work of 1668, for we find Hooke exploring the idea of possible alterations in the position of the earth's centre of gravity in more detail. Evidently he was dissatisfied with the manner in which he had left the question at the end of the 'Discourse', and the paper contains an interesting elaboration of the concepts previously outlined.

Hooke begins the paper in a somewhat amusing manner:

> I am not, I suppose, the only Person who hath heard some Persons (with what Reason I dispute not) ask what the Royal Society hath done for so

[16] Davies 1964, op. cit. (1), 495 and 496
[17] Waller, op. cit. (4), p. 316.
[18] Ibid.
[19] See p. 112, proposition 11.
[20] See note 1.
[21] Waller, op. cit. (4), pp. 321–2.
[22] See note 7 (a).

> many Years as they have met. And other Persons as confidently affirm that they have done just nothing.[23]

Of course, at that time and also in the eighteenth century, there was considerable discussion as to whether the early members of the Royal Society were or were not doing 'just nothing', or whether what they were doing was of any value.[24] It is not, however, my intention to become involved in this particular question, since for my purpose the interest lies in the fact that, in answer to this challenge, Hooke chose to give us his ideas on the methodology of science.

He suggests that science may be thought of as if it were a large building, with the observations, experiments, and trials of scientists (no doubt typified in the work of the Royal Society) acting as the stones and bricks of the edifice. For Hooke:

> The Structure aimed at, is a true and certain knowledge of the Works of Nature, and this is designed to be attained as fast as may be, and to be perfected as far as may be; or the end of the Inquisition is the promotion and increasing of Natural Knowledge.[25]

We note, incidentally, that he believed, as did many others in the seventeenth century, that the methods of science could provide 'true and certain knowledge of the Works of Nature', a dream now long recognized as being unattainable. By what methods did Hooke hope to realize this aim? To answer this question it is necessary to quote him at some length. He wrote:

> The methods of attaining this end may be two, either the Analytick, or the Synthetick. The first is the proceeding from the Causes to the Effects. The second from the Effects to the Causes: The Former is the more difficult, and supposes the thing to be already done and known, which is the thing sought and to be found out; this begins from the highest, most general and universal Principles or Causes of Things, and branches itself out into the more particular and subordinate.
>
> The second is the more proper for experimental Inquiry, which from a true information of the Effect by a due process, finds not[26] the immediate Cause thereof, and so proceeds gradually to higher and more remote Causes and Powers effective, founding its Steps upon the lowest and more immediate Conclusions.
>
> An Inquisition by the former Method is resembled fitly enough by that Example of an Architect, who hath a full comprehension of what he designs to do and acts accordingly: But the latter is more properly resembled to that of a Husbandman or Gardener, who prepares his Ground and sows his Seed, and diligently cherishes the growing Vegetable, supplying it continually with fitting Moisture, Food, Shelter, etc. observing

[23] Waller, op. cit. (4), p. 329.

[24] Cf. Thomas Sprat, *The history of the Royal Society of London* (London, 1667), p. 153: 'Yet I over-hear the whispers, and doubts of many, who demand, what they have done all this while? and what they have produc'd, that is answerable to these mighty hopes, which indeavour, to make the world conceive of their undertaking?'

[25] Waller, op. cit. (4), p. 330.

[26] This appears to be a misprint for 'out'.

and cherishing its continual Progression, till it comes to its perfect Ripeness and Maturity, and yields him the Fruit of his Labour.[27]

After further discussion of these metaphors[28] and reference to examples of them in his own astronomical work, and also in Newton's work, Hooke continues with a description of what is apparently a version of the 'hypothetico-deductive' method:

wherein from an Hypothesis being supposed or a premeditated Design, all the Phaenomena of the Subject will be *a Priori* foretold, and the Effects naturally follow as proceeding from a Cause so and so qualified and limited. And in truth the Synthetick way by Experiments Observations, etc. will be very slow if it be not often assisted by the Analytick, which proves of excellent use, even tho' it proceed by a false position; for that the discovery of a Negative is one way of restraining and limiting an Affirmative.[29]

He defines knowledge as 'a certainty of information of the Mind and Understanding founded upon true and undeniable Evidence'.[30] Further, in his view, true and undeniable evidence is 'afforded either immediately by Sense without Fallacy, or mediately by a true Ratiocination from such Sense.'[31]

There follows a somewhat circular statement of what he means by 'Sense without fallacy':

I call that Sense without Fallacy, where the fallacies of Sense being detected and known, the Evidence produced thereby is examined and found to be free and clear of all such Fallacies.[32]

Evidently we are to understand the phrase 'Ratiocination from such Sense' to mean a deductive process from the true knowledge given by the senses; he compares the process to the kind of reasoning employed in geometry.

However, he continues:

in Physical Inquiries, by reason of the abstruseness of Causes, and the limited Power of the Senses we cannot thus reason, and without many Inductions from a multitude of Particulars come to raise exact Definitions of things and general Propositions; yet by comparing of varieties of such Inductions we may arrive to so great an assurance and limitation of Propositions as will at least be sufficient to ground Conjectures upon, which may serve for making Hypotheses fit to be enquired into by the

[27] Waller, op. cit. (4), p. 330.

[28] The example of the architect is appropriate. One may envisage the *a priori* conception of the plans for a building, followed by the construction of the edifice according to this programme. Here Hooke appears to have rationalist, deductive elements in mind. However, it is not so clearly evident that the work of a gardener is analogous to attempts to build an edifice of knowledge by laborious inductive methods. The plants will grow of their own accord. Does induction proceed likewise? Doubtless Hooke's chosen metaphor was intended to emphasize the slow and gradual character of the development of the new science, in which daring flights of fancy were to be eschewed.

[29] Waller, op. cit. (4), p. 330.

[30] Ibid.

[31] Ibid., p. 331.

[32] Ibid.

> Analytick method, and thence to find out what other Experiments or Observations are necessary to be procured for the further progress in the Synthetick, which will questionless so far inform us of the general and universal progress of the Operations of Nature, that nothing but what is really the truth shall be proposed but the absurdity and insufficiency thereof will presently be detected and proved.[33]

It is interesting to note that Hooke apparently emphasizes the relevance of the *modus tollens* to the methodology of science, for he writes:

> tho' possibly we may not be able to produce a Positive Proof, yet we may attain to that of a Negative, which in many cases is as cogent and undeniable, and none but a willful or senseless Person will refuse his assent unto it.[34]

Hooke does not say very much about how he thinks hypotheses are generated, and this is perhaps scarcely surprising in view of the difficulty still found today in analysing the processes involved. It seems reasonable, however, to judge that he would have considered the role of analogy to be important. For example, there is a passage in the 'Discourse' (pp. 321–2) in which Hooke attempts to justify the hypothesis of 'pole wandering' (which, as will be shown below, he postulated in order to account for the alterations in the relative levels of land and sea and the consequent imprisonment of fossil shells in sediments) by claiming that such movements were comparable to the known wandering of the magnetic poles. Again, in the *Micrographia* we find Hooke arguing that if there were no analogy to be recognized between independent phenomena, 'our inquiries would be endless, our tryals vain, and our greatest inventions would be nothing but the mere products of chance, and not of Reason'.[35] However, Hooke does not use the word analogy itself in these discussions.

By the 'Analytick' method, Hooke appears to mean a deductive process, starting from principles which are taken to be true, *a priori*. Hence, for him, analysis is to be used for making deductions from the assumed hypothesis, thereby permitting the experimental tests of these conjectures. Conversely, he seems to mean by the 'Synthetick' method the inductive processes (usually associated with the name of Bacon) in which scientific knowledge is built up on the basis of a variety of experiments and observations.

Hooke does not seem to have any logical objections to his 'Synthetick' or inductive methods. Indeed, his description of the knowledge acquired by the senses being used as material for the process of ratiocination appears truly Aristotelian. It is, rather, that Hooke considers that the

33 Ibid.

34 Ibid.

35 R. Hooke, *Micrographia: or some physiological descriptions of minute bodies made by magnifying glasses. With observations and inquiries thereupon* (London, 1665), p. 14. Hooke touched on the methodology of science in his Preface to the *Micrographia*, but there he was principally concerned with the aids which instruments could offer to the senses. For his more detailed considerations, it is necessary to consult his *Posthumous works*.

scientific edifice, if erected purely by the 'Synthetick' method, will be painfully slow in rising from the ground. Therefore, he calls in the 'Analytick' method to his aid, not for logical reasons, but merely to expedite the whole process. His preferred method of scientific investigation can thus be summarized in the following manner:

1. Collect data and perform experiments, which give rise to propositions that can serve as a basis for the propounding of explanatory hypotheses.
2. On the basis of the information collected, and the propositions arising directly from this information, erect hypotheses to account for the phenomena under investigation.
3. Deduce the logical consequences which follow from these hypotheses.
4. Carry out further observations and experiments to test the consequences so deduced.
5. Hence eliminate some of the hypotheses, leaving only one remaining.[36]
6. Use the data that support the successful hypothesis as a basis for further investigations.[37]

If this method is followed repeatedly, we have, according to Hooke, a much more rapid method for reaching the desired goal, the certain knowledge of general propositions, than can be achieved by the use of the 'Synthetick' method alone.

There are several interesting points which arise from this account of Hooke's methodological views, which are shown diagrammatically in Figure 1. In the first place, its apparent similarity to some of the ideas in current circulation may be noted.

Secondly, we should consider the general relationship of Hooke's thinking to that of Bacon. Hooke's constant debt to Bacon is everywhere manifest, but is most obvious in his 'General Scheme' for conducting science, given in the first section of the *Posthumous works*.[38] Here we see Hooke at his most Baconian. The doctrine of idols is announced once more, and all the appropriate methods for the collection of data, which will serve as the 'solid basis for the superstructure of true philosophy', are described again. The tabulation of positive and negative instances is specified anew, and the subsequent processes of elimination are urged.

[36] Hooke does not seem to consider the possibility that none of the proposed hypotheses may be correct. Rather, like the sleuth of Baker Street, he seems to suggest that if all possibilities but one are eliminated, the remaining one must be correct.

[37] It might be thought that the further progress in the 'Synthetick' method would result from using the hypothesis which has survived experimental testing as a spring-board for further 'upward' movement. This does not, however, appear to be the intention of Hooke's method. He believes that the new data, which arise during the course of the testing of the successful hypothesis, provide the basis for the further ascent.

[38] See note 41 above, which gives the title of Hooke's methodological essay according to the table of contents at the head of the *Posthumous works*. Within the text, the essay is entitled: 'A general scheme, or idea of the present state of natural philosophy, and how its defects may be remedied by a methodical proceeding in the making experiments and collecting observations. Whereby to compile a natural history, as the solid basis for the superstructure of true philosophy.'

It seems that in this essay Hooke was thinking of the processes of scientific discovery as being largely mechanical, and he speaks of a 'Philosophical Algebra'[39] composed of two parts: the first dealing with the collection of data, the second with methods for dealing with the material thus collected. The second part, which would have been much the more interesting, was never published, so that we are merely left with his account of the Baconian methods of natural history. Hence it is the later lectures, published with the 'Discourse of earthquakes', to which we must turn

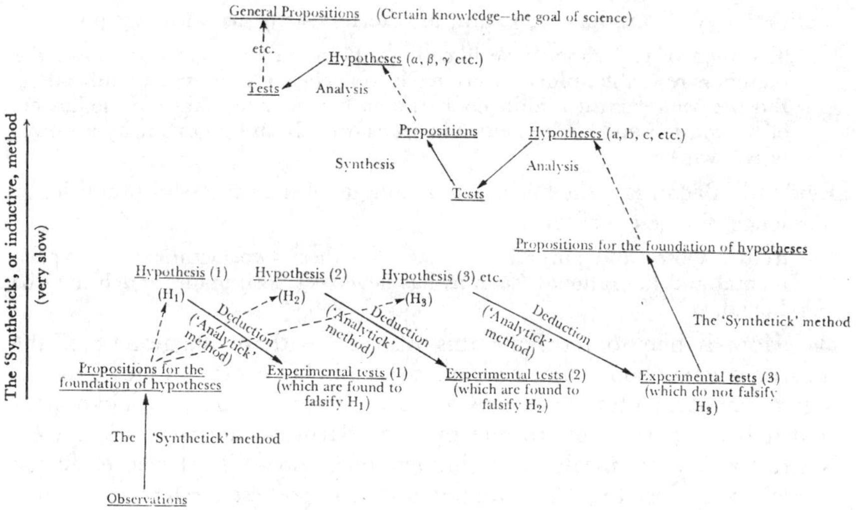


FIG. 1.

in order to understand Hooke's more mature methodology. Here there is less insistence upon the wholly mechanical nature of the processes of scientific enquiry, though it should be noted that Hooke viewed the general workings of the mind itself in a highly mechanical way, reminiscent of Cartesianism.[40]

To decide whether Hooke was a pure Baconian we must look again at the notions of Bacon himself. For many years the Lord Chancellor suffered opprobrium for his advocacy of what was held to be an utterly unworkable methodology of science. Bacon claimed:

> the course I propose for the discovery of sciences is such as leaves but little to the acuteness and strength of wits, but places all wits and understandings nearly on a level. For as in the drawing of a straight line or a perfect circle, much depends on the steadiness and practice of the hand, if it be

[39] See Hesse 1966, op. cit. (1).
[40] Waller, op. cit. (4), pp. 139–47.

> done by aim of hand only, but if with the aid of rule or compass, little or nothing; so it is exactly with my plan.[41]

So, we might be led to believe, the adoption of the Baconian methods—the tabulation of examples, the processes of looking for similarities and differences in the phenomena thus tabulated, the processes of exclusion and rejection, the drawing of the first vintage, etc.—would enable all to attain the fruits of knowledge. Diligence is required, but wit and sagacity are not pre-requisites.

But certain other passages in the *Novum organum* must be taken into consideration before we adopt the purely mechanical view of the Baconian methodology. Consider, for example, the following well-known passage:

> The men of experiment are like the ant, they only collect and use; the reasoners resemble spiders, who make cobwebs out of their own substance. But the bee takes the middle course: it gathers its material from the flowers of the garden and of the field, but transforms it and digests it by a power of its own.[42]

Evidently Bacon saw the bee as the analogue of the successful practitioner of science, for he concludes:

> from a closer and purer league between these two faculties, the experimental and the rational (such as has never yet been made) much may be hoped.

How is one to reconcile this passage, with its suggestion of the incorporation of both empiricist and rationalist elements into the methodology of science, with the previous quotation from Bacon? It would appear that it is not permissible to interpret the Baconian method as if it were based entirely on mechanical, inductive methods. To be sure, Bacon inveighed against the unwarranted use of hypotheses, referring to such anticipations of nature as 'rash and premature'. But we are not justified in discounting the role of hypotheses in Bacon's methodology entirely, and deduction does play some part in his scheme, even though the main emphasis is on the inductive ascent. For example, we read in the *Novum organum*:

> from the new light of axioms, which have been educed from those particulars by a certain method and rule, [and] shall in their turn point out the way again to new particulars, greater things may be looked for. For our road does not lie on a level, but ascends and descends; first ascending to axioms, then descending to works.[43]

So, one should ascend by induction, and then descend by a process of deduction—a procedure that enables one to test the axioms raised by

[41] F. Bacon, *Novum organum*, Book I, aphorism LXI (London, 1620), trans. J. Spedding, R. L. Ellis, and D. D. Heath, and published in the *Works of Francis Bacon, Baron of Verulam, Viscount St. Alban, and Lord High Chancellor of England* (14 vols., London, 1857–74), iv. pp. 62–3. Republished as: F. H. Anderson (ed.), *Francis Bacon, Baron of Verulam, Viscount of St. Albans. The new organon and related writings* (Indianapolis, New York, and Kansas City, 1960), pp. 58–9. Subsequent references will be made to this edition of 1960.

[42] Anderson, op. cit. (41), p. 93; Book I, aphorism XCV.

[43] Ibid., p. 97; Book I, aphorism CIII.

induction. The axioms, ostensibly obtained inductively with the aid of the tabulated information, are tested against fresh data, and, if they survive the testing, they may be used to induce higher level axioms.[44] It will be seen that Hooke's methodology, summarized in Figure 1, has a great deal in common with this Baconian scheme. The outstanding difference lies in the relative contribution allowed to hypotheses. Whereas Bacon's drawing of the 'first vintage' must indeed be accounted a process in which an explanatory hypothesis is formulated, the Lord Chancellor himself took this step most unwillingly, craving 'an indulgence of the understanding' from his readers at this point in the description of his method. By contrast, Hooke freely urged the formulation of hypotheses as an aid to the speedier prosecution of the scientific enquiry. This makes his whole method much more convincing than Bacon's, and one can envisage Hooke's method actually being put into practice. Indeed, we do see it in practice in the later lectures of the *Posthumous works*, as will be demonstrated below. By contrast, few have ever been convinced that Bacon could have arrived at the proposition that heat is a form of motion, by the mere application of the mechanical methods described in the *Novum organum*.

We conclude, then, that Hooke's methodology was essentially Baconian, except that he openly urged that hypotheses should be proposed, as part of the process of raising 'the exact Definitions[45] of things and general Propositions'. As we have seen, this gives Hooke's methodological views a modern ring, even though he purports to have used hypotheses for pragmatic reasons—merely to expedite his enquiries—rather than because of logical difficulties associated with inductive procedures. As a result of his recommendation to use hypotheses in science, wit and sagacity would undoubtedly be needed by those who attempted to follow his methodology, whereas Bacon, by contrast, had hoped that his 'plan' would place 'all wits and understandings nearly on a level'.

Further problems are raised by Hooke's use of the terms 'Analytick' and 'Synthetick'. From the context, it would appear that we should understand analysis to involve deduction, and synthesis to involve induction (using the terms deduction and induction in their modern sense),[46] and one may feel that there is some sort of correspondence with the meanings of the Kantian terms, analytic and synthetic. However, when we compare Hooke's usage of the terms with that of his contem-

[44] Ibid., p. 99; Book I, aphorism CVI.

[45] The influence of Aristotelianism on Bacon has been discussed in R. E. Larsen, *Journal of the history of ideas*, xxiii (1962), 435–50. It is interesting to note that here too Hooke reveals Aristotelian influence in his remark that science seeks to establish 'the exact Definitions of things'.

[46] It should be remembered, of course, that in the seventeenth century these terms did not have their present, fairly clearly defined meanings. Newton, for example, in a letter to Cotes, dated 31 March 1713, wrote: '. . . Experimental philosophy proceeds only upon Phenomena and deduces general Propositions from them only by Induction'; see J. Edleston, *The correspondence of Sir Isaac Newton and Professor Cotes* (London, 1850), p. 156.

poraries, it may be something of a shock to find that treatments are sometimes diametrically opposed. For example, in Hooke, we read:

> The Structure aimed at, is a true and certain knowledge of the Works of Nature, . . . The methods of attaining this end are two, either the Analytick, or the Synthetick. The first is the proceeding from the Causes to the Effects. The second from the Effects to the Causes . . .[47]

By contrast, Newton, in the 31st Query of the *Opticks*, wrote:

> By this way of Analysis we may proceed from Compounds to Ingredients, and from Motions to the Forces producing them; and in general, from Effects to their Causes and from particular Causes to more general ones, till the Argument end in the most general. This is the Method of Analysis: And the Synthesis consists in assuming the Causes discover'd, and establish'd as Principles, and by them explaining the Phaenomena proceeding from them, and proving the Explanation.[48]

This unfortunate confusion in terminology does not seem to have been confined to Hooke and Newton, and several other instances could be cited. As William Hamilton once wrote in the *Edinburgh review*:

> 'In one respect,' says Aristotle, 'the Genus is called a part of the Species; in another, the Species a part of the Genus . . .'
>
> In like manner, the same method, viewed in different relations, may be styled either Analysis or Synthesis. This, however, has not been acknowledged; nor has it even attracted notice, that different logicians and philosophers, though severally applying the terms only in a single sense, are still at cross purposes with each other. One calls Synthesis what another calls Analysis; and this both in ancient and modern times.[49]

To examine this terminological confusion in detail would be a task beyond the scope of the present paper, for it is a problem of considerable complexity. Nevertheless, just a few points about the matter may be made here. The term synthesis was used in the Greek geometrical tradition to denote a series of deductions from accepted axioms to theorems. The term corresponded with what today is generally called axiomatics. Conversely, geometrical analysis corresponded to the move (also deductive) from some unknown theorem (its truth being provisionally assumed) to some known theorem or axiom.[50] Furthermore, it was considered proper that geometrical analysis should be followed by synthesis to provide a complete account. These classical mathematical methods of analysis and synthesis had their counterparts in discussions of scientific method, in which the terms resolution and composition were generally

[47] Waller, op. cit. (4), p. 330.

[48] I. Newton, *Opticks: or, a treatise of the refractions, inflexions and colours of light* (4th edn., London, 1730), pp. 404–5.

[49] In an unsigned review of R. D. Hampden's article 'Aristotle's philosophy' in the seventh edition of the *Encyclopaedia Britannica*; see *Edinburgh review*, lvii (1833), 236. Here Hampden and Hamilton were concerned with the use of the terms analysis and synthesis in Aristotle's philosophy of definition, rather than in experimental science, but, as Crombie has shown (op. cit. [51], pp. 61–74 and *passim*), the distinction between analysis and synthesis has some of its roots in the Aristotelian notions of genus and species.

[50] T. L. Heath, *The thirteen books of Euclid's Elements translated from the text of Heiberg with introduction and commentary* (2nd edn., 3 vols., New York, 1956), i. 137–42.

used.[51] These again were paralleled in the rhetorical tradition, where the terms invention and judgement were applied.[52]

In the Paduan methodological writings of the sixteenth century, and earlier, the process of resolution was used to denote a movement from observed effects to presumed causes, while composition referred to the deductions of effects from causes previously established by resolution. The mathematical and methodological traditions are found united in Galileo[53] and Newton, and although their terminology is an evident development of earlier writings, we do not find any unexpected reversal of meaning, for they continue to associate synthesis with deduction. Why, then, should Hooke have chosen to alter the customary usages?

The difficulty probably centres around the word analysis rather than synthesis, for it seems to have been a term with a multiplicity of meanings. Descartes certainly used the word ambiguously[54] and may have provided a warrant for Hooke's usage. As was noted above, Hooke referred to his movement from *assumed* hypotheses to testable consequences as the 'analytick method'. Here he was doubtless following the mathematical tradition, in the sense that the Greek geometers, in their process of analysis moved from *assumed* theorems to known propositions. As M. B. Hesse has pointed out,[55] this would fit in with a 'philosophical algebra', employing the deductive testing of hypotheses. Synthesis, then, for Hooke, would correspond to the slow ascent from 'a true information of the Effect' to 'the immediate Cause thereof', and so on 'to higher and more remote Causes and Powers effective'.

If this interpretation of the position is correct, then we can understand the rationale for the manner in which Hooke employed the traditional terms, though the incongruity of the opposition between the usages of Hooke and Newton remains.

I come now to a consideration of the actual example that Hooke

[51] For discussion of the usages of the terms resolution and composition see, for example: A. C. Crombie, *Robert Grosseteste and the origins of experimental science 1100–1700* (Oxford, 1953), pp. 27–8, 75–9, 81–3, and *passim*; J. H. Randall Jr., 'The development of scientific method in the school of Padua', *Journal of the history of ideas*, i (1940), 177–206; L. M. Régis, *Epistemology* (New York, 1959), pp. 443–65; and N. W. Gilbert, *Renaissance concepts of method* (New York, 1960), *passim*.

[52] For a discussion of the relationship between the rhetorical tradition and the development of scientific method, see R. McKeon, 'Philosophy and the development of scientific methods', *Journal of the history of ideas*, xxvii (1966), 3–22.

[53] For a statement of Galileo's position, see his letter to Pietro Carcavy (Arcetri, 1637) in G. Barbera (ed.), *Le opere di Galileo Galilei* (20 vols., Florence, 1966), xvii. 88–93.

[54] R. Descartes, *Objections faites par des personnes très doctes contre les précédentes méditations avec les réponses de l'auteur: Réponses . . . aux secondes objections recueillies . . . par le R. P. Mersenne*, §§121–2, in *Oeuvres philosophiques*, 2 vols. (Paris: Editions Garnier Frères, 1967), ii. 581–3. The editor of this edition, Ferdinand Alquié, has drawn attention here to the great difficulty in interpreting this portion of the text. For a general discussion of Descartes's usage of the terms analysis and synthesis, see L. J. Beck, *The method of Descartes: a study of the Regulae* (Oxford, 1952), pp. 155–71. G. Buchdahl, in *Metaphysics and the philosophy of science. The classical origins, Descartes to Kant* (Oxford, 1969), pp. 118–47, has also treated the problem at length, and has remarked on the fluidity of the Cartesian terminology. He has found it necessary to offer some codification of the different Cartesian uses of the term analysis, in order to clarify the philosopher's intentions.

[55] Hesse 1966, op. cit. (1), 82

gave to illustrate his methodology. The example was a most interesting one, which arose from his simple investigations of fossilized shells, but which soon led him to consider fundamental problems concerned with the figure of the earth and its motion in space. The following account is based upon five of the lectures on geological topics that Hooke gave to the Royal Society, after his presentation of the 'Discourse of earthquakes'.[56]

First, Hooke considers the problem of fossils. They undoubtedly *look* like the remains of once living marine organisms. They have all the properties such remains would reasonably be expected to have. All tests confirm the supposition that they are the remains of former living creatures; therefore, this supposition *must* be accepted. Presumably, this would be an example of 'True and undeniable Evidence . . . afforded either immediately . . . or mediately by a true Ratiocination from such Sense'. (In support of this contention, he discusses the problem of how one would know that a metal resembling gold is genuinely gold.[57] His treatment is reminiscent of Locke's discussion in the *Essay concerning human understanding*.[58]) Hence one is led to the conviction that fossils are indeed the remains of marine organisms that were once alive. The presence of such shells many miles from the sea can only be accounted for, Hooke argues, by the supposition that in the past there have been marked changes in the relative levels of land and sea. Now, by his 'Synthetick' method, he has reached a position where hypotheses may be proposed to account for the phenomena.

Several hypotheses immediately present themselves to Hooke's mind:

1. The fossils were placed in their present positions by Noah's flood.
2. There has been a gradual swelling of the earth owing to a subterranean fermentation.
3. Erosion, with consequent deposition of sediment on the organisms, has occurred.
4. In the past there have been changes in the earth's axis of rotation, so that, for example, England may at one time have been in the 'Torrid Zone' (thereby accounting for the giant ammonites of the Portland and Purbeck beds). Further, such changes may have led to a variety of marine transgressions and regressions.[59]

In my view, this is a most attractive approach to the problem, reminiscent of the justly famed 'method of multiple working hypotheses'

[56] See note 7(a)–(e).

[57] Waller, op. cit. (4), p. 332.

[58] J. Locke, *Essay concerning human understanding*, (London, 1690), Book 3, chapter 6, section 35, and *passim*. Locke, of course, was primarily concerned here with establishing the distinction between the real essence and the nominal essence of objects. Hooke does not make this distinction.

[59] Waller, op. cit. (4), pp. 341–3. The third hypothesis was intended to be independent of the fourth; that is, in the third case, he considered the situation which would arise if the erosion and sedimentation occurred without accompanying changes in the polar axes.

of T. C. Chamberlin,[60] though it must be admitted that Hooke is evidently unduly attached (reprehensibly so, from Chamberlin's standpoint) to the fourth hypothesis, and the first three are perhaps put in for the sake of appearances, for they are rapidly dismissed. If Noah's flood were responsible, Hooke asked, would not the shells all be small ones? How could they grow to their observed sizes in, say, two hundred days? If a ferment were to produce a swelling of the earth, would not the shells be found on the upper surface of the ground and not buried in subterranean depths? If burial under sediment were the sole cause of the present situation of fossils, how could 'the Mountain come to be placed on top of them'? Evidently it was the fourth hypothesis that required further consideration. Hooke took up the question in his next lecture.

In this lecture he accepted the Copernican system, which was by then standard, and also the notion of the regular motion of the earth's axis, which accounted for the precession of the equinoxes. In addition to this well-known motion, however, he postulated a motion of the position of the poles over the earth's surface. This polar wandering was described in the following terms:

> I suppose . . . further, that the Axis of the Diurnal Rotation of the Earth hath also had a progressive motion, and hath, in process of time, been chang'd in position within the Body of the Earth, and consequently that the Poler [*sic*] Points upon the Surface of the Earth, have alter'd their Situation; so that the present Polar Points have formerly been distant from those Poles that were then; and consequently that those former Polar Points are now remov'd to a certain distance from the present, and move in Circles about the present.[61]

The postulates of Hooke's theory may be clarified by reference to Figure 2. Here, the dotted lines represent the *initial* paths which objects would follow if they were free to fall towards the interior of the earth. Hence, for Hooke, the *effective* centre of gravity of the earth was to be taken as a section of a line coincident with the axis of rotation, rather than a point at the centre of the earth.[62]

[60] T. C. Chamberlin, 'The method of multiple working hypotheses', *Journal of geology* (1897), 837–48. Originally read before the Society of Western Naturalists, 1892.

[61] Waller, op. cit. (4), p. 346. It is not exactly clear what Hooke envisaged when he wrote 'and move in Circles about the present'. The meaning would be clearer if these words were omitted, thus making it easier to reconcile the whole passage with others (e.g. p. 347) where he speaks of alterations in latitudes.

[62] In a subsequent lecture (Waller, op. cit. [4], p. 356), Hooke argued that bodies would fall in the directions indicated by the dotted lines of Figure 2, because of the different centrifugal effects at different latitudes. The effect is, of course, a real one, but scarcely equal to the task of producing earthquakes as the latitude of a place on the earth's surface changes owing to pole wandering. Thus calculation reveals that an object at the latitude of North 45° falls initially towards a point about ten miles to the south of the geometrical centre of the earth. It should be noted that there appear to be two distinct elements in Hooke's cyclic geological theory: 1. the rise and fall of the earth's crust due to earthquakes; 2. alterations in sea level resulting from the presumed oblately spheroidal shape of the earth's envelope of water and a postulated pole wandering. The earthquakes of '1' are presumed to result from the pole wandering of '2', since a given portion of the earth's crust experiences different gravitational effects, depending on the latitude.

It would follow from these assumptions, Hooke argued, that, as the latitude of a point on the earth's surface gradually altered, the direction of the resultant of the forces acting upon it would change correspondingly. This, he believed, would result in a 'sliding, subsiding, sinking and changing of the Internal Parts of the Earth, as well as External',[63] with the consequent production of earthquakes. Further, since the rotation of the earth gives an oblately spheroidal form to the oceans, the waters are generally deepest near the equator, so that as a point on the earth's surface gradually moves into the higher latitudes, a marine regression will necessarily occur (see Figure 2). Conversely, a point in the southern

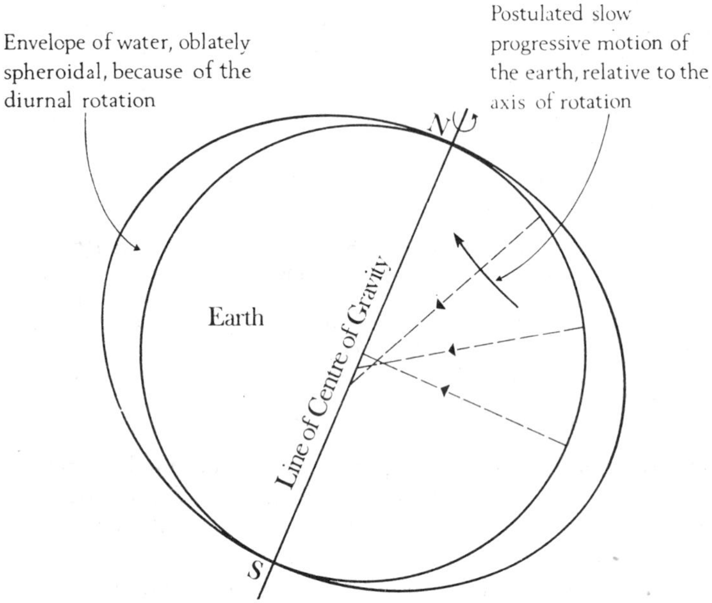


FIG. 2.

hemisphere (on the right-hand side of Figure 2) gradually approaching the equator, will suffer a marine transgression and will consequently become covered in layers of sediment, with associated shells. On the other hand, as the lands in the northern hemisphere (again on the right-hand side of Figure 2) are gradually exposed to the action of erosion, they become wasted and decayed. Simultaneously, the lands entering the seas (for example, those in the southern hemisphere on the right-hand side of the figure) become repaired by the deposition of layers of sediment. As Hooke put it:

> Hence also it will follow, that the Earth itself doth, as it were, wash

[63] Waller, op. cit. (4), p. 347.

and smooth its own Face, and by degrees to remove all the Warts, Furrows, Wrinckles and Holes of her Skin, which Age and Distempers have produced.[64]

Here, then, we have the efficient cause for Hooke's cyclic theory. The hypotheses were of considerable subtlety, as might be expected for a man of Hooke's genius. If granted, they would account most satisfactorily for many field observations, including the extraordinary sizes of some of the English fossils. For if, for example, England had once been situated in deep water near the equator, it was to be expected that tropical shells such as the giant ammonites and nautili would be incorporated in the sediments, later to be revealed by erosion when England rose out of the waters. Perhaps similar ammonites might be found, in the fullness of time, in the deeper waters of the present tropical seas![65]

Hooke believed that there was good evidence for the influence of the diurnal rotation of the earth upon the weights of bodies. He argued that the effective force of gravity becomes weaker as one approaches the equator, since, with the high velocity at the equator (compared with the almost stationary poles), there would be a detectable diminution of the acceleration due to gravity in the lower latitudes. Was it not necessary to shorten the pendulum in tropical countries to make it keep 'due Time'?[66] It may be noted that in this lecture Hooke did not consider the second major factor which contributes to variations in the acceleration due to gravity: the oblately spheroidal figure of the earth. It is true that in the subsequent lecture he did consider whether the earth (as well as its envelope of water) was spherical or otherwise, but the methods that he proposed for testing this question need not be dealt with here.

There remained, then, the problem of finding some experimental method for testing the supposition that the poles were wandering. Even now we have considerable difficulty in finding a method with sufficient precision to determine a rate of continental drift. For Hooke the detection of slow changes in latitude and longitude was technically impossible. Nevertheless, his fertile brain suggested at least three methods which might have been feasible in principle. At this point we move on to a consideration of his lectures of 2 and 9 February 1686/7.

It occurred to Hooke that if such pole wandering movements occurred, then the examination of ancient buildings, constructed for some reason so that they were originally aligned with the meridian, might reveal the movements. Hooke's commitments at the time were such that he could not contemplate a visit to Egypt in order to determine the present alignments of the Pyramids.[67] Nevertheless he examined the recorded

[64] Ibid., p. 348.
[65] Ibid., pp. 340–3 and 433–5.
[66] Ibid., p. 349.
[67] For a modern discussion of the influence of continental drift on the alignments of the Pyramids, see A. L. Du Toit, *Our wandering continents* (Edinburgh, 1937), pp. 299–301.

observations of his contemporaries, but was disappointed to find that they ignored the question of the alignment of the Pyramids, being so engrossed with the determination of their sizes. As Hooke rightly pointed out:

> This shews how useful Theories may be for the future to such as shall make Observations; nay, tho' they should not be true, for that it will hint many Inquiries to be taken notice of which would otherwise be not thought of at all.[68]

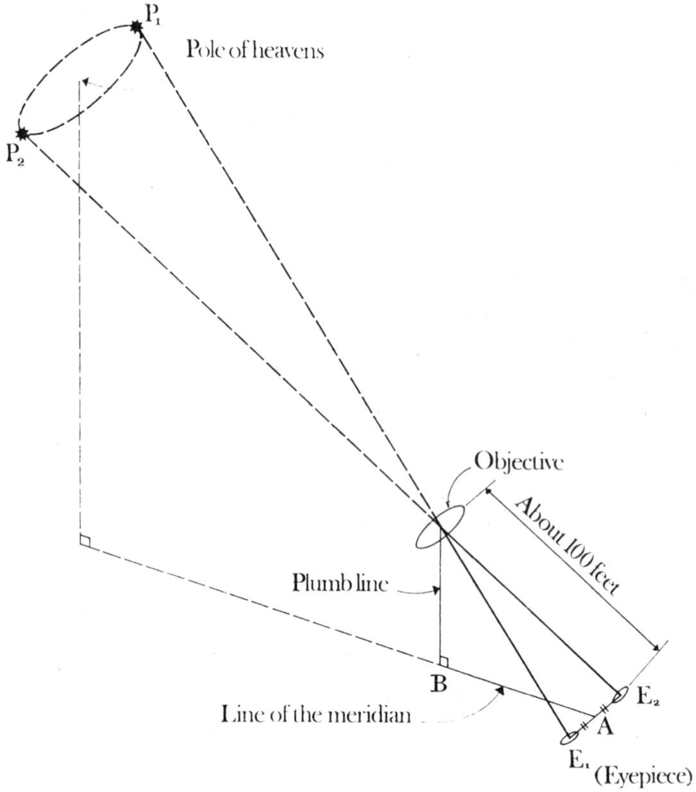


FIG. 3.

Disappointed by the Pyramids, Hooke also considered the possibility of examining the alignments of the cathedrals, but we have no record of his actually putting the matter to the test.

Finally, then, we come to Hooke's two audacious astronomical methods for determining alterations in the direction of the meridian, methods which, so he supposed, would provide crucial experiments for testing his whole chain of reasoning.

[68] Waller, op. cit. (4), p. 353.

For his first method a telescope, about 100 feet long, was to be constructed, with the objective lens placed at the top of a building and the eyepiece near the ground. The most easterly and westerly excursions of the pole star were to be determined on a given date, and hence the direction of the meridian would be ascertained with considerable accuracy. The arrangement is shown diagrammatically in Figure 3.

To perform the experiment, the most easterly position of the pole star ($P_1$) is observed from $E_1$ and the western station is observed from $E_2$. $E_1E_2$ is bisected at A. B is the point perpendicularly below the centre of the objective. If, then, the telescope is directed horizontally along the line of the meridian AB, so determined, towards a distant object, it is possible to draw what is seen in the field of view; and if the experiment is repeated a few years later, any shift in the direction of the meridian should become evident. To our knowledge, Hooke never put these notions to the test, although a good deal of time was devoted to the determination of the meridian by other methods during the later part of his life.

His second method also relied upon the determination of the direction of the meridian. A piece of glass with concentric circles inscribed on it was to be placed within a long telescope, which was to be directed to the pole of the heavens. Here one would see a constellation, invisible to the naked eye, called the English Rose. By suitable adjustments of the telescope one might bring it to such a position that its axis would be pointing directly to the North pole of the heavens. The adjustment would be guided by the fact that when the telescope was correctly aligned in the direction of the meridian, and at the correct altitude,[69] the stars of the English Rose would appear to move in circles parallel to the lines marked on the plate in the telescope. The line of the meridian on the ground could then be determined by suitable plumb lines suspended from the telescope.

Such, then, was the final stage of Hooke's line of reasoning. The fossil shells suggested changes in the relative levels of land and sea. This could be explained by the hypothesis of the slow movement of the surface of the earth relative to the direction of its poles in space, within an oblately spheroidal envelope of water. If this hypotheses were true, he believed that earthquakes could be accounted for, and the presence of apparently

[69] It should be noted that although the precession of the equinoxes does not have any effect on either of the two methods which Hooke proposed for the determination of changes in a line of meridian upon the surface of the earth (and he does not concern himself with precession in his discussions of these proposals), any alterations in the obliquity of the ecliptic would result in changes in the altitude of the pole of the heavens in the sky, though they would not affect the direction of a line of meridian on the surface of the earth. Justifiably, therefore, Hooke did not consider the effect of changes in the obliquity of the ecliptic. (I am indebted to Professor J. B. Thornton for his thought concerning the null effect of any possible changes in the obliquity of the ecliptic on Hooke's scheme.)

tropical shells within English strata became intelligible. A check on the whole theory could be made by means of astronomical observations.

Hooke's daring cyclic earth theory may have seemed absurd to his contemporaries, and, as Carozzi has recently observed,[70] apart from R. E. Raspe, Hooke had almost no geological followers. Yet, to the modern philosopher of science, Hooke's notions have all the hallmarks of a genuine scientific theory. The postulates of this theory were not in any way outrageous. The Huttonian theories of heat[71] and the Wernerian concept of a universal ocean had many more dubious hypotheses, yet both attracted a wide following. The whole of Hooke's methodological discussion is mature and convincing to the modern reader, despite his prolix style with its frequent repetitions and digressions. We may quibble that he believed in the possibility of science achieving some kind of certainty, a claim which, in many cases, we have reluctantly had to relinquish. One may point to the logical *lacunae* in his arguments—the niceties of the Duhem-Quine thesis[72] were evidently not apparent to him—but it cannot be questioned that Hooke's approach to the methodology of scientific investigation has much in common with modern accounts. We may accept it as a specimen of what could be achieved if a man of more than ordinary sagacity attempted to pursue a version of an essentially Baconian methodology.

*Acknowledgements*

I am most grateful to my colleagues, Mr W. H. Leatherdale and Dr R. M. Gascoigne, for their comments on the first draft of this paper; to Professor G. Seddon, for his remarks made in his capacity as commentator on the paper when it was presented at the Brisbane meeting of the Australasian Association for the History and Philosophy of Science in August 1971; and to an anonymous referee for a number of most helpful suggestions.

[70] Carozzi, op. cit. (1).

[71] The question of some of the more subtle assumptions of Hutton's theories has been discussed recently in P. A. Gerstner, 'James Hutton's theory of the earth and his theory of matter', *Isis*, lix (1968), 26–31. A slightly different point of view is adumbrated in my paper: 'The Vulcanist-Neptunist dispute reconsidered', *Journal of geological education*, xix (1971), 124–9.

[72] That is, the thesis that experiments and observations can never falsify an individual hypothesis, forming part of a theoretical system. When a hypothesis is 'put to the test', experiments and observations may not accord with predictions made on the basis of the hypothesis, This certainly shows that 'something is wrong somewhere', but it does not necessarily show that the hypothesis has been falsified, for it may be that it was the 'background information'—the 'auxiliary hypotheses' assumed when the 'test implications' were being deduced from the hypothesis—which was somewhere at fault. Moreover, any hypothesis may be saved from falsification if sufficient adjustments are made in the 'background information'. Hooke, of course, is not concerned with such subtleties. He is *quite sure* that evidence shows that fossils were not emplaced by Noah's flood. For further discussion of the thesis, see I. Lakatos and A. Musgrave (eds.), *Criticism and the growth of knowledge* (Cambridge, 1970), pp. 184–8.

# XI

# FROM RENAISSANCE MINERAL STUDIES TO HISTORICAL GEOLOGY, IN THE LIGHT OF MICHEL FOUCAULT'S *THE ORDER OF THINGS*

W. R. ALBURY AND D. R. OLDROYD

*Introduction*

IN this paper we examine the study of minerals from the Renaissance to the early nineteenth century in the light of the work of Michel Foucault on the history of systems of thought. In spite of a certain number of theoretical problems, Foucault's enterprise opens up to the historian of science a vast terrain for exploration. But this is the place neither for a general exegesis nor for a general criticism of his position; our aim here is the more modest one of taking certain points from Foucault's study, *The order of things*,[1] and seeing how far they can be extended into an area not explicitly considered in that work.[2]

In *The order of things* Foucault argues that the study of plants and animals has undergone two radical reorganizations since the sixteenth century. These 'transformations', which twice completely altered the conceptual system of the discipline, Foucault situates around the middle of the sixteenth century and at the beginning of the nineteenth century, thus giving demarcations which correspond approximately to the familiar categories of 'the scientific revolution' and 'the rise of historicism'. However these categories, as commonly used, implicitly depend upon a number of presuppositions that are alien to Foucault's analysis; and we shall therefore not make use of them in our discussion. Instead, we shall follow Foucault in referring to the interval between the two transformations (c. 1660–1800) as the 'classical period'.

Foucault, then, distinguishes three different systematic approaches to the study of living things: in the Renaissance this study took the form of 'histories' of plants and animals; in the classical period it was organized as 'natural history'; and in the nineteenth century it became the science of biology. In addition, Foucault shows that the study of language and the study of wealth were systematically organized according to the same conceptual structure as that underlying the study of living things in each of these three periods. The existence of this relationship between such disparate empirical field leads us to ask whether a similar relationship may not be found between fields which have been more overtly connected. In particular, we should like to know the extent to which Foucault's

The body of the text of this paper was prepared through the cooperation of the two authors The Appendix was written by W. R. Albury alone, but the suggestions put forward there have the support of his co-author.

interpretation of the study of plants and animals since the Renaissance is applicable to the study of minerals during the same span of time. We are moved to undertake this enquiry not by a conviction that one uniform system of thought, a *Weltanschauung*, must prevail throughout a culture at any given time, but rather by a desire to explore further a definite series of concrete historical relations existing between the study of minerals, plants, and animals—e.g., the relations between geology and biology in the nineteenth century, or the relations between studies of the animal, vegetable, and mineral 'kingdoms' in eighteenth-century natural history. Let us begin this exploration, then, by summarizing the distinctions that Foucault makes between Renaissance 'histories' and classical 'natural history'.

Foucault's analysis of Renaissance 'histories' of plants and animals reveals that these works were composed according to a system of knowledge quite different from that of the 'natural histories' of the classical period. The role of the Renaissance scholar, when giving an account of a particular plant or animal, was to compile all the information about it that was available at the time. This information included not only a description of external characteristics of appearance and behaviour, but also a discussion of such things as magical and medicinal powers, uses in heraldry and mythology, 'correspondences', sympathies and antipathies, and especially a summary of all that had previously been said or written about the object in question. This last point was particularly important, for the written descriptions of a thing were thought to form part of its very 'nature'. It was the task, then, of the 'historian' to display all such information, organized according to the principle of 'resemblance' or 'similitude'. This principle formed the basis for order in Renaissance 'histories'; hence the importance of the 'doctrine of signatures' in these works.

By contrast, the discipline of 'natural history', as it arose in the mid-seventeenth century, turned its back on this complex web of similitude. It concentrated, instead, upon giving an exact description of the external (and particularly the visible) structures of natural objects, and upon formulating a correspondingly exact nomenclature for them. As Linnaeus later wrote:

> He may call himself a naturalist [a natural historian], who well distinguishes the parts of natural bodies by sight and describes and names all these rightly in agreement with the threefold division. Such a man is a lithologist, a phytologist or a zoologist.[3]

This reliance upon visible structure, Foucault argues, depended upon the principle of 'representation', which replaced 'similitude' as the basis for order in classical natural history. The notion that the 'essences' of bodies were faithfully 'represented' by their external features, and that these features themselves were subject to exact analysis, made possible the establishment of ordered, hierarchical tables of classification based upon the systematic comparison of visible structures. The works of Linnaeus himself, particularly in botany, best exemplify the tabular order which natural history sought to discover in nature.

This brief and necessarily oversimplified outline of the distinction which Foucault has made between Renaissance 'histories' of plants and animals and classical 'natural histories' of the animal and vegetable kingdoms, should now permit us to determine whether an equivalent distinction may be found between Renaissance studies of minerals and classical 'natural histories' of the mineral kingdom.

*Renaissance studies of the mineral kingdom*

We may begin our consideration of this question by turning to the mineral writings of two Renaissance scholars who are also noted for their works on plants or animals, Conrad Gesner and Ulisse Aldrovandi. In his *De rerum fossilium, lapidum et gemmarum* (1565),[4] Gesner compiled for minerals the same sort of information as typically appeared in contemporary 'histories' of plants and animals, all of which he organized according to the doctrine of signatures or correspondences. Thus his fifteen categories of mineral substances were based upon the resemblances between minerals and a variety of geometrical forms, celestial objects, animals, plants, and artefacts. Likewise, Aldrovandi, in a posthumously published collection of writings (1648),[5] treated minerals according to the following headings: synonyms, definition, origin, nature and properties, varieties, mode and place of occurrence, uses, historical references, sympathies and antipathies, mysteries, miracles, moralia, mythology, dreams, symbols and lapidati. This list may be compared with an equivalent series of headings cited by Foucault from Aldrovandi's *Historia serpentum et draconum*.[6]

The mineral writings of Gesner and Aldrovandi, then, display features similar to those which Foucault has shown to be characteristic of Rennaissance 'histories' of plants and animals. For 'historians' such as Gesner and Aldrovandi, minerals occupied the same conceptual domain as plants and animals, and were studied and described according to the same system of knowledge. But the question may be raised whether a scholar of the period such as Georgius Agricola, who approached minerals from a background of mining and metallurgy, may not have studied minerals according to a different conceptual system altogether.

In its accounts of individual minerals, Agricola's *De natura fossilium* of 1546[7] does, in fact, have characteristics very similar to those we have seen in the mineral writings of Gesner and Aldrovandi. Agricola included, for example, descriptions of minerals based upon 'imitation' and 'similitude', as when he wrote:

> Several minerals imitate the color of living things. For example, the gem *icterias* is similar in color to saffron; hematite, to blood; *prasius*, to the leek; *aspilates*, to fire.[8]

In addition, his accounts of the uses of minerals contain discussions of such matters as religious rituals[9] and legendary medicinal properties.[10]

But what of the organization of the work as a whole? An examination of the text of *De natura fossilium* reveals what at first glance appears to be an unsystematic collection of such descriptions and literary references. Nevertheless the book *is* underpinned by a general taxonomic scheme; and it is, perhaps, for this reason that Agricola was later hailed as a beacon pointing the way for the future development of the science of minerals. His underlying classification was based partly on a theoretical account of the genesis and composition of minerals (details of which may be found in his *De ortu et causis subterraneorum* of 1546)[11] and partly on a consideration of certain observable properties. 'Thus', he wrote, 'minerals have differences which we observe by color, taste, odour, place of origin, natural strength and weakness, shape, form, and size'.[12] Most of these features, however, as well as theoretical considerations upon the origin of mineral forms, were to be rigorously excluded as taxonomic characters by the natural historians of the classical period. Furthermore, Agricola did not systematically make use of his classificatory scheme in ordering his discussions of mineral substances. In his chapter on earths, for example, he briefly indicated his method of classification, but then went on to describe the various earths, first according to the art or trade to which they were most useful,[13] and second, according to the meanings of their names.[14] We see, then, that Agricola's text did not represent a fundamental departure from the Renaissance mode of studying minerals, and it is only by anachronism that this work can be considered as a 'natural history of the mineral kingdom'.

Agricola did not set out his classification of minerals in tabular form. However other Renaissance scholars who wrote on the subject of minerals did construct taxonomic tables of these substances. Thus we should consider whether such tables were ordered according to the same principles as underlay the tabular classifications of classical natural history. To this end we shall take as our example the work of Boethius de Boodt, *Gemmarum et lapidum historia* (1609),[15] which not only contains the kind of information about minerals which we have found in the writings already examined, but also includes two different tables presenting alternative systems of mineral classification.[16]

Although de Boodt's two mutually inconsistent classificatory schemes were largely divorced from his theoretical concepts regarding the structure, composition and generation of minerals, they nevertheless took as taxonomic characters a wide range of features alien to 'natural history', such as aesthetic appeal ('beautiful', 'ugly') and place of origin. Moreover, the variety of taxonomic characters used in each of these two tables shows the difference between de Boodt's principle of classification and that underlying the 'artificial systems' of the later 'natural historians'. It was indeed possible for the natural historian to construct alternative 'artificial systems' of classification (rather than a single 'natural method' of taxonomy).[17] But artificial systems were always formed on the basis of an exhaustive description of one specific part or feature of the body in question (for example, the Linnaean system of botanical classification concentrated upon the sexual organs of plants), not by taking a wide range of characteristics into account as did de Boodt. We may conclude, then, that the appearance of tables in *Gemmarum et lapidum historia* does not signal the beginnings of a natural history of minerals, since the organizing principle of these tables was quite different from that of classical taxonomies, both 'artificial' and 'natural'.

Just as in the study of plants and animals, so also in the study of minerals, it was not the appearance of something new (such as tables) that signalled the change from Renaissance 'histories' to classical 'natural histories', but rather the disappearance of much that was old. In the late seventeenth century, as Foucault has pointed out,[18] authors who wrote about plants and animals began deliberately to discard information concerning the secret 'virtues' of substances, their literary and heraldic significances, and other associated items of Renaissance lore. These writers did not necessarily know a great deal more than their predecessors; in fact they often gave less information. But there was (by the new standards) an increasing 'clarity' of exposition, arising from the elimination of 'superfluous' data and the fixing of attention upon external features for the purpose of identification. Thus, just as Buffon, in the eighteenth century, was to express astonishment at finding in the

> [zoological] work of a naturalist like Aldrovandi such an inextricable mixture of exact description, reported quotations, fables without commentary, remarks dealing indifferently with an animal's anatomy, its use in heraldry, its habitat, its mythological values, or the uses to which it could be put in medicine or magic,[19]

so too, by the 1670s, John Webster was to write of one of the typical mineral works of the late Renaissance:

> The Jesuite *Bernardus Caesius* writ a Voluminous Piece of Mineralogie, or Natural Philosophy; wherein, though he expatiated too far to fetch in all things that might seem any way of kindred to that kind of knowledge; and that it was but a meer Collection and heap stoln from other Authors, and hardly anything except notions; yet is there something in it (especially concerning the signs of discovering Mines and Ores) that may advantage such a Reader, as either hath the skill, or will take the pains to sever the Tares from the Wheat, and separate the gold from the dross.[20]

## *'Classical' studies of the mineral kingdom (c. 1660–1800)*

Foucault sees the turning-point in the approach to the study of living things in Jonston's *Historia naturalis de quadripedibus* of 1657.[21] In this work, Foucault notes, the question of 'animal semantics'—the literary and symbolic histories of animals so important to Renaissance scholars—was entirely omitted, thereby producing a considerable restriction in the type of information dealt with. It is interesting to note, therefore, that the same restricted approach can be found in Jonston's brief handbook of the mineral kingdom, published in 1661,[22] where a simple system of classification, based primarily on external characteristics, is presented.

A much more elaborate mineral classification which nevertheless had many features in common with Jonston's taxonomy of 'fossils', was that published by John Woodward in 1704 as a contribution to the *Lexicon technicum* of John Harris.[23] Here again, the classification was based almost exclusively on visible, external characteristics. It may also be noted that Woodward's ideas on the 'history' of the globe[24] did not have any marked effect on his taxonomic scheme, for these speculations were reflected only peripherally in a conspicuously unelaborated taxon of 'adventitious' fossils 'reposited in the Earth by the Universal Deluge', forming an insignificant component of the total classification.

It seems, however, that Woodward was not entirely satisfied with a classification based solely upon external characteristics, for in a letter to Newton, published in 1728, he indicated that he favoured the use of chemical tests (principally the action of the furnace) to help 'discover the nature of such parts in . . . [mineral] bodies, as do not immediately fall under the senses.'[25] Woodward's interest in chemical tests is noteworthy, even though the mineral classification which he published in 1728 differed only slightly from that of 1704 and contained few chemical taxa; for we do not find plant and animal taxonomists of the classical period relying, to any significant extent, upon information about the internal composition of living bodies (whether chemical[26] or anatomical and physiological) as a basis for their classifications. Rather, as Foucault explains, the natural history of the animal and vegetable kingdoms during this period

> is provided by surfaces and lines, not by functions or invisible tissues. The plant and the animal are seen not so much in their organic unity as by the visible patterning of their organs . . . Natural history traverses an area of visible, simultaneous, concomitant variables, without any internal relation of subordination or organization.[27]

In the case of the mineral kingdom, however, we find that many classical natural historians had recourse to chemical criteria in establishing their taxonomies.

Even Linnaeus, the prime exponent of restriction to a limited set of visible characteristics in botanical classification,[28] relied partially upon chemical criteria to establish his mineral taxonomy, as can be seen in his classification of the mineral kingdom, in the first edition of the *Systema naturae* (1735).[29]

It is evident from this classification that Linnaeus, like Woodward, took some account of the results of the action of heat on mineral substances. His tripartite division of stones, for example, into the orders Apyrous (unaffected by heat), Calcareous (forming a calx when heated) and Vitrescible (forming a glass when heated), was clearly founded on this basis. In addition, he adopted a number of the familiar chemical categories of his time, such as salts, sulphurs and mercuries, to serve as taxa.

On the other hand, most of the Linnaean genera, once established, could have been recognized as such largely by ocular inspection; and it seems likely that the chemical tests used in the classification did not proceed much further than the action of heat (possibly with fluxes and reducing agents in the case of the metallic ores, according to the traditional methods of the assayers[30]) and extraction with water. 'Schist', for example, is listed by Linnaeus as a genus of calcareous stone on the basis of the action of heat. Obviously, treatment with acid was not used as a diagnostic test.

It must be granted, then, that although Linnaeus organized his mineral system in a manner entirely characteristic of classical natural history—arranging his taxa in classes, orders, genera and species according to the method that was followed for the plant and animal kingdoms—the mineral system itself was based upon a mixture of visible (external) and chemical (internal) criteria. And despite successive refinements, Linnaeus was never able to do away entirely with the use of chemical criteria in his system of mineral taxonomy. He did, however, succeed in reducing their role so that, for example, in the twelfth edition of the *Systema naturae* (1766–8) the Apyrous and Vitrescible taxa of the first edition are replaced by the orders *Argillaceae* (clays) and *Arenatae* (sandy materials).

Also in the twelfth edition of the *Systema naturae* we find a system for classifying and naming crystalline minerals that was, *par excellence*, a system based on visible structure—with attention directed to the surface of things. This new method rested on the mental 'unfolding' of the crystal surfaces, and the counting and naming of the resultant two-dimensional geometrical figures. Thus, for quartz, Linnaeus gave the following diagrams:

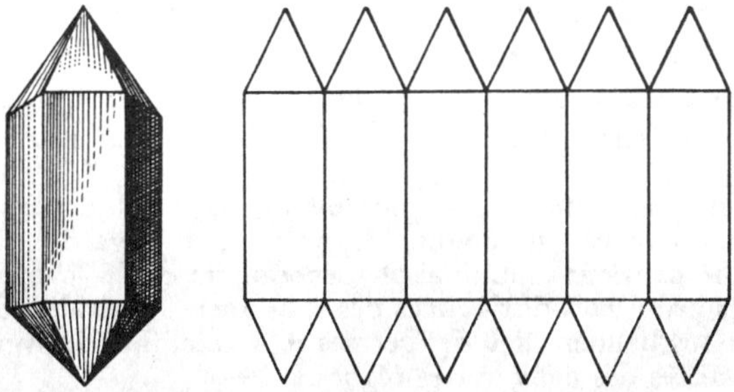

From these systematic description could be given:

*Columna* oblonga hexaëdra aequalis
*Pyramides* hexaëdra aequalis;[31]

and

*Figura Latera* 18-edrum
*Plana* Rectangulus 6
*Trigonis* 12.[32]

The method enabled one to describe and classify simply by counting edges or surfaces, just as in the sexual system for plants one counted the number of male or female organs per flower. It was immaterial what the internal chemical composition of the quartz crystals might be, so far as their description and classification by this method were concerned.

The Linnaean approach to crystals, and the classification of minerals by their visible characteristics, was carried forward in France by J. B. Romé de l'Isle. He offered[33] diagrammatic representations of crystals along the lines proposed by Linnaeus; and his general crystallographical scheme[34], in which crystals were mentally constructed by various bevellings and truncations of certain simple geometrical solids, was clearly one based on consideration of the surfaces of things, not on internal, chemical, criteria.

In addition, Romé de l'Isle offered a list of three 'external' criteria which would, he thought, if considered in conjunction with each other, enable one to recognize a particular mineral kind without ambiguity.[35] His diagnostic features were: crystalline form (with determinate interfacial angles), specific gravity, and hardness. Two different crystals might well have the same crystalline form, the same hardness, or the same density, but no two different kinds would have all these features in common. Romé de l'Isle, then, was attempting to erect a system of mineral taxonomy based on the determination of a restricted range of characteristics. Other features such as colour could be wholly ignored. Thus it would seem that his procedures were very close to those of Linnaean botany, with its restricted criteria for the determination of vegetable species.

It should be noted, however, that Romé de l'Isle, writing in 1784, gave considerable thought to the problem of the relationship between internal composition and external characteristics. He believed that there was a one-to-one 'representative' relationship between these two levels, with the externals serving as 'le vrai *cachet de la Nature*'.[36] In his view:

> [Les] principes élémentaires, plus ou moins simples, mais doués d'une figure qui leur est inhérente, & qu'on doit regarder indestructible, deviennent par leur combinaison générateurs de la forme cristalline & des autres propriétés distinctives des substances minérales, & conséquemment ils nous tiennent lieu des germes reproductifs exclusivement attachés aux substances des deux autres régnes.[37]

Nevertheless his crystallographic method offered no empirical means, whether chemical or geometrical, for investigating these *principes élémentaires*. Thus for Romé de l'Isle the connexion between the internal and external levels of crystals remained a purely speculative one. In the slightly later system of R. J. Haüy, however, a connexion between the external forms of crystals and their internal geometrical structure was established on an empirical basis.

In brief, Haüy attempted to determine the geometrical character of the 'primitive forms' of crystals by using, where possible, the natural cleavage planes as guides. For example, he claimed that all calcite crystals could be cleaved so as to yield rhombs of facial angles 101° 32′ 13″ and 78° 27′ 47″.[38] These were the 'primitive forms', or 'nuclei [*noyeaux*]', of calcite, as determined by direct experimental investigation. Haüy believed that his experiments revealed that six kinds of 'primitive forms' underlay all crystals.[39] Below the level of the 'primitive forms' there were three kinds of 'integrant molecules'[40] which were themselves made up of their various chemical components, or 'premiers principes ou les élémens des corps'.[41] Physical analysis could lead one to the 'primitive forms'. By further mental analysis, one might hope to penetrate to the level of the 'integrant molecules'. By a separate process—chemical analysis—one might reveal the natures of the 'constituent molecules'. Haüy was not, however, much concerned with these. He was a crystallographer rather than a chemist.

Having once penetrated to the level of the 'integrant molecule', Haüy was able to suggest ways in which the various known forms of a crystal of a particular kind (say calcareous spar) might be (mentally) synthesized by the various juxtapositions of the 'integrant molecules', according to different mathematical laws of decrement.[42] Thereby, each crystal 'species' might be determined by the particular angular characteristics of its 'primitive forms', thus resolving the problem of the existence of numerous 'secondary forms' of crystals having the same chemical composition.

Unlike Romé de l'Isle, then, Haüy offered a system of crystallography which established an empirical connexion between the external 'secondary' forms of crystals and their internal 'primitive forms' and 'integrant molecules'. This step was a decisive one which we may compare with the approximately contemporary development of the new basis for animal and plant classification to which Foucault calls attention in *The order of things*. Citing the work of such figures as A. -L. de Jussieu, Vicq-d'Azyr and Lamarck—all of whom placed primary emphasis upon the connection between the organic structure of living beings and their external features—Foucault writes as follows of the change introduced into taxonomy in the last decades of the eighteenth century:

> To classify . . . will no longer mean to refer the visible back to itself, while allotting one of its elements the task of representing the other; it will mean, in a movement that makes analysis pivot on its axis, to relate the visible to the invisible, to its deeper cause, as it were, then to rise upwards once more from that hidden architecture towards the more obvious signs displayed on the surfaces of bodies.[43]

This description, we suggest, applies to Haüy's crystallography quite as well as it does to Jussieu's botany or to Vicq-d'Azyr's comparative anatomy.

There was, however, an important aspect in which Haüy's theory of crystals was lacking. The hoped-for correlation of the external properties of crystals with their 'constituent molecules', which was to have been provided by chemical analysis, was not forthcoming owing to difficulties over isomorphism. But this problem speaks as much for the refinement of the techniques of chemical analysis in the last third of the eighteenth century as it does for the inadequacy of Haüy's theory; and it is to this technical refinement, especially as it bears upon mineralogy, that we now wish to turn our attention.

In the earlier part of the eighteenth century, as we have seen, Woodward and Linnaeus made a limited use of chemical criteria in establishing their predominantly 'external' systems of mineral classification. By the middle of the century, however, there were chemists such as Axel Cronstedt who openly attacked the use of external features in mineral taxonomies and attempted to devise purely chemical systems of classification. By means of tests with the blowpipe, Cronstedt distinguished nine 'simple earths', which he took to be the basic constituents of mineral bodies and used as the foundation of his taxonomic system. But these tests were not, in fact, true chemical analyses of mineral substances as Cronstedt thought; instead they merely served to provide additional criteria for the characterization of these substances.[44]

It was not, then, by 'fiery' techniques, such as were employed by Cronstedt, that the chemical analysis of minerals was achieved in the eighteenth century. Rather it was through the introduction of 'humid' methods by Torbern Bergman in the 1770s[45] that chemists came to analyze minerals with reasonable success. Bergman himself emphasized the value of these new methods, [46] and he was the first to attempt a chemical classification of minerals on this basis. He claimed that the only proper way to arrive at the 'principles' of bodies was by 'humid' analysis. A 'slight agreement with other known bodies' (by which, it may be presumed, he was referring to the external criteria of the natural historians) was not deemed sufficient.[47]

Bergman believed that five simple earths,[48] when combined in twos, threes, fours, or all five together, could provide sufficient variety to account for the various 'earthy' species of the mineral kingdom. He designated his five earths by 'p[onderous]', 'c[alcareous]', 'm[agnesian]', 'a[rgillaceous]', and 's[iliceous]'.[49] Then his 'double' species were, for example, 'pc', 'pm', 'pa', 'ps', 'cm', 'ca', etc. However, 'ps' and 'sp', were held to represent different substances, 'ps' indicating an excess of ponderous earth, and 'sp' an excess of siliceous earth. Working in this way, Bergman predicted the existence of three hundred and twenty five different combinations which, he thought, were more than sufficient to account for the range of mineral substances encountered in nature, if the contributions of the various metals were also taken into account.

Besides gaining a much clearer idea of the chemical composition of minerals as a result of the applications of his methods of analysis, Bergman could also put forward an associated scheme of mineral classification and nomenclature. If, for example, the substance 'xy' contained more 'x' than 'y', then it should belong to the genus 'x', and its species should be determined by 'y'. However, this rule might require modification if a particularly valuable component (say 'z') were found in conjunction with 'xy'. Then one might have to say that the specimen, 'xyz' belonged to the genus 'z'. The wide variety of siliceous minerals also presented a difficulty, and Bergman decided that the generic name should be determined by the next heaviest component after silica—otherwise too many disparate substances would be placed under the siliceous head.

Bergman extended his system beyond the earths in the following way. He adopted the four venerable taxa of Avicenna[50]—now termed salts ('s'), earths ('t'), metals ('m'), and phlogistic substances ('i')—as his higher category of *classes*. Members of the four classes could supposedly be combined in twos, threes of fours, giving yet another range of mineral substances, and since, as it will be recalled, there were some three hundred and twenty five kinds of earths, and also many divisions of the salts, as well as several different inflammables, there could be an enormous variety of 'Fossils mixed mechanically'.[51] However, these higher combinations were not thought to be chemical in nature, whereas the combinations of the primitive earths were explicitly conceived as involving chemical unions.

Although Bergman hoped that he had 'in some measure laid the foundations of a general system of mineralogy',[52] he nevertheless had not presented a completely developed systems. For example, although minerals could be composed of the same earths but with quantitative differences, Bergman's scheme did not make much allowance for such variations. It recognized minerals that might be designated by 'sp' or 'ps' but was forced to overlook intermediate ratios of 's' and 'p'. Also, Bergman did not attempt to integrate his extensive knowledge of chemical affinities with his system of mineral classification. Thus, unlike Haüy's crystallography, Bergman's mineralogy failed to establish a connexion between internal components and external characteristics. The simple earths appeared as the last term of analysis, but there was no understanding of how they united together to form particular mineral combinations—no chemical law comparable to Haüy's crystallographic 'law of decrement' which showed how crystals were built up from their 'primary forms'. Thus while Bergman's system made it possible, for the first time, for an adequate chemical mineralogy to be developed, it did not really represent as significant a step in the chemical approach to minerals as did Haüy's system in crystallography.

Both of these systems, however, were concerned with the internal make-up of minerals and in this respect they were on a par with the analyses of organic structure which appeared in the natural histories of plants and animals at the end of the classical period. Bergman's chemical investigation of mineral composition did not represent a radically new approach, for, as we have seen, chemical criteria had been used in one fashion or another in a number of mineral natural histories since the beginning of the eighteenth century.[53] But in the system of Haüy, like that of Vicq-d'Azyr, Lamarck, and Jussieu, the turn toward internal structure represented a very significant departure from past methods of analysis of the three kingdoms of nature.

Foucault argues[54] that the new concern with the organic structure of plants and animals at the end of the eighteenth century had as one of its consequences the introduction of a radical distinction between living and non-living things. Naturalists began to speak of two kingdoms of nature rather than three, and the science of biology became possible as the study of the phenomena of living things. Foucault sees the earliest clear elaboration of the new biological point of view of the nineteenth century in the work of the zoologist Georges Cuvier, whose comparative anatomy presented 'Life' as a hierarchy of physiological functions carried out in the various classes of animals by different anatomical systems.[55] The visible surfaces of animals, which had been the most important features for the classical natural historian, now were of only subsidiary significance. Furthermore, as Foucault suggests,

> . . . by substituting anatomy for classification, organism for [external] structure, internal subordination for visible character, the series for tabulation, [biology] was to make possible the precipitation into the old flat world of animals and plants, engraved in black on white, a whole profound mass of time to which men were to give the renewed name of *history*.[56]

This *possibility* of a history of life, however, 'becomes real history only by means of the external world in which [organisms] live'.[57] Thus it is historical geology, the record of the changing conditions of this external world and of the organic forms associated with them, to which we must now turn our attention. With the radical separation of the living from the non-living at the end of the classical period, mineralogy diverged from botany and zoology to become increasingly a branch of the nineteenth-century physical sciences. Thus we can not expect to find a transformation of the system of Haüy which would introduce the possibility of a history of minerals in just the same way that Cuvier's transformation of Vicq-d'Azyr's comparative anatomy made possible a history of life. It was the geological history of the earth as a whole, rather than a history of individual mineral forms, that was a necessary adjunct to nineteenth-century biology. On the other hand, we should not rule out in advance the possibility of finding some other type of relationship, not precisely analogous to the one between the work of Vicq-d'Azyr and that of Cuvier, which would link Haüy's mineralogy to both the geology and biology of the nineteenth century.[58]

Before turning to the nineteenth century, however, we must consider once more the last quarter of the eighteenth century—the end of the classical period—and attempt to locate the transformation in mineral studies which made the historical geology of later decades possible. To begin with, we shall ask whether this transformation was produced by the apparently historical mineral systems of Buffon and Lamarck, both of which presented a temporal sequence for the development of various minerals.

In his *Histoire naturelle des minéraux* (1783–1788),[59] Buffon proposed that the globe was initially molten, and that on cooling five primitive 'glasses' separated out (quartz, jasper, feldspar, schorl,[60] and mica) though two of these (jasper and mica) might, he admitted, be derivatives of the other three, and possibly all might ultimately be derived from quartz alone. As this glassy matter settled out on cooling, 'pure air' was left, but containing a certain quantity of fire. With water, this 'air' gave rise to 'aerial acid', which acting on vitreous matter, gave 'vitriolic acid'. Waters originally covered the earth, except for the highest hills, and in these fecund conditions many shells were produced. Rain falling on the hot, vitreous material caused it to break up into fragments, and the subsequent action of the water reduced it to clayey matter, which then formed schists and slates as it dried out. Tiny fragments broken off by waters from the primitive 'glasses' would become reunited and aggregated to form stalactites and also various mineral crystals and gems:

> C'est ainsi que le quartz, pénétré & dissous par l'eau a produit par exsudation, les cristaux de roche blancs & les cristaux colorés, tels que les améthystes, cristaux topazes, chrysolites & aigues-marines, lorsqu'il s'est trouvé des matières métalliques, & particulièrement du fer dans le voisinage ou dans la route de l'eau chargée de ces molécules quartzeuses.[61]

Similarly, feldspar supposedly gave rise to sapphire, cat's eye, aventurine, opal, etc.; schorl (sometimes with quartz and felspar also) gave emerald, beryl, peridot, garnet, hyacinth, tourmaline, etc.; mica gave talcs, jade, serpentine, Briançon chalk, etc.; calcareous matter (from shells) produced stalactites, Iceland spar, marbles, gypsum, etc.; a calcareous '*suc*' and a vitreous '*suc*' together could give substances such as lapsis lazuli, zeolites, flints; and, finally, vegetable earth derived from the detritus of plants and animals, and retaining a portion of fire from the original organisms, would give things like phosphorescent bodies, pyrites and diamonds.

It is evident that Buffon's account of the origin of minerals, upon which he erected a system of taxonomy, was temporally-based. The same thing can be said of the several accounts produced by Lamarck a few years later.[62] Lamarck proposed his scheme of the development of minerals as an alternative to the analytical schemes offered by contemporary chemists, whose conclusions he seriously distrusted. Retaining in his own system many features of earlier chemical theories that most of his contemporaries had discarded, Lamarck asserted that decayed matter from animals and plants served as the starting points for two main chains of mineral conversions, both ultimately leading to rock crystal.[63] But Lamarck did not offer a clear indication of mechanisms for the supposed changes, though there was the suggestion that, starting from organic debris, they involved a gradual stripping away of extraneous matter from the pure elementary 'earth'.[64]

Given the importance of temporal development in these two mineral systems, can they be seen as introducing the possibility of an historical study of the earth? For various reasons we would not subscribe to this interpretation. For one thing, like the earlier speculative 'theories of the earth', the system of Buffon and Lamarck were not results of, nor instruments for, empirical research. Also, as Foucault writes,

> For eighteenth-century thought, chronological sequences are merely a property and a more or less blurred expression of the order of beings; from the nineteenth century, they express in a more or less direct fashion, and even in their interruptions, the profoundly historical mode of being of things and men.[65]

The 'eighteenth-century' transformationist theory of Lamarck's *Zoological philosophy* (1809), for example, expresses the temporal unfolding of the chain of organized beings rather than an historical evolutionary scheme.[66] The inevitable 'complexification' of life results from an internal principle rather than from the contingent relations between beings and their environment which merely affect the forms which that complexity takes. Similarly, Lamarck's chains of minerals express the inexorable sequence of 'decomplexification' through which the products and remains of life must pass. And for Buffon as well, although the process is more involved, the future development of a mineral is, so to speak, 'programmed' into it. In both systems there is a supposed sequence of events that a mineral will naturally follow after its initial creation. Such an approach to the past was very different from that of the nineteenth-century stratigrapher, who, by examining strata and their spatial relationships, together with their fossil and mineral contents, attempted to work out the contingent geological *history* of the globe, rather than an inevitable sequence of mineral development.[67]

*Historical geology*

Can one find in late eighteenth-century accounts of minerals some general scheme that *did* display a significant historicist orientation, thereby producing the transformation which led to the historical geology of the nineteenth century? Let us look at the mineralogical and geological work of Werner with this question in mind.

Werner's early mineral system[68] was based almost exclusively on external characteristics, and although he failed to achieve a tightly ordered system such as that of Linnaeus—for he did not limit himself to the consideration of a severely restricted range of characters—this early work was characteristic of the eighteenth-century natural historian. But when we examine aspects of the later celebrated Wernerian theory itself we find a significantly different approach. Consider, for example, the following passage by Werner, which indicates what he had in mind when he was writing his important *Kurze Klassifikation* of rock masses (1786):

> Since *observation of nature's products* shows that, even though nature now and then in widely separated periods was similar, it gradually changed very much during the various successive and in part enormously great time spans, which perhaps far exceed our imagination of the time of development of the rock masses, *I had to be guided completely in the classificatory presentation or tabulation of these masses by the discoverable time sequences of the particular formations* if I wanted to remain true to my plan to sketch through this classification a foundation for a complete canvass of the universal formation of these masses.[69]

Werner's intention, then, was to build up, by 'observation of nature's products', a classification of rocks based upon 'the discoverable time sequence of the particular formations'. The temporalization of mineral natural history, as a matter of empirical research, is at once evident in this programme. Werner's conception of the stratigraphical column[70] was that of a record of contingent events. And the 'Neptunist' Wernerian theory itself was an attempt to link particular rock formations in this record with particular periods in the earth's past.

Werner was not, of course, the first person to consider the stratigraphical relations of the rocks of the earth's crust.[71] But he was, as his contemporaries acknowledged,

> . . . the first to draw the attention of geologists, explicitly, to the *order of succession* which the various natural families of rocks are found in general to present, and . . . [he] himself developed that order to a certain extent, and with a degree of accuracy which, before his time, was unattainable, from the want of sufficient methods of discriminating minerals and their compounds.[72]

In Werner's view, this order of succession ought to be found in all parts of the globe, allowing the correlation of similar strata in different regions according to a single historical time-scale. To those interested in the study of the earth, the Wernerian geognosy[73] held out, for the first time, the promise of a general method of enquiry, to be used in researches all over the world—indeed it was the very generality of this method, rather than Werner's Neptunist explanation of the development of the earth's crust, that gave the Wernerian system its particular significance in the eyes of so many of his contemporaries. Followers such as Leopold von Buch and Alexander von Humboldt were willing to accept a much greater role for the action of heat in the development of the earth than Werner ever countenanced, but they took pains to emphasize the value of Wernerian geognosy as a basis for world-wide research into the earth's history.[74]

In the geognosy of Werner, then, we have the transformation of classical mineralogy into a new science which would make possible the historical investigation of the earth. The link between this historical geology and the biology of the nineteenth century lay in the relations drawn between organic fossils and geological strata. And although Werner himself did not go as far as William Smith, who suggested that fossils rather than mineral composition ought to be used to identify strata, 'he never failed, in his course of lectures, to fix the attention of his pupils on the relations that exist between certain fossils and [geological] formations of different ages'.[75] But, in addition to this obvious link, at a deeper conceptual level there was an important parallel between geognosy and biology, a parallel involving also the crystallography of Haüy.

In order to exhibit this parallel we must recall the consequences, for the natural history of plants and animals, of the introduction of organic structure into taxonomy by Vicq-d'Azyr and others at the end of the eighteenth century. Prior to this event classification was a tabulation of visible identity and difference. With the anatomical investigation of organic structure, however, the identity of parts found in various individuals could no longer be based upon what was immediately visible—not because some parts were internal, but because individual living things of widely differing organization and complexity were now being compared. Vicq-d'Azyr avoided the full consequences of this problem by restricting himself to the examination of fairly closely-related animal forms, but the comparative anatomy of Cuvier faced the problem in its most general terms. How was it possible to identify the anatomical 'elements' from one species to the next and trace their variations throughout the entire animal kingdom?

There were, as Foucault points out,[76] two alternative answers to this question, one of which was put forward by Cuvier and the other by Etienne Geoffroy St-Hilaire. Cuvier's solution to the problem was based on physiological function; for the chief functions, such as respiration, digestion, and circulation, persist through all variations of form, and the different organs for performing these functions (later named *analogous parts*) can be compared with one another (e.g. for respiration: lungs, gills, tracheae, etc.). Geoffroy's solution, on the other hand, was founded upon topological considerations. For Geoffroy, an anatomical element was identified not by its functions but by its spatial relations—its connexions—with other elements.[77] A series of parts, in Geoffroy's view, might be variously enlarged or diminished from species to species; their forms might be altered; some of them might even be absent in certain species, but they would never appear in a different order of connexion relative to one another.[78] For Geoffroy, then, identical anatomical elements (later named *homologous parts*) were to be recognized by use of the 'principle of connexions'.[79]

Now if we return to the geognosy of Werner we can see how its relation to biology can be understood in terms of the comparative anatomy of Geoffroy St-Hilaire. The Wernerian geognostic programme, like comparative anatomy, required 'assembling, analyzing, relating and comparing a large number of distant facts which do not have the slightest apparent relation to one another'.[80] This programme raised for classical mineralogy the same problem of the criteria of identity based upon visibility as Vicq-d'Azyr's introduction of organic structure did for classical zoology. And like Vicq-d'Azyr, Werner was at first spared the full consequences of this problem by the fact that he confined his investigations to fairly closely-related objects—in this case the statigraphy of a somewhat restricted area of Europe. The stratigraphical 'elements' compared throughout this region comprised not simply individual minerals but groups of associated minerals, called by Werner 'formations'; and the identity of formations from place to place was defined by three criteria: the extent of the deposit, the structure and texture of its mineral composition, and relative position in the stratigraphical column.[81] Werner applied this concept of formation in his own geognostic research, but he attempted to draw conclusions about the globe as a whole from his study of formations in eastern Europe. As von Humboldt wrote:

> In the first period of the science, the *type of formations* was founded on a small number of observations, and resembled too much the description of the places where it originated. The mineral masses which in other countries are but subordinate or accidental beds, were mistaken for independent formations; the existence of those formations that are important in equatorial America, and in the north and west of Europe, was unknown . . .[82]

Cuvier and Geoffroy could have whole specimens brought to them in Paris for anatomical comparison; but for the programme of geognostic comparison to be carried out the investigation of stratigraphical relationships had to be undertaken by field workers in all parts of the globe.[83] Thus it was former students of Werner such as von Buch and von Humboldt who took geognosy beyond 'the first period' by their extensive travels. Humboldt expressed this point when he wrote that Werner

> . . . was the first who taught the art of observing and distinguishing the formations; and it is by the further application of that art that geognosy has become a positive science . . . It is only in extending our observation to a greater part of the globe, that the type of formations can be enlarged and simplified. It has thus been made to correspond better with the geognostic constitution of continents considered in a general point of view.[84]

But this extension of geognostic research to distant parts of the globe necessitated a modification of Werner's original criteria for identifying formations, since it was discovered that formations could vary greatly from one region to another both in thickness and in composition. Thus for von Buch and von Humboldt, 'The essential character of the identity of an independent formation is its relative position, or the place which it occupies in the general series of formations'.[85] Mineral composition and organic fossils were used in a manner subsidiary to this principle of position in determining the equivalence of formations in places remote from one another—the term 'identical formations' being applied to those which were the same both in relative position and mineral content; while formations occupying the same relative position in the stratigraphical column but differing in mineral composition were called 'parallel'. 'My *parallel* formations', wrote von Humboldt, 'are *geognostic equivalents*; they are the rocks that represent each other'.[86] Finally, since an extensive formation found in one region could be insignificant or entirely lacking in another one, thickness or extent of deposit was no longer a criterion of identity or equivalence between formations in different parts of the earth (although it remained as an indicator of the mode and duration of the process by which the deposit was formed).

For the Wernerians, then, the geognostic study of formations was guided by a principle analogous to Geoffroy's principle of connexions: a formation was 'essentially' defined by relative position. As W. H. Fitton wrote, the mature Wernerian doctrine held

> . . . *that the exterior of the earth consists of a series of these formations, laid over each other in a certain determinate order.* Not that the whole series is anywhere complete; but that the relative place of its several members is never departed from. Thus in the series A, B, C, D, it may happen that B or C, or both, may be occasionally wanting, and consequently D be found immediately above A; but the succession is *never* violated, nor the order inverted, by the discovery of A above the formations B, or C, or D, nor of B above those that follow it, &c.[87]

This order of superposition revealed the relative age of the formations in the stratigraphical column because it reflected their order of deposition. Similarly for Geoffroy the anatomical order of the organs directly reflected the process of their development from the nutritive fluids of the body. 'The essential point', he wrote,

> . . . is that each subdivision of the principal [circulatory] branch deposits a part of the fluid which it contains and exactly yields its diverse products in an order of superposition which is the same as the order in which these subdivisions are attached to the principal branch.[88]

Finally, just as Geoffroy was able to account for such apparent anomalies as monstrous deformations by means of his principle of connexions, showing how these deviations arose from the fusion, exaggerated development, or disappearance of certain parts, so too the Wernerian geognosts sought to account for stratigraphical 'monstrosities' such as the extraordinary local increase of one formation so that it envelops a second formation which ought normally to succeed it, or the local suppression of a particular formation owing to the over-development of another formation. According to von Humboldt,

> These effects of the alternation and unequal development of rocks are so much the more worthy of attention, as the study of them may throw light on some apparent deviations from a type of superposition generally received, and may serve to refer to a common type the series of positions observed in countries very distant from each other.[89]

Thus in teratology, as in morphology and in organogenesis, Geoffroy's biology with its principle of connexions was paralleled by the Wernerian geognosy based upon the stratigraphical principle of position. But these parallels can be extended to Haüy's crystallography as well,[90] for in place of the principle of connexions or position, one finds as the basis of Haüy's work the constancy of angular relations between the molecular components of the various crystals. This constancy permitted Haüy to calculate the mathematical laws of decrement according to which these molecular elements are layered upon one another (cf. organogenesis, successive deposition), and to define precisely the resultant crystalline form (cf. morphology, stratigraphical typology). Finally, Haüy showed that even crystals which had previously been regarded as anomalous could be understood as modifications of regular forms, thus bringing them under the general laws of crystal structure (cf. teratology, explanation of stratigraphical anomalies).

We see, then, that there was a significant relationship between the work of Haüy and the subsequent development of nineteenth-century geology, even though it was different from the relationship between the work of Vicq-d'Azyr and the subsequent development of nineteenth-century biology. The problem set for comparative anatomy be Vicq-d' Azyr was analogous to the one set by Werner for geognosy at approximately the same time, and the solution to this problem adopted by Geoffroy St-Hilaire in his biological studies was paralled by the one which the followers of Werner adopted in their geological researches. Both these solutions, however, depended upon the application of the same kind of spatial relational principles as those which underlay Haüy's crystallography.

Haüy's importance, then, for geology lies not in his having defined the field of inquiry (a step taken by Werner), but in the logic of his investigations into the related field of mineralogy. This logic contributed to the development of Wernerian geognosy into the historical geology of the nineteenth century, and hence to the production of a 'geological record', not only of fossil organic remains but also of the environmental history of the earth—a record of the conditions to which living things had been exposed throughout the course of their existence. And this understanding of geology, in turn, made possible the delineation by Darwin of the real history of living things.

*Conclusion*

In conclusion, let us assess how far Foucault's account, in *The order of things*, may be extended from studies of the animal and vegetable kingdoms to studies of the mineral kingdom. It appears that this extension can be made quite satisfactorily for most of the period that has been considered in the present paper, though, as has been pointed out above, the early chemical studies of the mineral kingdom had no exact counterpart in the contemporary studies of the animal and vegetable kingdoms by the natural historians of the classical period. But in other respects, until the latter part of the classical period, the similarities that may be discerned in studies of the three kingdoms of nature are most marked. This, of course, is not in itself unexpected, for in the Renaissance and the classical period these three kingdoms were viewed as three comparable subdivisions of the continuum of nature.

In the latter part of the eighteenth century significant parallels may still be found, though there were increasing divergences. Thus crystallographical studies, like the studies of living bodies, came to place emphasis upon internal structure and composition, while external characteristics were explained in terms of internal structures. But this emphasis upon the internal structure of plants and animals brought the functions of these organisms into prominence, and led to the introduction of a rigid distinction between living and non-living things. The concept of organic function had no counterpart in the inorganic realm—indeed it became one of the defining characteristics of life itself. Thus in the nineteenth century the biological interest of Cuvier in function and the correlation of parts found no exact parallel in contemporary studies of the mineral kingdom, though important comparisons can be drawn between the anatomical work of Geoffroy St-Hilaire and the geognostic investigations of the Wernerian school. Of greater significance was the advent, at the end of the eighteenth century, of theoretical conceptions that made possible the historical investigation of all three kingdoms of nature. With respect to the mineral kingdom, this development is revealed by the establishment of geology (or geognosy) as an independent science, separate from mineralogy, at about the same time that the natural histories of the animal and vegetable kingdoms were transformed into the new science of biology.

In closing, we should emphasize that we are not arguing for specific causal connections to account for the parallels that we have discussed in this study. At this stage we are only seeking to call attention to these parallels, without entering into the difficult researches that would be needed in any attempt to explain them. It is only where we have found obvious differences between the study of the mineral kingdom on the one hand, and the study of plants and animals on the other, that we have offered suggestions as to why and how these differences may have arisen.

There remains, then, the perplexing question of the nature of the causal connexions that may, through further investigations, be found to explain the many similarities that have been identified in the study of the three realms of nature. The importance of this question has, we hope, been heightened by showing that an analysis in the manner urged by Foucault, can reveal additional systems of relations, beyond those already established in *The order of things* between theories of language, of wealth, and of plants and animals. This may emphasize still more acutely, for philosophers and students of intellectual history, the many fertile problems raised by Foucault's investigations. In addition, students of the history of the earth sciences may find some satisfaction in the interesting pattern of their subject matter that is revealed when it is examined in the light of *The order of things*.

## Appendix by W. R. Albury

In natural histories of the plant and animal kingdoms it was not until the last quarter of the eighteenth century that the internal parts of living things —their organization or 'organic structure'—became taxonomically important. Why was it, then, that the programme of investigating the internal constituents of bodies as an aid to their classification appeared so much earlier among natural historians of the mineral kingdom? This question is one which cannot be answered definitively at the present state of research, but the following considerations are relevant and may offer some general indications of a solution.

1. *The development of analytic techniques.* It does not seem that the methodological difference between the natural history of minerals and that of plants and animals can be accounted for by a difference in the degree of development of the appropriate analytic techniques. At the beginning of the eighteenth century animal dissection, for example, was as advanced an art in the hands of the physicians as mineral analysis was in the hands of chemists and assayers. And as we have seen, even the considerable improvement in analytic technique introduced by Bergman in the 1770s

could not, of itself, bring about an adequate system of chemical mineralogy.

2. *The nature of the objects studied.* In classical natural history the plant or animal species, the basis of all taxonomy, was seen as a collection of similar individuals which maintained its identity through time by the reproduction of additional individuals similar to their parents. In the case of the mineral kingdom, however, it was not at all clear what constituted the mineral 'individual' (was it a homogeneous mass or a heterogeneous mass of a given pattern?), nor could the reproductive link of like arising from like be brought to bear on the problem. Thus, by the middle of the eighteenth century the whole question of the mineral 'species' was at issue.[91] Toward the end of the century the crystallographers would resolve this difficulty by taking the '*molécule intégrante*' as the mineral 'individual' and substituting the fixity of chemical composition for the criterion of reproduction. Thus for Haüy the mineral species was '*une collection de corps dont les molécules intégrantes sont semblables, et composées des mêmes élémens unis en même proportion*'.[92] We should note, however, that the problem of the mineral species seems not to have been raised until after the work of Linnaeus had appeared,[93] and that the use of chemical criteria by Linnaeus himself and by other mineral taxonomists before him may not, therefore, have been directly related to it.

3. *The location of the objects studied.* Plants and animals were, for the most part, presented directly to the sight of the naturalist in their various habitats. They were dispersed over the surface of the globe but not concealed within it; and when they were partially concealed and partially visible, the concealed parts (e.g. the roots of plants) were systematically neglected in favour of the more obvious ones. Minerals, on the other hand, were normally concealed within the earth; they were 'fossils' (i.e., things dug up). The mineral kingdom was thus defined, in the *Encyclopédie* for example, as 'l'assemblage total des corps qui appartiennent à la terre, & qui se forment dans son sein',[94] and Axel Cronstedt excluded such substances as soot and tartar from his mineral system because they were 'never to be met with under the surface of the earth'.[95] Thus the natural location of minerals, and the activities of digging up, uncovering and separating which were associated with them, may have gone some way toward overcoming the kinds of ethical, religious and epistemological attachments which bound classical natural history to the visible exteriors of plants and animals.[96] It is worthy of note, in this connexion, that even those mineralogists who sought to avoid any use of chemical criteria in their taxonomic systems had recourse to a number of 'external' characteristics of minerals (e.g., hardness, specific gravity) that were not directly visible and could only be determined by subjecting the specimen in question to certain tests. Nevertheless, the location of minerals within the earth, if it did contribute to the early acceptance of chemical criteria

into mineral natural histories, would have done so more as a negative factor (in helping to remove an obstacle) than as a positive one.

4. *Relations between associated fields.* It was chiefly the medical disciplines which furnished the concepts and techniques (wholeness of the organism, anatomical dissection, physiological investigation, correlation of internal and external phenomena) that transformed zoological and botanical taxonomy at the end of the eighteenth century. Thus it may be instructive to compare the relationship between medicine and the natural history of the plant and animal kingdoms during the classical period, with the contemporaneous relationship between chemistry and the natural history of the mineral kingdom.

In contrast to chemistry's traditional concern with a wide variety of minerals, the focus of medical theory and practice was not living things in general but man alone, with animal dissection and experimentation being undertaken by physicians and surgeons primarily for the light which could thereby be thrown upon the human body and its functions. The status of the medical professions, with their valued social role and established positions in both universities and scientific academies, was such that their members had no need to demonstrate the worth of their disciplines by proposing medical concepts and techniques as a basis for the reform of botanical and zoological natural history. On the contrary, those physicians who devoted themselves to natural history (e.g., Boerhaave,[97] Linnaeus) did so without any sign of wishing to bring medical theory and practice to bear upon their systems of plant and animal classification.

The social position of chemistry, however, was somewhat different from that of medicine at the beginning of the eighteenth century. Although represented in the scientific academies of the day, the study of chemistry was carried out in the universities primarily as an appendage to the study of pharmacy in the medical faculties—there were few chairs for chemists *per se*. And the value to society of an independent discipline of chemistry was by no means clear. In these circumstances chemists as a group had much to gain in putting forward the concepts and methods of their discipline as the basis for a reform of mineralogy, especially as they saw a growing public interest in natural history developing early in the century. It was not to aristocratic collectors, with their 'cabinets d'histoire naturelle', that the chemists addressed their works, however, but to those individuals with practical interests. The early chemical system of Pott, for example, which was based upon 'the assaying of stones by fire', brought its author 'considerable honour', according to Cronstedt, because of its value to 'Miners and other manufacturers'.[98] A taxonomy based on visible features may have been satisfactory for the arrangement of a collection meant only for display, but to the technologically-oriented bourgeoisie the contemplation of natural objects was less important than their utilization.[99] To this group the chemists' claim to offer '*the only method of obtaining any knowledge of the Mineral Kingdom*' and their dismissal of mineralogists who were '*entirely taken up with the surface of things*'[100] had considerable appeal, as the following avowal from the *Encyclopédie*, that great propaganda-piece of bourgeois practicality, makes clear:

> . . . le coup d'oeil extérieur ne suffit point pour nous faire connoître les corps du règne minéral . . . il faut avoir une connoissance des qualités internes des substance minérales, pour savoir les usages auxquels ils peuvent être employés; et ce n'est que la chimie qui puisse procurer cette connoissance.[101]

In proving the utility of chemical mineralogy for 'Miners and other manufacturers' the chemists stood to win more than 'considerable honour'; for by the beginning of the eighteenth century the mining and metallurgical industries had already attained a position of central importance in the economies of western European states, and the mercantilist theories of the time encouraged active governmental support of such key industries. Thus Cronstedt was careful to point out that chemistry would have to enjoy the patronage of 'those to whom the heavy burthen of governing human societies is allotted' if the full value of chemical mineralogy were to be realized:

> It is from their care we are to expect the compleat tribunal where all disputes in this matter can be accurately decided, and all things be rendered trulv useful; I mean the institution of a Laboratory . . .[102]

Indeed Cronstedt went on to suggest that in such laboratories the 'chemical experiments upon vegetables' which had been carried out in the seventeenth century by the *Académie royale des sciences* in Paris, might 'at some future time be repeated with advantage, when more knowledge in that matter is obtained'.[103] Thus he held out to potential patrons of chemistry the prospect of a chemically-based natural history of plants as a complement to the chemical systems of mineralogy already being devised.

As we have noted above, however, it was medicine rather than chemistry that provided the basis for the reorganization of animal and vegetable taxonomy at the end of the eighteenth century. This development was made possible by two important changes in medical orientation which occurred around the middle of the eighteenth century and which were highlighted by definite events in the early 1760s.

One such event was the publication in 1761 of J. B. Morgagni's *De sedibus et causis morborum*,[104] the first systematic study of pathological anatomy. From Morgagni on, medicine became increasingly concerned with the establishment of correlations between external sysmptoms of disease and internal lesions as determined by post-mortem dissection. This growing importance of pathological anatomy reflected the acceptance by physicians of 'the surgical point of view'[105] at a time when surgery, although still regarded as distinct from medicine, had freed itself from its traditional guild association with unlettered barbers and become established as a learned profession.[106] The fact that surgery involved manual skills no longer relegated it to a status inferior to that of medicine, as the older aristocratic values were supplanted by those of the bourgeoisie, for whom 'social distinctions must be founded on usefulness to society'.[107]

Together with this development of the surgical orientation in medicine, with its concern to correlate the external and internal phenomena presented by the patient, there also occurred a change in the relationship between medicine and animals necessitated by the increasing economic importance of the latter as sources of food and clothing for the growing urban populations. This change was institutionalized in the early 1760s with the opening of the first veterinary schools;[108] and in the following decade, when the *Société royale de médicine* was established by the French government, it was given the responsibility of studying not only epidemic diseases of humans but those of livestock as well.[109] Thus animals became medical objects in their own right, rather than as surrogates for humans; and the peculiarities of their organization became the subject of special study. Consequently, at the end of the century, when surgically-oriented physicians such as Vicq-d'Azyr (founder of the *Société royale de médicine* and author of works on the epidemic diseases of cattle[110]) turned to natural history, they saw animals (and, by extension, plants[111]) as beings to which the concepts and techniques of medicine were wholly appropriate.

To summarize our argument in this section, then, we may say that we find a combination of economic and social factors underlying the relationship of natural history to chemistry and medicine. On the economic side there was the relatively earlier development of the mining and metallurgical industries, as compared with the large-scale cultivation of livestock; while on the social side there was, first, the relatively lower status of chemistry as compared with medicine, and second, the rising status of surgeons to a position approaching that of physicians. Such factors, we suggest, interacting in the manner outlined above, contributed largely to the methodological difference which we have found between mineral natural histories and those of plants and animals, with the former introducing chemical criteria from the early 1700s while the latter maintained a rigorous adherence to the visible exteriors of living things until the end of that century.

NOTES

1 M. Foucault, *Les mots et les choses: une archéologie des sciences humaines* (Paris, 1966); *The order of things: an archeology of the human sciences* (London, 1970).

2 Also relevant to this paper is Foucault's clarification of his methodology in: *L'archéologie du savoir* (Paris, 1969); *The archeology of knowledge* (London, 1972).

3 C. Linnaeus, *Systema naturae* (1735); facsimile of the first edition with an introduction and a first English translation of the Observations by Dr. M. S. J. Engel-Ledeboer and Dr. H. Engel (Nieuwkoop, 1964), p. 19.

[4] C. Gesner, *De rerum fossilium, lapidum et gemmarum maximè, figuris & similitudinibus liber: non solum medicis, sed omnibus rerum naturae ac philologiae studiosis, utilis & iucundus futurus* (Zurich, 1565).

[5] U. Aldrovandi, *Musaeum metallicum in libros IIII distributum* (Bologna, 1648).

[6] M. Foucault, op. cit. (1, 1970), p. 39.

[7] G. Agricola, *De natura fossilium, lib. X* (Basle, 1546); M. C. Bandy & J. A. Bandy, *De natura fossilium* (*Textbook of mineralogy*) by Georgius Agricola: translated from the first Latin edition of 1546 . . . for the Mineralogical Society of America (New York, 1955).

[8] G. Agricola, op. cit. (7, 1955), p. 6. Italics in original.

[9] Ibid., pp. 39–40.

[10] Ibid., p. 121. 'Horus writes that [yellow quartz] is a remedy for mongoose bite when burned and pulverized'.

[11] G. Agricola, *De ortu et causis subterraneorum lib. V* (Basle, 1546).

[12] G. Agricola, op. cit. (7, 1955), p. 15.

[13] Ibid., p. 23. 'I shall now take up the forms of earths, those cultivated by farmers and those used by potters, fullers, painters, carpenters, and other artisans.'

[14] Ibid., p. 29. 'I shall now discuss earths which have taken their names either from the locality in which they are found or from their color . . .'

[15] A. Boethius de Boodt, *Gemmarum et lapidum historia, qua non solum ortus, natura, vis & precium, sed etiam modus quo exiis, olea, salia, tincturae, essentiae, arcana & magisteria arte chymica confici possint, ostenditur. Opus principibus, medicis, chymicis, physicis, ac liberalioribus ingeniis utilissimum. Cum variis figuris, indeceque duplici & copioso* (Hanover, 1609).

[16] Ibid., facing page 3 and facing page 7.

[17] M. Adanson, *Familles des plantes* (2 vols., Paris, 1763), enumerated sixty-five different artificial systems of plant taxonomy. For a discussion of the difference between the 'natural method' and 'artificial systems' in classical natural history, see M. Foucault, op. cit., (1, 1970), pp. 138–45.

[18] M. Foucault, op. cit., (1, 1970), pp. 128–32.

[19] Ibid., p. p.39.

[20] J. Webster, *Metallographia: or, an history of metals* (London, 1671), p. 30. Italics in original.

[21] J. Jonston, *Historia naturalis de quadripedibus* (Amsterdam, 1657).

[22] J. Jonston, *Notitia regni mineralis, seu subterraneorum catalogus, cum praecipuis differentiis*, Leipzig, 1661.

[23] J. Harris, *Lexicon technicum: or, an universal English dictionary of arts and sciences: explaining not only the terms of art, but the arts themselves* (London, 1704), article entitled 'Fossils'.

[24] For an account of these ideas, see: V. A. Eyles, 'John Woodward, F. R. S., F. R. C. P., M. D. (1665–1728): a bio-bibliographical account of his life and work', *Journal of the Society for the bibliography of natural history*, v (1971), 399–427.

[25] J. Woodward, *Fossils of all kinds, digested into a method, suitable to their mutual relation of affinity; with the names by which they were known to the antients, and those by which they are at this day known; and notes conducing to the setting forth the natural history and the main uses of some of the most considerable of them. As also several papers tending to the further advancement of the knowledge of minerals, of the ores of metalls, and of all other subterranean productions* (London, 1728), p. 4.

[26] There were, during the classical period, attempts by *chemists* to introduce chemical criteria into plant and animal taxonomy, but the results of these projects were inconclusive and no significant use was made of them in *natural history*. See F. L. Holmes, 'Analysis by fire and solvent extractions: the metamorphosis of a tradition', *Isis*, lxii (1971), 129–48; and D. C. Goodman, 'The application of chemical criteria to biological classification in the eighteenth century', *Medical history*, XV (1971), 23–44.

[27] M. Foucault, op. cit. (1, 1970), p. 137.

[28] Linnaeus believed that only the number, form, proportion, and situation of the parts should be considered in plant taxonomy: C. Linnaeus, *Philosophia botanica in qua explicantur fundamenta botanica cum definitionibus partium, exemplis terminorum, observationibus rariorum, adjectis figuris aeneis* (Stockholm, 1751), pp. 116 and 257.

[29] C. Linnaeus, op. cit. (3).

[30] See: F. Greenaway, 'The historical continuity of the tradition of assaying', *Proceedings of the tenth international congress of the history of science: Ithaca 26 VIII 1962–2 IX 1962*, (2 parts, Paris, 1964), part ii, pp. 819–23.

[31] C. Linnaeus, *Systema naturae per regna tria naturae, secoundum classes, ordines, genera, species, cum characteribus & differentiis* (12th edn., 3 vols., Stockholm, 1766–8), iii, 213.

[32] Ibid., p. 215.

[33] J. B. Romé de l'Isle, *Essai de cristallographie, ou description des figures géométriques, propres a différens corps du regne minéral, connus vulgairement sous le nom de cristaux, avec figures et développemens* (Paris, 1772). This volume contains (following p. 384) a '*Tableau cristallographique*' and ten plates of figures giving representations of common crystal forms.

34 For an account of this, see J. G. Burke, *Origins of the science of crystals* (Berkeley and Los Angeles, 1966), pp. 62–7 and 69–76.

35 J. B. Romé de l'Isle, *Des caracteres exterieurs des minéraux, ou response à cette question: existe-t-il dans les substances du règne minéral des caracteres qu'on puisse regarder comme spécifique; & au cas qu'il en existe, quels sont ces caracteres? Avec un apperçu des différens systêmes lithologiques qui ont paru depuis Bromel jusqu'à présent. Suivi de deux tableaux synoptiques des substances pierreuses et métalliques, pour servir de suite à la cristallographie* (Paris, 1784), pp. 10 and 55.

36 Ibid., p.55. Italics in original.

37 Ibid., pp. 39–40.

38 R. J. Haüy, *Essai d'une théorie sur la structure des crystaux, appliquée a plusieurs genres de substances crystallines, par M. l'Abbé Haüy*, (Paris, 1784), p. 97.

39 R. J. Haüy, *Traité de minéralogie, par le Cen. Haüy* (5 vols., Paris, 1801), i, 283–4.

40 Ibid., pp. 31 and 284.

41 R. J. Haüy, op. cit. (38), p. 49.

42 For details of these, see J. G. Burke, op. cit. (34) pp. 94–100.

43 M. Foucault, op. cit. (1, 1970), p. 229.

44 See: D. R. Oldroyd, 'A note on the status of A. F. Cronstedt's simple earths and his analytical methods', *Isis*, lxv (1974), 506–12.

45 Bergman's method was described in his paper: 'Disquisitio chemica de terra gemmarum', *Nova acta Regiae Societatis Scientiarum Upsaliensis*, iii (1777), 137–70. For an English translation, see: T. O. Bergman, *Physical and chemical essay; translated from the original Latin of Sir Torbern Bergman, by Edmund Cullen, M. D.* (2 vols., London, 1784–8), ii, 27–56.

46 T. O. Bergman, op. cit. (45, 1784–8), i, p. xxvii. ('*Analysis should chiefly be conducted in the humid way*'. Italics in original.)

47 Ibid., p. xxv.

48 By the 1770s, most chemists were agreed that there were several individual, empirically-separable, simple earths, rather than a single 'terrene principle', such as had earlier been envisaged by P.-J. Macquer. There was some uncertainty as to the precise number of the simple earths, but commonly silica, lime, magnesia, alumina and baryta were considered.

49 T. O. Bergman, *Meditationes de systemate fossilium naturali* . . . (Florence, 1784); *Physical and chemical essays;* translated from the original Latin of Sir Torbern Bergman (3rd vol., Edinburgh, 1791), pp. 205–316.

50 E. J. Holmyard and D. C. Mandeville, *Avicennae de congelatione et conglutinatione lapidum* (Paris, 1927), pp. 33–4.

51 T. O. Bergman, op. cit. (49, 1791), p.285.

52 Ibid., p. 314.

53 We disagree, on this point, with the view commonly expressed in the secondary literature, that the application of chemical criteria to mineral classification was introduced 'in the last quarter of the eighteenth century'; F. D. Adams, *The birth and development of the geological sciences* (2nd edn., New York, 1954), p. 200. In the Appendix, below, some suggestions are offered as to why the programme of investigating the internal constituents of bodies arose in mineral taxonomy so much earlier than it did in the classification of plants and animals.

54 M. Foucault, op. cit., (1, 1970), pp. 226–32.

55 Ibid., pp. 263–79.

56 Ibid., p. 138. Italics in original.

57 Ibid., p. 293.

58 We shall return to the question of this relationship below.

59 G. L. L. de Buffon, *Histoire naturelle des minéraux* (5 vols., Paris, 1783–8).

60 In the eighteenth century, several distinct minerals were conflated under the name 'schorl', but in general it referred to members of the tourmaline group.

61 G. L. L. de Buffon, op. cit. (59), iv (1786), 441.

62 J. B. Lamarck, *Encyclopédie méthodique. Botanique*, vol. ii (Paris, 1786), 34; *Recherches sur les causes des principaux faits physiques* (2 vols., Paris, Year 2), ii, facing p. 366; *Mémoires de physique et d'histoire naturelle* (Paris, Year 5), facing p. 349; *Hydrogeology*. Translated by Albert V. Carozzi (Urbana, 1964), p. 117.

63 In the version of the Year 5 (1797), the table was headed: 'Tableau des corps bruts, ou exposition des principales substances minérales, disposées dans un ordre relatif au progrés des altérations qu'ont subi les dépouilles des corps vivans, et successivement leur différens produits'. In a footnote to the table, Lamarck stated that it described 'le véritable ordre natural des minéraux'.

64 J. B. Lamarck, op. cit. (62, 1794), p. 355 and passim.

65 M. Foucault, op. cit. (1, 1970), p. 276.

66 J. B. Lamarck, *Zoological philosophy; an exposition with regard to the natural history of animals*. Translated by H. Elliot (New York and London, 1963), p. 179.

67 It may be noted here that in the opening of his *Époques de la nature* of 1778, Buffon did indicate his wish to unveil the Earth's past by a variety of empirical means, including an examin-

ation of 'les monumens qu'on doit regarder comme les témoins de ses premiers âges'. But, in practice, his method was primarily that of a cosmogonist, rather than an earth historian.

[68] See: A. G. Werner, *On the external characters of minerals*. Translated and edited by A. V. Carozzi (Urbana, 1962).

[69] A. G. Werner, *Short classification and description of the various rocks*. Translated with an introduction and notes by Alexander M. Ospovat (New York, 1971), p. 20. Emphasis added.

[70] Ibid., pp. 128–9.

[71] Bergman, for example, took an interest in this subject. See: H. D. Hedberg, 'Influence of Torbern Bergman (1735–1784) on stratigraphy', *Acta Universitatis Stockholmiensis, Stockholm contributions in geology*, xx (1969), 19–47. This paper is also useful for its account of several earlier stratigraphical investigations. We do not, however, find that any of these earlier works offered the general programme for stratigraphical enquiries, with the explicit objective of determining the history of the earth, that is to be found in the Wernerian system. But Hedberg (p. 30) quotes Bergman's opinion (1766) that fossils 'are actually medallions of a sort, which were laid down on the originating earth surface, whose layers are archives older than all annals, and which, appropriately investigated, give much light on the natural history of this our dwelling place'.

[72] [W. H. Fitton,] 'Art. IV. Transactions of the Geological Society, established November 1807. Vol. III. 4to. pp. 444. W. Phillips, London, 1816', *Edinburgh review*, xxix (1817), 70–94 (p.71). Italics as in original.

[73] It is noteworthy that Werner saw his work as the establishment of a new science, quite distinct from both mineral natural history (which he called 'oryctognosy') and speculative theorizing on the origin of the earth ('geogony'). In order to mark this distinction, he coined the name 'geognosy' for his new science shortly before Lamarck and others, with similar intent, adopted the term 'biology' for the new science of living things. In addition to geognosy and oryctognosy, Werner recognized three other branches of mineralogy: mineralogical chemistry, mineralogical geography and economic mineralogy.

[74] This is apparent, for example, in: L. von Buch, 'Uber den Gabbro, mit einigen Bermerkungen uber den Begriff einer Gebirgsart' *Der Gesellschaft naturforschender Freunde zu Berlin: Magazin für die neuesten Entdeckungen in der gessammten Naturkunde*, v (1810), 128–49; and A. von Humboldt, *A geognostic essay on the superposition of rocks, in both hemispheres* . . . translated from the original French (London, 1823), pp. 3, 23 and passim.

[75] A. von Humboldt, op. cit. (74), p. 67. Jameson also emphasized that Werner recognized the importance of organic fossils in stratigraphical investigations ('Professor Jameson's notice of Werner's views of the natural history of petrifactions', *The Edinburgh new philosophical journal*, vii (1829), 78–9).

[76] M. Foucault, 'La situation de Cuvier dans l'histoire de la biologie–II', *Revue d'histore des sciences*, xxiii (1970), 63–69.

[77] E. Geoffroy Saint-Hilaire, *Philosophie anatomique* . . . (2 vols., Paris, 1818–22), i, xxv–vi: 'Or il est évident que la seule généralité à appliquer dans l'espèce est donnée par la position, les relations et les dépéndances des parties, c'est-à-dire, par ce que j'embrasse et ce que je désigne sous le nom de connexions'.

[78] Ibid., p.xxx: 'un organe est plutôt altéré, atrophié, anéanti, que transposé'.

[79] Ibid., p. xxviii.

[80] J. F. d'Aubuisson de Voisins, 'Suite de la lettre de M. d'Aubuisson, a M. Berthollet, sur les travaux de M. Werner, en minéralogie', *Annales de chimie*, lxix (1809), 225–48 (p. 229). Our translation.

[81] A. M. Ospovat, in: A. G. Werner, op. cit. (69), pp. 97–8.

[82] A. von Humboldt, op. cit. (74), p. 81. Italics as in original.

[83] Ibid., pp. 4–5.

[84] Ibid., pp. 82–3.

[85] Ibid., p. 9. Von Buch, in his discussion of the meaning of the term '*Gebirgsart*', took the same view. Comparing rock formations to the numbered houses of a street, he maintained that it was natural to base one's classification on positions in an ordered sequence, and he asserted that any other procedure would fail to reveal the great laws which nature followed in forming *Gebirgsarten*. (L. von Buch, op. cit. (74), pp. 132–3.)

[86] A. von Humboldt, op. cit. (74), p. 17. Italics as in original (Cf., also, p. 56, and J. F. d'Aubuisson de Voisins, *Traité de géognosie* (2 vols., Strasbourg & Paris, 1819), ii, 255.)

[87] W. H. Fitton, op. cit. (72), 71. Italics as in original.

[88] E. Geoffroy Saint-Hilaire, op. cit. (77), i, xxix. Our translation.

[89] A. von Humboldt, op. cit. (74), p. 43. Also worthy of note is the parallel between Geoffroy's experimental study of the deforming effects of variations in temperature and mechanical agitation upon a developing embryo, and the study by the Wernerian stratigraphers of the role of extreme heat, volcanic eruptions, and earthquakes in producing deviation from the normal sequence of formations.

[90] F. Dagognet, 'Valentin [sic] Haüy, Etienne Geoffroy Saint-Hilaire, Augustin P. de Candolle: Une conception d' ensemble mais aussi un ensemble de conceptions' *Revue d'histoire*

*des sciences*, xxv (1972), 327–36. Dagognet, as his title indicates, attempts to extend the system of parallels to the botanical work of A. P. de Candolle also; but since de Candolle drew as much upon the functional principles of Cuvier as he did upon the topological principles of Haüy, we have not included a consideration of his work in our present discussion.

91 See: S. H. Mauskopf, 'Minerals, molecules and species' *Archives internationales d'histoire des sciences*, xxiii (1970), 185–206.

92 R. J. Haüy, op. cit. (39), i, p. 162. Italics as in original. Cf. the statement of Romé de l'Isle, cited above (note 37).

93 S. H. Mauskopf, op. cit. (91), p. 199.

94 [P. H. T. d'Holbach,] 'Minéral regne', in Diderot *et al.* (eds.), *Encyclopédie, ou dictionnaire raisonné des sciences, des arts et des des métiers*, . . . (17 vols. and supplements, plates and tables, Paris, 1751–80), x (1765), 534.

95 A. F. Cronstedt, *An essay towards a system of minerology . . . Translated from the original Swedish, with notes, by Gustav von Engestrom . . .* (2nd English edn., London 1772), p. xix.

96 For a discussion of some aspects of these attachments to visible exteriors in the context of medicine, see D. E. Wolfe, 'Sydenham and Locke on the limits of anatomy', *Bulletin of the history of medicine*, xxxv (1961), 209–17.

97 See, for example, H. Boerhaave, *Index alter plantarum quae in horto academico Lugduno-Batavo aluntur* (Leyden, 1720); and T. A. Sprague, 'Boerhaave as a botanist' in *Memorabilia Hermann Boerhaave optimi medici* (Haarlem, 1938), pp. 91–4.

98 A. F. Cronstedt, op. cit. (95), p. ix.

99 D. Longeot, 'La genèse de la crystallographie et de la minéralogie scientifique', *Dix-huitième siècle*, iii (1971), 253.

100 A. F. Cronstedt, op. cit. (95), p. xii. Italics as in original.

101 [P. H. T. d'Holbach,,] 'Minéraux, mineralia', in D. Diderot, op. cit. (94), p. 544.

102 A. F. Cronstedt, op. cit. (95), p. x.

103 Ibid., pp. x–xi. For a discussion of the seventeenth-century experiments in question see F. L. Holmes, op. cit. (26).

104 J. B. Morgagni, *De sedibus et causis morborum* (Venice, 1761).

105 O. Temkin, 'The role of surgery in the rise of modern medical thought', *Bulletin of the history of medicine*, xxv (1951), 255.

106 T. Gelfand,'From guild to profession: the surgeons of France in the 18th century', *Texas reports on biology and medicine*, xxxii (1974), 125–7.

107 This phrase was cited as a motto by a group of naval surgeons in their *Principes de réclamations pour les chirurgiens de la marine* (Brest, 1790). See T. Gelfand, op. cit. (106), p. 131.

108 F. Dagognet, *Le catalogue de la vie* (Paris, 1970), p. 68.

109 C. C. Hannaway, 'The Société Royale de Médicine and epidemics in the ancien régime', *Bulletin of the history of medicine*, xlvi (1972), 257–73; M. Foucault, *The birth of the clinic* (London, 1973), pp. 26–8.

110 F. Vicq–d'Azyr, *Recueil d'observations sur les différentes méthodes proposées pour guérir la maladie épidémique qui attaque les bêtes à cornes, sur les moyens de la reconnaître par-tout où elle pourra se manifester et sur la manière de désinfecter les étables.* (Paris, 1775); *Exposé des moyens curatifs et preservatifs qui peuvent être employés contre les maladies pestilentielles des bêtes à cornes* (Paris, 1776).

111 See, for example, F. Vicq-d'Azyr, 'Troisième discours sur l'anatomie', in *Oeuvres recueillies de Vicq-d'Azyr*, J. –L. Moreau (de la Sarthe), ed., (6 vols., Paris, 1805), iv, 229 ff.

# HISTORICISM AND THE RISE OF HISTORICAL GEOLOGY, PART 1

## INTRODUCTORY REMARKS : HISTORICISM

It is commonly recognized that an intellectual shift of considerable importance occurred towards the end of the eighteenth century, when many aspects of man's culture came to be viewed from an historical perspective. According to Collingwood, for example, the rise of history as a scientific discipline occurred at about this time, and the work of Herder contributed very significantly to the establishment of an historical outlook in the late eighteenth century.[1] In a similar vein, Butterfield, in his well-known *Man on his past*,[2] drew attention to the work of the Göttingen school of history in the latter part of the century, and stated that it marked the advent of the professionalization of historical studies.

Needless to say, the influence of the new historical outlook was not just confined to the discipline of history. One need only point to the changing of architectural styles in the period, with the beginnings of the Gothic revival,[3] to realize that men were developing a new interest in, and sympathy with, the past. An appreciation of the extent of the general change in outlook is further reinforced by an examination of contemporary literature, for although the historical novel was primarily a nineteenth century phenomenon, one can readily identify early instances of the *genre* in the latter part of the preceding century. Indeed, it might be said that the very rise of the novel in the eighteenth century marked a growing awareness of the notion of the flux of time.

In music, one finds performers of the early nineteenth century turning back to the music of the past with awakened interest,[4] and the rise of musicology in the late eighteenth century as a distinct intellectual discipline, manifested in the writings of the Göttingen scholar, J. N. Forkel,[5] marked the addition of a kind of temporal dimension to the study of music. Again, sciences such as archaeology, with an obvious historical concern, made their appearance towards the end of the eighteenth century.[6]

My intention, then, in the present paper, will be to examine the rise of historical geology in the late eighteenth and early nineteenth centuries, and to attempt to connect the appearance of this discipline with the more general changes in the Western intellectual tradition that were occurring at about this time, which have been adumbrated above. For I do not think it merely coincidental that historical geology appeared when it did. Rather, it will be claimed that the rise of historical geology was a concomitant of other, more broadly-based, changes, and meshed with them. In addition, I shall discuss the general intellectual commitments of some of the leading geologists of the period under consideration, in so far as they were concerned with this question of 'thinking historically'.

In the exposition that follows, it will be convenient to use the term 'historicism' to refer to the view that was commonly espoused by those who exhibited the new 'historical consciousness' in its many different aspects. The word has previously been used by some writers with respect to the history of geology,[7] and thus there is some precedent for my usage. It has not, however, been the subject of a particular investigation in the history of geology. And writers who have used the term 'historicism' with respect to the history of geology have not generally done so in a manner which accords satisfactorily with the customarily-accepted usage by students of intellectual history.

The term historicism is a slippery one, and has been given a wide range of interpretations.[8] I shall, however, try to use the word in a manner consonant with the meaning which it held in the early nineteenth century (in its German form, *Historismus*),[9] such as was intended by the followers of the German historical school of law. Thus, unlike G. G. Simpson who employs the term in reference to progressionist or developmentalist viewpoints,[10] or Popper who uses it in reference to those who attempt to make predictions into the future on the basis of past happenings,[11] I shall use the word 'historicism' to refer to a belief in the efficacy of offering *explanations* of the nature of things or phenomena by means of their history. It is in this sense that Runes's *Dictionary of philosophy* defines historicism as

> ...the view that the history of anything is a sufficient explanation of it, that the values of anything can be accounted for through the discovery of its origins, that the nature of anything is entirely comprehended by its development . . . or, [in general, it is] the doctrine which discounts the fallaciousness of the historical fallacy.[12]

However, to use the term in this way still leaves open the question of the meaning of the "history of anything". I maintain that this does *not* mean simply a knowledge of the antecedents of a thing or event, which are in some manner 'causally connected' with it. For the term also has methodological connotations.

For example, a Laplacian determinist might claim that, given a knowledge of some initial set of circumstances and the general 'laws of nature', he could *explain* the occurrence of some particular event or phenomenon. Such a mode of explanation would certainly be concerned with the past as a means of explaining the present. But it would not be an *historical* explanation, and it would not correspond to the approach favoured by the historicist. For an historian approaches his task, not by appeal to general laws and 'boundary conditions',[13] but by rummaging in libraries and archives, selecting from the information there discovered, so as to piece together an account of a sequence of events that can serve as a means of accounting for certain subsequent events or phenomena. It is the interest in the unique historical events, rather than general historical laws, that is the hallmark of the historian, and which characterizes the historicist attitude. Let us, therefore, distinguish between historical explanations, and those that require knowledge of a set of antecedent circumstances plus certain laws of change or development. This latter mode of explanation is, of course, appropriate in some circumstances, more especially in the physical sciences. Let us refer to it as the *genetic* mode of explanation, emphasizing that it is substantially different from that which is characteristically employed by the historian, and which is advocated by the historicist.[14] The distinction intended here is of importance in the argument that follows.

The use of the term 'Historismus' may be traced to the eighteenth century,[15] but the rise of historicism as an intellectual movement is usually linked with the so-called 'historical school of law', which came to the fore in Germany in the early years of the nineteenth century under the leadership of F. C. von Savigny.[16] In his pamphlet, *Ueber den Beruf unserer Zeit für Gesetzgebung und Rechtswissenschaft,*[17] for example, he wrote:

> The historical spirit . . . is the only protection against a species of self-delusion, . . . namely, the holding that which is peculiar to ourselves to be common to human nature in general. Thus, in times past, by the omission of certain prominent peculiarities, a natural law was formed out of the Institutes, which was looked upon as the immediate emanation of reason. There is no one now who would not regard this proceeding with pity; and yet we meet with people daily, who hold their juridical notions and opinions to be the offspring of pure reason, for no earthly reason but because they are ignorant of their origin. When we lose sight of our individual connection with the great entirety of the world and its history, we necessarily see our thoughts in a false light of universality and originality. There is only the historical sense to protect us against this . . . .[18]

Von Savigny's main argument was directed against the establishment of legal codes, such as that of Napoleon, in Germany. Nevertheless, we can see clearly enough in this passage evidence for the nineteenth century turning away from the natural law doctrines of the Enlightenment in favour of historically-based justifications of legislation, emphasizing individuality and the uniqueness of events, rather than their law-like character. Calvin Rand has put the matter well in the following words:

> . . . the nature of . . . [the historicist's] world appears to be [one of] flux and change, a process and development of newly emerging *individuals,* each in its unique position in time and place, rising and falling and nonrecurring. . . . [History] moves to no goal, only individual ends and values appear momentarily within the process.[19]

The rise of the historical school of law does not, of course, stand as an isolated phenomenon in the history of Western thought. Clearly, it may be correlated with the growth of the Romantic Movement and *Naturphilosophie,* the rise of nationalism and the political events that occurred in Europe following the French Revolution, and many other factors. And the whole German movement, of which it formed a part, may be seen in the special context of the intellectual, cultural and political characteristics of that country. To discuss these matters is not the business of this paper. Rather, it is my intention to show that the ideas and methods of the early practitioners of historical geology bore considerable analogy to the attitudes of the adherents of the historical school of law, or, rather, the historicist movement as a whole. By this means one may show how the rise of historical geology fitted naturally into the intellectual context of the late eighteenth and early nineteenth centuries.

One must first consider, however, whether it is correct to accept the view that the historical age was a mark of the *post*-Enlightenment period, and that the thinking of the Enlightenment was essentially ahistorical or even anti-historical. Also, can instances of historical approaches to the study of the Earth be found in periods well before the turn of the nineteenth century—in the seventeenth century, for example?

Collingwood, it will be recalled, represented the period of the late eighteenth and early nineteenth centuries as that of "the threshold of scientific history".[20] Yet something that might reasonably be referred to as 'history' was most certainly being written in the Enlightenment period, and indeed much earlier. The critical use of sources, for example, was not an invention of the nineteenth century. Voltaire was something other than a mere chronicler.[21] And Cassirer has stated categorically that "the common opinion that the eighteenth century was an 'unhistorical' century, is not and cannot be historically justified".[22] Nevertheless, I subscribe to the view that the period of Enlightenment was not one that preferred *historical* studies as guides to an understanding of man. Condorcet, for example, set up an *idealized* account of man's past, illustrating a supposed gradual growth and spread of rationality. This picture was not derived from an examination of the details of individual items of historical information concerning different people, races and nations, with this information being woven together so as to give a history of mankind in all its complexity and individuality. Rather, one should select facts "from the history of different peoples, compare them and combine them in order to extract the *hypothetical* history of a single people and to compose the picture of its progress".[23] And where physical explanations proffered in the Enlightenment invoked an unfolding of nature in the course of time (as, for example, in Kant's theory of the heavens),[24] they may be regarded as genetic rather than historical in character. The same holds true for Linnaeus's early accounts of the development of the surface of the globe through the gradual diminution of the seas.[25] On the other hand, in Germany movements towards an historical study of mankind (in a fashion customarily associated with nineteenth century historicism) were beginning to be developed from the middle years of the eighteenth century, at a time when French thought was still consciously anti-historical.[26] This should be borne in mind if one is concerned to identify links between geological thought and opinions concerning the history of mankind.

A further difficulty may seem to be presented by the fact that studies that may with some justification be regarded as constituting early examples of 'historical geology' were in fact being undertaken in the seventeenth century.[27] For instance, as is well known, in Steno's *Prodromus* there is the fundamentally important stratigraphical concept now known as the 'principle of superposition', which enables one to forge a link between observations of strata and presumed temporal sequences.[28] And Hooke, in 1688, spoke to the Royal Society about the possibility of using fossils "to raise a Chronology out of them, and to state the intervalls of the Times wherein such, or such Catastrophies and Mutations have happened".[29] But Hooke did not, in fact, actually write a history of the globe on the basis of fossils or observations of strata, and neither did any of his contemporaries. Moreover, his interest in this discourse to the Royal Society was a means towards providing secure empirical support for the traditions of catastrophes indicated by ancient mythologies—for *man's* history in other words—rather than as a means whereby the geological history of the *Earth* might be elucidated. And in his earlier statement of geological doctrine (1668) we have a cyclic theory which Hooke did not seek to justify by examination of the details of individual strata, though he thought that the presence of fossils within the strata could be accounted for by means of his theory.[30] We must, however, note Hooke's interest in providing a link between studies of the Earth's past and mankind's past.[31] His geological work meshed with the concern with man's past that was evinced in his day. But this concern was with ancient mythology, Biblical records, and millenarian doctrines—confirming the unfolding of history according to some inspired plan—rather than with the understanding of man's past through the painstaking piecing together of the secular record of man's affairs.

Steno also, was concerned to offer theoretical reasons to account for the sequence of events which he believed to have taken place in the Earth's past.[33] He was prepared to some degree to generalize the inferences which he made on the basis of his observations of the strata of Tuscany,[34] and he sought a unification of the testimony provided by the observations of strata and the reading of the holy scriptures.[35] But he did not achieve or even contemplate the writing of a complete history of the globe on the basis of his Tuscan investigations. There was no piecing together of discrete pieces of evidence respecting strata in order to build a composite historical account, which would, *qua* history, stand as a means of explaining the Earth's present conditions. The intentions of Hooke and Steno, then, offered parallels to the aims of the antiquaries, chroniclers and natural historians of the seventeenth century.[36] But Hooke and Steno did not provide examples of Earth histories, except in so far as their accounts intertwined with scriptural or mythological versions of the Earth's past. Hooke had a *theory* of the Earth; but this was not historical in character.

In a recent paper, W. R. Albury and I have discussed some of the eighteenth century 'genetic' accounts of the Earth's minerals, such as those of Buffon and Lamarck, so it is not necessary that this ground be retraced here, beyond saying that we have not found such theories to have been historical in intent.[37] But did any other eighteenth century figures manifest an historical or historicist outlook in their investigations of the Earth? Let us look at the work of the Italian investigators Moro and Arduino with this question in mind.

### INVESTIGATIONS IN ITALY: INTIMATIONS OF HISTORICAL GEOLOGY

Moro's thesis was not typical of eighteenth century schemes, for it envisaged a sequence of 'Plutonic' events, rather than the formation of strata by typical sedimentary processes. Taking the formation of the new volcanic island of Mea Kaumen in 1707 as the key to his understanding of the Earth's past, Moro employed an actualistic methodology. He divided the material of the Earth's crust into gneissic Primary mountains elevated by Plutonic forces, and Secondary mountains produced by further igneous activity and supplemented by the addition of sedimentary material.[38] The superposed strata represented for Moro the chronological sequence of the formation of the Earth, and could give an indication of its past circumstances.[39] Thus, as early as 1740 there was in Moro's work something approaching an historical attitude towards a study of the Earth, despite the fact that it was linked with a particular *theory,* and also attempted a union with the traditional Judaeo/Christian history of Genesis. Moro, I suggest, was not promulgating a purely 'genetic' scheme. The Earth's exact and particular sequence of volcanic eruptions could not be known *a priori,* even if the general principles of the 'Plutonic' theory were accepted. On the other hand, the broad patterning of events (Primary and Secondary) does not seem to arise wholly and directly from the evidence of observations.

The main contribution of Arduino was to publish a general four-fold classification of the Earth's strata, the four divisions coinciding closely with the four modern major divisions of the stratigraphical column.[40] His thinking was also actualistic in character:

> It is only by the guide of practical knowledge of those physical agents which we see actually operative in the works of nature, as well as that of the corresponding effects, that we can, with reasonable principles, find out what might have been the effective causes in times which are remote.[41]

Moreover, Arduino perceived the directionalist tendencies of the fossil record, and in the following passage he revealed what may appear as an historicist temper:

> He who wishes to grasp, without being deceived, the great history of our planet, and the different epochs of the great catastrophes and changes undergone by it, cannot dispense with attentively contemplating these signs and monuments; and the diversity, number, circumstances, materials, relations, and all the other phenomena of earths, sands, rocky strata, both on the mountains and on every part of the earth.[42]

At first sight this passage may seem to offer the same theoretical position as that presented by Hooke. But the two are not precisely the same. Arduino is concerned with "signs and monuments" for the purpose of grasping "the great history of our planet", not thinking of this as simply an adjunct to a knowledge of human history. Thus, it would seem, Arduino is taking up an historical attitude towards the study of the Earth, for its own sake, so to speak. Certainly his view is catastrophist, but this is not incompatible with an historicist view of the Earth, though from the nineteenth century standpoint (Lyell) it came to be felt necessary to smooth away the discontinuities (catastrophes) in the Earth's history so that an evenly flowing narrative of its past might be provided.

It may be noted that although Arduino did not produce a geological map in order to illustrate his two important letters to Vallisnieri, he *did* produce most interesting manuscript sketches, a number of which have been reproduced by Stegagno. One of these sketches is a remarkable vertical section of the mountains up the valley of the Agno from Montecchio Maggiore to the high peaks of the Alps above Recoaro.[43] It gives a graphic representation of Arduino's subdivisions of the Earth's strata and understanding of the stratigraphy of northern Italy. It would seem that his work contained many features that we have come to associate with nineteenth century geology.

But Arduino's work was of uncertain influence. He certainly established no general school, though he was subsequently appointed to a chair of mineralogy at Padua. And as is well known, it was in Germany that studies of the Earth succeeded in establishing their first major paradigm —that of the Wernerian school. So let us turn to consider some of the features of the work of Werner and his disciples.

### GERMANS AND GEOGNOSY

I shall not attempt here to trace the earliest roots of the Neptunist doctrines of the Freiberg school.[44] For our present purposes, it is not necessary to go back further than the work of J. G. Lehmann,[45] for long recognized as a precursor of Werner, and one of the important founders of the science of stratigraphy.

Lehmann believed in a 'catastrophist' theory of the Earth, and he spoke of the Deluge as an important geological agent. He referred to himself unequivocally as an "historian" of the Earth,[46] and was prepared to offer a graphic discription of the earlier states of affairs of our planet. Moreover, he was able to show that there was a general patterning of the strata over wide areas of Germany. He asserted boldly that "Nature has always acted in a uniform manner in the formation of beds",[47] and described a typical section of thirty-one strata in the county of Hohenstein and up to Mansfeld, illustrating the account with a section of the strata. Some of the terms that he employed (e.g., *Zechstein*) are in use to this day. Lehmann claimed that no one before him had treated the layered mountains in his way.

Lehmann's general theory proposed a tri-partite division of rocks, into those that were contemporary with the formation of the globe (*Gange-Gebürge*), those that were formed from these in a general "revolution" (the *Flötz-Gebürge*), and those that were formed by purely local "revolutions".[48] His diagrammatic section showed the chief subdivisions of the *Flötz-Gebürge*.[49]

The point to be noted here is that Lehmann believed that his observations enabled him to give an *universal* history of the Earth. It is true that claims of this kind had been made previously. We have, for example, already noted the tentative universalism of the theory of Steno. But the *Prodromus* offered a sequence of supposedly universal *events,* rather than a sequence of *things* (strata) which might be thought universal and which consequently might be *tested* by subsequent investigators in areas far distant from the part of Germany where the scheme was originally proposed.

A work by G. C. Füchsel[50] shows a considerable similarity to the scheme of Lehmann, and Füchsel explicitly refers to his countryman in his text. The list of strata proposed as a general schematization of the geology of Thuringia was:

Muschelkalk
rothe Gypslager
Sandgeburge
Glühsand, Hüttensandlager
Meelbatzen, Brennkalchgeburge
graue Gyps oder Alabasterlager
Floetze, Floetzgeburge
Sandfloetz
weiß Geburge, weißlich Schalgeburge
rothe Geburge, roth Schalgeburge
schwarz-blau Schalgeburge, Dach-Schiefergeburge
Alaunschiefer, schwarz Alaunlager, Vitriolisches Geburge
Steinkohlicht Geburge, Steinkohlenfloetze
rothe tode Lager
Grundegeburge, Gangegeburge [=Ambitus fundamentus]

Füchsel accompanied his memoir with several maps, the first of which (T.♀) shows a kind of "perspective plan"; the second and third (T.☽ and T.♁) show combinations of cross-sections and plans.

T.♀ has been described as the "first published geological map",[51] and it is true that we do have here perhaps the first serious attempt to portray the history of strata by means of a map. However, from this point of view, a section is really much more significant than a map, and in terms of principle Lehmann had already shown what might be done with the help of a geological section.

The first part of the *Historia terrae et maris* is chiefly concerned with the geometrical disposition of the layers of rock which Füchsel observed in Thuringia. He introduces the terms *stratum* (*Schicht,* bed or stratum), *situs* (*Lager,* bed or stratum), *positus* (*Stand,* position), and *series montana* (*Gebürge,* chain of mountains, formation). The concept of *situs* differed from that of *stratum* in that it held the connotation of being adjacent to particular neighbours in a stratigraphical series. A mountain might have the same stratigraphical status, or *situs*, as some other mountain, but yet occupy a different geographical location (*positus*).

Füchsel emphasized that the strata had not simply settled out in the order of their specific gravities, and in his second section he began to unravel the stratigraphical history of the Thuringia district: "One may combine the preceding observations both with each other and with others, and deduce events or a history from them."[53] Here certain speculative elements came into play, but his basic conception of the deposition of sediments from an overlying sea was sound: "The bodies of strata lying on themselves are produced in as many measures of time as there are strata."[54] Also, a particular age might conveniently be denominated by the strata to which it gave rise.[55] Füchsel discussed the possible conditions of deposition for the formation of strata; for example, in terms of streams carrying sediments in particular directions at particular times, or windstorms producing particular sedimentary materials in the overlying oceans at given periods. However, his time-scale was restricted to the order of hundreds of thousands of years. He speculated on the marine transgressions and regressions as having been due to "shakings" of the globe—possibly due to disturbances produced by the gradual accumulation of sediments.

Before considering the ideas of Werner and his followers, a few words should be said about the work in Russia of the great traveller and natural historian, P. S. Pallas, for he also contributed to the establishment of a general description of the layers of the Earth's surface. In his celebrated memoir of 1777,[56] Pallas noted that Swedish and German scholars had previously expressed ideas on "the order that Nature has followed in forming these elevations of the globe and in the arrangement of its beds".[57] And he believed that his observations in Asia confirmed these earlier generalizations.

Broadly Pallas's theory was similar to that of Lehmann and Füchsel, but he was also drawing on ideas expressed by Linnaeus and Bergman.[58] He maintained that the central cores of the chief mountain ranges of the Earth were formed of granite, flanked by schistose and sedimentary strata. The strata that he observed in Russia were divided into Primitive, Secondary and Tertiary categories.

Pallas wrote metaphorically:

> These two orders of mountains [Secondary and Tertiary] offer the most ancient chronicle of our globe, least liable to falsifications and at the same time more legible than the writing of the primitive ranges. They are Nature's archives, prior to even the most remote records and traditions that have been preserved for our observant century to investigate, comment on and bring to the light of day, and which

will not be exhausted for several centuries after our own.[59]

The character of this passage provides further evidence for the eighteenth century view that the work of the natural historian was analogous to the work of the literary historian, and the belief that one could, in principle, establish an account of the Earth's past by means of empirical enquiry. But we have here a history based in large measure upon a particular directionalist *theory,* rather than piecing together of individual items of stratigraphical evidence. There were strong 'genetic', as opposed to 'historical', elements in Pallas's approach.

Let us turn now to the work of A. G. Werner and the school of geognosy that he established at Freiberg. In his early work of 1774, *On the external characters of minerals,*[60] Werner chose to give a classification using chiefly external characteristics as taxonomic criteria—a method that accorded well with typical eighteenth century mineral taxonomies such as that of Linnaeus. But in Werner's later *Short classification* of 1787[61] an altogether different programme of enquiry was put forward, stemming from the stratigraphical investigations of men such as Lehmann and Füchsel, discussed above.

In the *Short classification,* minerals (or rather rocks) are ranged in what is taken to be their chronological sequence, which was to be determined by recourse to the examination of strata in the field. A. M. Ospovat has published a version of Werner's sequence of the younger strata in Germany.[62] Other sequences illustrating Werner's stratigraphical column may be found in various publications.[63] For example the Primitive mountains (*Urgebirge*) were supposedly represented by the following sequence:[64]

1. Granite.
2. Gneiss.
3. Micaceous schist.
4. Argillaceous schist.
5. Primitive limestone.
6. Trap.
7. Porphyry.
8. Syenite.
9. Serpentine.
10. Topaz rock.
11. Quartz.
12. Siliceous schist.

Such lists were in some measure direct records of field observations, interpreted in the light of the very general principle of superposition. But they were also underpinned by the speculative 'genetic' theory of the precipitation of a set sequence of mineral substances from a hypothetical universal ocean.

From Werner's personal notes, unpublished in his lifetime, Ospovat has published the following significant remark, from a manuscript giving a clarification by Werner of what he was seeking to achieve in writing his *Short classification*:

> I had to be guided completely in the classificatory presentation or tabulation of these [rock] masses by the discoverable time sequence of the particular formations if I wanted to remain true to my plan to sketch through this classification a foundation for a complete canvass of the universal formation of these masses.[65]

Werner, therefore, was looking for a "discoverable time sequence" as a basis for the classification of rocks, which sequence was to be determined by empirical investigations, with spatial distributions of strata serving as indicators of a temporal sequence of events. And, like Füchsel, Werner attempted to read backwards from the observed present conditions of rocks to their conditions at their time of deposition.

All this may suggest that Werner's geognosy was basically 'historical', or 'historicist' in character, and Ospovat has laid particular emphasis upon Werner's role as a "historical geologist".[66] He and his students could use the stratigraphical sequence of Saxony as the key to the understanding of strata in all parts of the globe, thereby providing a paradigmatic basis for a fruitful programme of geological (or geognostical) enquiry. His pupils, such as von Buch, von Humboldt and Jameson, constantly emphasized this as the underpinning of their inquiries, giving them a basis for investigation in all parts of the globe.[67] Werner himself clearly considered that his programme was one of the highest significance, leading to the establishment of a science worthy of being dignified by the use of the new term 'geognosy'. Under the aegis of the Wernerian paradigm the old 'flat' mineral tables of the earlier part of the eighteenth century were given a depth of perspective by the addition of the dimension of time.[68] The earlier studies of the natural history of the mineral kingdom, coupled with the cosmogonical systems of the eighteenth century, became transmuted into geognosy, or—as we might say—into historical geology, meshing well with the intellectual temper of the end of the century, thereby accounting very satisfactorily for the remarkable success of the Freiberg school.

But a number of cautionary remarks must be added if we are to avoid a simple-minded acceptance of an apparently plausible schema. It has been suggested by Nieuwenkamp, for example, that "chronology was of little importance for the Wernerians, and indeed was wilfully neglected by them".[69] And in fact it may be that in *practice* geognosy meant primarily the working out of spatial relations and the geometrical ordering of rock suites, rather than the elucidation of the globe's *history*. A fair amount of evidence can be brought forward in support of such a contention.

For example, Werner visited Delamétherie in Paris in 1802, and the editor of the *Journal de physique* promptly took the opportunity of informing his readers of his understanding of the German master's theoretical opinions. Thus we have the following statement:

> Geognosy (which we call geology), according to [Werner], is that part of mineralogy which, considering minerals as portions of the globe, chiefly deals with their layering, and their spatial relationships, one to another.[70]

It is true that Delamétherie continued immediately:

> Observations with this objective [in mind] can lead us to certain ideas concerning the formation of minerals and even of the terrestrial globe [itself].[71]

But this remark is promptly qualified by the statement:

> But everything that is hypothetical is absolutely banned. Geognosy is nothing but generalized facts, and the immediate consequences which he [Werner] believes one cannot but draw from them.[72]

From this, we may be led to conclude that the objective of the followers of the geognostic paradigm was primarily the elucidation of the spatial ordering of the strata, thus providing a new means for establishing a system of taxonomy.

Further evidence on the question may be sought by examining some mineral catalogue arranged according to avowed Wernerian principles. The so-called Leskean collection, arranged by Werner's pupil, D. L. G. Karsten, and still held in Dublin to this day, provides a useful source for English readers.[73]

Werner thought that there were five broad divisions of mineralogy: a knowledge of external characters (oryctognosy), chemical mineral classification, the Earth's internal structure (geognosy or geology), mineral geography, and economic mineralogy. A cabinet, therefore, should display specimens illustrating all these facets of the science. In the Leskean catalogue, the third collection is described as being one in which "minerals are arranged according to their position and relative situation in the internal structure of the Earth".[74] This does not stress the temporal aspect of the scheme, and in any case it is stated that the second part of the total collection—that concerned with mineral classification—was "the most important of all".[75] We are told that "in this Collection the articles are disposed in genera and species according to the nature and proportion of the ingredients in their composition, but the varieties of each species are disposed according to external characters".[76] There is no question of an historical element entering into the taxonomy at this point. Rather, Füchsel's concept of *'situs'* or *'Lager'* seems to have provided the main focus of attention.

As a further consideration we may note what P. Neill, the translator and editor of d'Aubuisson de Voisins's work on the basalts of Saxony, had to say about "geognostic character". He wrote:

> Although Werner has depended principally on the external characters, he does not, for the determination of his species, confine himself to them alone; but likewise employs, particularly what he terms *geognostic character,* that is the relation in respect to position, which the mineral bears to the accompanying substances.[77]

By such an approach, one might hope to evade the difficulties raised by the revelation of von Buch and d'Aubuisson that certain basalts were indubitably produced by the action of volcanoes.

The point that emerges from these discussions is that although there was certainly a temporal component to Wernerian thinking it was by no means the sole, or even the most important, consideration. Werner displayed the typical eighteenth century passion for order, and perhaps his chief intention was to determine a natural taxonomy of the mineral kingdom. This, in one of its aspects, involved a classification with an essentially historical base, but this was by no means the exclusive or even the principal concern. It is also worth noting that Goethe, who studied for a time under Werner and adopted a theory much beholden to the Neptunist doctrine, did *not* see the Earth as a record of historical development. Goethe attempted to display the unity and continuity of the mineral kingdom, and thinking in static, archetypal terms he constructed a series of "transitions", from granite, through porphyry and pudding-stone to sandstone, and also from granite through gneiss to micaschist. But in the opinion of G. A. Wells, Goethe's "geological thinking clearly lacked the crucial . . . [concept] of time".[78]

Pursuing the tradition of the Wernerian school, one should consider the ideas of perhaps his most celebrated pupil, Alexander von Humboldt, who studied with Werner for just eight and a half months in 1791/92. It was from von Humboldt that Cuvier acquired much of his information respecting Werner when he prepared his *Éloge;* and Geikie seemingly drew heavily on this source when writing his *Founders of geology.* Consequently, it might be said that von Humboldt has become the chief source of information about Werner for non-professional English-speaking readers of the history of geology.

The text of von Humboldt's manuscript was published some years ago by William Coleman. The following passage is relevant to our concerns.

> The most brilliant part of M. W[erner's] work is without doubt that which concerns the structure of the Earth, and the history of rocks and of formations. Geology teaches two distinctly different things: the way in which rocks are superposed one on another and the causes which have brought about this order of things. This latter more problematical and uncertain part constituted, so to speak, geology before M. W[erner's] time. It was a succession of day-dreams on comets which caused deluges and conflagrations of the surface of the globe. M. W[erner] was the first to bring home the fact that before trying to know the origin of things it is necessary to know the arrangement that they display in all the most distant countries, the superposition of beds and their relative ages.[79]

Again, we may note the emphasis on a knowledge of the structural ordering of strata as a prerequisite for eliciting an understanding of the Earth's past.

But the idea which particularly appealed to von Humboldt was that of the *universality* of the Wernerian methodology:

> It is to M. W[erner] that we owe the fundamentally important idea of the identity of certain formations over the whole surface of the globe, and the regular succession according to which the different rocks are to be found emplaced.[80]

It was this theme that von Humboldt explored at length in his *Geognostic essay on the superposition of rocks.* He noted that miners had customarily been in possession of local knowledge of the successions of particular formations and had believed that "each portion of the globe differed in geological constitution".[81] However, von Humboldt maintained it was the recognition of a *general* structure to the constitution of the Earth's crust that marked the transition from the art of the miner to the science of geognosy. It was the search for a *general* order of succession of rock formations, and the attempted confirmation of this order in all parts of the globe that was the hallmark of the Wernerian programme of inquiry, as construed by von Humboldt.

The task which von Humboldt and Werner's other disciples undertook was certainly daunting. Von Humboldt deliberately chose not to use palaeontological criteria in his ordering of the Earth's strata, for such a system would necessarily entail the omission of the many unfossiliferous formations that the Wernerians had for long considered of the greatest importance. And because it was not possible in practice to identify a wholly uniform sequence of formations throughout the globe, in certain instances von Humboldt had to attempt to identify what he called '*parallels*' or '*geognostic equivalents*'.[82] Moreover, the guiding principle that had underpinned the earlier investigations of Werner—that all formations of a particular kind or constitution were produced at a particular epoch—could no longer be sustained in the light of the research conducted in the early years of the nineteenth century. Nevertheless, the broad principle of a single very general order of formations was retained, and was considered by the writers of the period to be of essential significance. The matter was stated clearly by W. H. Fitton in the following words:

> The exterior of the earth consists of a series of . . . formations, laid over each other in a certain determinate order. Not that the whole series is anywhere complete; but that the relative place of its several members is never departed from. Thus in the series A, B, C, D, it may happen that B and C, or both, may be occasionally wanting, and consequently D may be found immediately above A; but the succession is *never* violated, nor the order inverted, by the discovery of A above the formations B, or C, or D, nor of B above those that follow it, &c.[83]

It is noteworthy that von Humboldt had hopes of reducing his scheme to a comprehensible system in three ways, either by presenting a simple written list of the strata (as was given, for example, in his *Geognostic essay*),[84] by presenting a tabular idealized cross-section of the Earth's strata,[85] or by means of a system of symbols—which he referred to as a "*pasigraphical method*".[86]

In this list, the main formations were represented in the following manner:

| | |
|---|---|
| alpha | Granite |
| beta | Gneiss |
| gamma | Mica-slate |
| delta | Clay-slate |
| omicron | Ophiolite, etc. |
| sigma | Syenite, etc. |
| pi | Porphyry |
| tau | Calcareous and gypseous formations |
| chi | Clastic rocks |
| xi | Coal |
| zeta | Rock-salt |

Various species of these principal types might be indicated by appropriate superscripts (e.g., $\tau^{o}$=oolitic limestone). The total stratigraphical sequence might then be represented in the compass of a few lines:[87]

$$\alpha,\ \alpha\beta,\ \beta+\pi,\ \beta\gamma,\ \gamma+\tau,\ \alpha,\ \gamma,\ \delta,\ \alpha,\ \beta,\ \delta,\ o\ ||\ \varkappa^{s},\ \tau',\ \delta\tau',\ \delta',$$
$$\delta'+\pi,\ \gamma,\ \tau',\ \sigma\pi,\ \sigma+\alpha,\ \sigma\pi,\ o\ ||\ \pi\varkappa^{a}+\xi,\ \tau^{a}+\vartheta,\ \varkappa^{n},\ \tau^{m},\ \varkappa^{q},\ \tau^{o},$$
$$\varkappa^{1},\ \tau^{c}\ ||\ \varkappa^{z1},\ \tau^{p}.\ .\ .\ .$$

Such a system had the great advantage of making appropriate acknowledgement of various transition types, or recurrences of formations at various points in the ordered arrangement, and it also allowed the ready comparison of successions in various parts of the world.

So far as I am aware, von Humboldt's system of pasigraphy did not achieve any wide degree of acceptance. It may be said, however, that it was a precursor of the general system of representation of rock units on geological maps by means of letters and subscripts or superscripts, as well as by colours. Moreover, it is interesting to note that the scheme paralleled the closing phases of the search for a general philosophical language, which was still being sought in the late eighteenth century.[88] It may be said, then, that von Humboldt's geognostic nomenclature was a clear reflection of the general linguistic and philological preoccupations of his contemporaries.

Let us consider now the extent to which a scheme such as that of von Humboldt should be construed as being historical in character, and representative of an historicist outlook. Evidently, there was a temporal and directionalist approach to the scheme. However, in so far as it represented the view that there was a unique system of universal events —linked to the theory of hypothetical oceans—there was still present an element of the genetic thinking of the eighteenth century, with an attempt to display a temporalized order of nature. The regular succession of strata seems to have represented an expression of the law-like characteristics of the Earth's structure and history. As such, it does not cohere with or exactly parallel the multiplicity and individuality of human characteristics as viewed by the nineteenth century historicists.

But whether it arose from pressures exerted by the historicist *Weltanschauung* of the nineteenth century, the effects of the greatly expanded field of vision brought about by the many geological travels and explorations of the period, or a combination of both, it seems that in the attempt to piece together the order of strata in various places by the identification of parallel formations and other devices, one was, in fact, moving closer to an historicist approach. The temporal element drawn from the eighteenth century cosmogonies was modified to yield an historical picture of the Earth's crust, this being achieved by the painstaking accumulation and assemblage of items of evidence from all quarters of the globe—as if one were solving an historical jig-saw puzzle. And this process seems to have grown quite naturally from its predecessor, the Wernerian geognostic system.

*(Part 2 of this article begins on p. 227 of our December issue.)*

REFERENCES

1. R. G. Collingwood, *The idea of history* (Oxford, 1946), 88-93.
2. H. Butterfield, *Man on his past: The study of the history of historical scholarship* (Cambridge, 1955), 39-61, and *passim.*
3. On this, see for example, K. M. Clark, *The Gothic revival: An essay in the history of taste* (3rd edn, London, 1962).
4. Mendelssohn's revival of the compositions of Bach is well known, but one may readily identify earlier examples of a rising enthusiasm for music of bygone times. It was, for example, evidenced by the series of so-called "Ancient concerts", presented in London in the late eighteenth century. See: C. Mackeson, "Ancient concerts", in E. Blom (ed.), *Grove's dictionary of music and musicians* (5th edn, 9 vols, London and New York, 1954), i, 144-5.
5. See V. Duckles, "Johann Nicolaus Forkel: The beginnings of music historiography", *Eighteenth century studies,* i (1968), 277-90. But Forkel's position can not be regarded as that of an historicist. He was, as Duckles puts it, "a true man of the Enlightenment".
6. See G. Daniel, *The origins and growth of archaeology* (New York, 1968).
7. E.g., F. C. Haber, *The age of the world: Moses to Darwin* (Baltimore, 1959), ch. 2 ("Historicism and the scientific revolution"); G. G. Simpson, "Uniformitarianism. An inquiry into principle, theory, and method in geohistory and biohistory", in M. K. Hecht & W. C. Steeres (eds), *Essays in evolution and genetics in honor of Theodosius Dobzhansky* (New York, 1970), 43-96 (republished in C. C. Albritton (ed.), *Philosophy of geo-history: 1785-1970* (Stroudsburg, 1975), 256-309).
8. For a discussion of the various usages, see D. E. Lee & R. N. Beck, "The meaning of historicism", *American historical review,* lix (1965), 568-77.
9. For a sketch of the etymology of the terms 'historicism', 'historism' and '*Historismus*', see G. G. Iggers, *The German conception of history: The national tradition of historical thought from Herder to the present* (Middletown, 1968), 287-90.
10. G. G. Simpson, *op. cit.* (ref. 7, 1975), 279-85.
11. K. R. Popper, *The poverty of historicism* (London, 1957).
12. J. K. Feibleman, in D. D. Runes (ed.), *Dictionary of philosophy* (Paterson, 1956), 127.
13. I do not wish to maintain dogmatically that the covering-law theory never finds application in historical explanation. But I find the description of historians' work given by John Passmore ("Explanation in everyday life, in science, and in history", in G. H. Nagel (ed.), *Studies in the philosophy of history* (New York, 1965), 16-34), emphasizing the historian's use of unique and individual events in their explanations, more appropriate.
14. There are published precedents for the distinction that I wish to make between 'historical' and 'genetic' explanation. (See, for example, I. F. Knight, *The geometric spirit: The Abbé de Condillac and the French Enlightenment* (New Haven and London, 1968), 28.) Nevertheless, many writers do not make a clear-cut distinction between the two terms.

15. See G. G. Iggers, *op. cit.* (ref. 9), and also his article "Historicism", in P. P. Wiener (ed.), *Dictionary of the history of ideas: Studies of selected pivotal ideas* (5 vols, New York, 1973-74), ii (1973), 456-64.
16. See H. Kantorowicz, "Savigny and the historical school of law", *The law quarterly review,* liii (1937), 326-43.
17. Berlin, 1814
18. F. C. von Savigny, *Of the vocation of our age for legislation and jurisprudence, translated from the German . . . by A. Hayward* (London, 1831), 134-5.
19. C. G. Rand, "Two meanings of historicism in the writings of Dilthey, Troeltsch, and Meinecke", *Journal of the history of ideas,* xxv (1964), 503-18, p. 511.
20. R. G. Collingwood, *op. cit.* (ref. 1), pt iii, 86-133.
21. This view is maintained, for example, in J. H. Brumfitt, *Voltaire: Historian* (Oxford, 1958).
22. E. Cassirer, *The philosophy of the Enlightenment* (Princeton, 1951), 197.
23. A. -N. de Condorcet (tr. J. Barraclough), *Sketch for a historical picture of the progress of the human mind* (London, 1955), 9—emphasis added (1st French edn, 1795). (We may note here also a remark of J. S. Mill: "The disrespect in which history was held by the *philosophes* is notorious; one of the soberest of them, D'Alembert we believe, was the author of the wish that all record whatever of past events could be blotted out" (*The London and Westminster review,* xxxiii (1839), 277).
24. I. Kant, *Allgemeine Naturgeschichte und Theorie des Himmels* (Königsberg, 1755).
25. C. Linnaeus, *Oratio de telluris habitabilis incremento . . .* (Leyden, 1744).
26. See P. H. Reill, *The German Enlightenment and the rise of historicism* (Berkeley, 1975).
27. The title of a well-known paper by C. J. Schneer ("The rise of historical geology in the seventeenth century", *Isis,* xlv (1954), 256-68) is suggestive in this context.
28. N. Steno, *The prodromus of Nicolaus Steno's dissertation concerning a solid body enclosed by process of nature within a solid . . .* (New York & London, 1968), 230.
29. R. Waller (ed.), *The posthumous works of Robert Hooke, M.D. S.R.S. Geom. Prof. Gresh., etc., containing his Cutlerian Lectures, and other discourses, read at the meeting of the illustrious Royal Society* (London, 1705; facsimile, New York & London, 1969), 411.
30. For discussions of this, see D. R. Oldroyd, "Robert Hooke's methodology of science as exemplified in his 'Discourse of earthquakes' ", *The British journal for the history of science,* vi (1972), 109-30; and A. J. Turner, "Hooke's theory of the Earth's axial displacement: Some contemporary opinions", *The British journal for the history of science,* vii (1974), 166-70.
31. A. P. Rossiter, writing in 1935, believed that Hooke's later work, with its increasing interest in Biblical chronology, compared unfavourably with the "Discourse" of 1668, and he ascribed the change in Hooke's interests and literary style to "personal misfortunes" (A. P. Rossiter, "The first English geologist: Robert Hooke (1635-1703)", *Durham University journal,* xxix (1935), 172-81). I do not find this suggestion particularly convincing.

32. On millenarian doctrines in Britain in the seventeenth century, see particularly C. Webster, *The intellectual revolution of the seventeenth century* (London, 1974), and M. C. Jacob, *The Newtonians and the English revolution* (London, 1976).
33. E.g., N. Steno, *op. cit.* (ref. 28), 266.
34. *Ibid.*, 263.
35. *Ibid.*, 263-70.
36. See, for example W. Camden, *Camden's Britannia, newly translated into English, with large additions and improvements, publish'd by Edmund Gibson* (London, 1695).
37. W. R. Albury & D. R. Oldroyd, "From Renaissance mineral studies to historical geology, in the light of Michel Foucault's *The order of things*", *The British journal for the history of science*, x (1977), 187-215. However, some further remarks about Buffon appear below.
38. A. L. Moro, *Dei crostacei e degli altri corpi marini che si trovano sui monti* (2 vols, Venice, 1740), ii, 428.
39. *Ibid.*, 430-32.
40. G. Arduino, "Due lettere del sig. Giovanni Arduino sopra varie sue osservazioni naturali", *Nuova raccolta d'opuscoli scientifici e filologici*, (n.s.) vi (1760), 97-132, 133-80.
41. Quotation from G. Stegagno, *Il Veronese Giovanni Arduino e il suo contributo al progresso della scienza geologica* (Verona, 1929), 24.
42. *Ibid.*, 29. (This passage is from an unpublished memoir from Arduino to Ferber.)
43. The original of this is held in the Verona Public Library, where the Arduino papers are to be found.
44. For a discussion of some of the theoretical antecedents of the Wernerian school, see my paper, "Some phlogistic mineralogical schemes, illustrative of the evolution of the concept of 'earth' in the 17th and 18th centuries", *Annals of science*, xxxi (1974), 269-305.
45. J. G. Lehmann, *Versuch einer Geschichte von Flötz-Gebirgen* (Berlin, 1756); French translation (which I have used), *Essai d'une historie naturelle des couches de la terre, dans lequel on traite de leur formation, de leur situation, des minéraux, des métaux & des fossiles qu'elles contiennent; avec des considérations physiques sur les causes des tremblemens de terre & de leur propagation . . .* (Paris, 1759). This formed the third volume of J. G. Lehmann, *Traités de physique, d'histoire naturelle, de minéralogie et de métallurgie* (3 vols, Paris, 1759).
46. J. G. Lehmann, *op. cit.* (ref. 45); *Essai*, 304.
47. *Ibid.*, 326.
48. *Ibid.*, 212-13.
49. See F. D. Adams, *The birth and development of the geological sciences* (New York, 1954), 376; or W. B. N. Berry, *Growth of a prehistoric time scale based on organic evolution* (San Francisco & London, 1968), 31.
50. G. C. Füchsel, "Historia terrae et maris ex historia Thuringiae, per montium descriptionem, eruta . . .", *Acta Academiae electoralis moguntinae scientiarum utilium, quae Erfordiae est*, ii (1761), 44-208; and "Eiusdem usus historiae suae terrae et maris", *ibid.*, 209-54.

51. B. Hansen, "Füchsel, Georg Christian", in C. C. Gillispie (ed.), *Dictionary of scientific biography,* v (New York, 1972), 205. V. A. Eyles ("Mineralogical maps as forerunners of modern geological maps", *The cartographic journal,* ix (1972), 133-5) has pointed to examples, such as those of Guettard, that were somewhat earlier than those of Füchsel, but strictly they were mineralogical rather than geological maps.
52. For a discussion of the meaning of this term, see A. M. Ospovat, "Reflections on A. G. Werner's 'Kurze Klassifikation' ", in C. J. Schneer (ed.), *Toward a history of geology* (Cambridge, Mass. & London, 1969), 242-56, p. 250.
53. G. C. Füchsel, *op. cit.* (ref. 50), §43.
54. *Ibid.,* §50.
55. *Ibid.,* §63.
56. P. S. Pallas, "Observation sur la formation des montagnes & les changemens arrivés au globe, particulièrement à l'égard de l'Empire de Russe; par M. P. S. Pallas lues le 23 Juin 1777 à l'assemblée de l'Académie Impériale des Sciences, que Monsieur le Comte de Gothland daigna illustrer de sa présence", *Acta Academiae scientiarum imperialis Petropolitanae pro Anno MDCCLXXVII. Pars prior* (St Petersburg, 1778), 21-64.
57. *Ibid.,* 24.
58. For an English account of Linnaeus's geological work, see A. G. Nathorst, "Carl von Linné as a geologist", *Annual report of . . . the Smithsonian Institution . . . for the year ending June 30 1908* (Washington, 1909), 711-43. On Bergman, see H. D. Hedberg, "Influence of Torbern Bergman (1735-1784) on stratigraphy", *Acta Universitatis Stockholmiensis, Stockholm contributions in geology,* xx (1969), 19-47. Linnaeus favoured a generalized five-fold division of the Earth's strata. Bergman's four-fold division is regarded as one of the chief sources of Werner's better known stratigraphical taxonomy.
59. P. S. Pallas, *op. cit.* (ref. 56), 46.
60. A. V. Carozzi (trans. and ed.), *On the external characters of minerals by A. G. Werner . . .* (Urbana, 1962).
61. A. G. Werner, *Short classification and description of the various rocks, translated with an introduction and notes by Alexander M. Ospovat* (New York, 1971).
62. *Ibid.,* 128-9.
63. E.g., *Des Herrn Nathaniel Gottfried Leske . . . hinterlassenes Mineralienkabinet systematisch geordnet und beschreiben auch mit vielen wissenschaftlichen Anmerkungen und mehreren äussern Beschriebungen der Fossilien begleitet, von Dietrich Ludwig Gustof Karsten . . .* (Leipzig, 1789); J. F. Daubuisson, "De la classification orictognostique des minéraux, par Werner . . .", *Journal de physique,* liii (*an* 9), 448-75; J. F. Daubuisson, "Tableau de la classification des minéraux; par M. Werner, Conseiller des mines de la Saxe, etc.: accompagné de quelques observations sur les travaux de ce minéralogiste; par M. d'Aubuisson", *Journal de physique,* lx (*an* 13), 171-8, 329-39.
64. This version is extracted from "Division et classification des montagnes et roches d'après le conseiller Werner; rédigée par MM. Hoffman et Eslinger, élèves de Werner", *Journal de physique,* i (*an* 8), 473-6.
65. A. M. Ospovat, *op. cit.* (ref. 61), 20.

66. A. M. Ospovat, "The work and influence of Abraham Gottlob Werner: A re-evaluation", *Proceedings of XIIIth International Congress of the History of Science, Moscow, August 18-24, 1971*, viii, 123-30, p. 127.

67. L. von Buch, "Ueber den Gabbro, mit einigen Bemerkungen über den Begriff einer Gebirgsart", *Der Gesellschaft naturforschender Freunde zu Berlin: Magazin für die neuesten Entdeckungen in der gesammten Naturkunde*, v (1810), 128-49; F. W. K. H. A. von Humboldt, *A geognostic essay on the super-position of rocks, in both hemispheres . . . translated from the original French* (London, 1823), 3, 8, and *passim;* R. Jameson, *System of mineralogy, comprehending oryctognosy, mineralogical chemistry, mineralogical geography, and economical mineralogy* (3 vols, Edinburgh, 1804-8), iii, 67-73. Jameson entitled his fourth chapter: "General account of the different formations, in regard to their succession and stratification, and this is illustrated by a short description of the Harz and the Saxon Erzgebirge."

68. For further discussion of this point, see W. R. Albury & D. R. Oldroyd, *op. cit.* (ref. 37).

69. W. Nieuwenkamp, "Buch, [Christian] Leopold von", in C. C. Gillispie (ed.), *Dictionary of scientific biography*, ii (New York, 1970), 552-7, p. 552.

70. J. C. Delamétherie, "Idées de Werner sur quelques points de la géognosie. Extraits de ses conversations; par J. -C. Delamétherie", *Journal de physique*, lv (*an* 10), 443-50, p. 444.

71. *Ibid.*

72. *Ibid.*

73. It is described in *A description of the minerals in the Leskean Museum. By Dietrich Ludwig Gustavus Karsten . . . . Translated by George Mitchell . . . . Vol. I. Containing the characteristic and systematic collections . . . Vol. II. Containing the geological, geographical and economical collections* (London & Dublin, 1798).

74. D. L. G. Karsten, *op. cit.* (ref. 73), vi.

75. *Ibid.*, v.

76. *Ibid.*, v.

77. J. F. d'Aubuisson, *An account of the basalts of Saxony, with observations on the origin of basalt in general . . . . Translated, with notes, by P. Neill, . . . secretary to the Wernerian Natural History Society* (Edinburgh & London, 1814), xi-xii.

78. G. A. Wells, "Goethe, Johann Wolfgang von", in C. C. Gillispie (ed.), *Dictionary of scientific biography*, v (New York, 1972), 442-6, p. 444. For discussions of Goethe's geological work, see R. Magnus, *Goethe as a scientist* (New York, 1949), 151-65; M. Semper, *Die geologischen Studien Goethes* (Leipzig, 1914); G. A. Wells, "Goethe's geological studies", *Publications of the English Goethe Society,* (n.s.) xxxv (1964-65), 92-137; R. Trümpy, "Goethes geognostisches Weltbild", in *Eidgenössische technische Hochschule, kultur und staatswissenschaftliche Schriften*, no. 127 (1968), 1-37.

79. W. Coleman, "Abraham Gottlob Werner vu par Alexander von Humboldt, avec des notes de Georges Cuvier", *Sudhoffs Archiv für Geschichte der Medizin und der Naturwissenschaften*, xlvii (1963), 465-78, pp. 473-4.

80. *Ibid.*, 474.

81. F. W. K. H. A. von Humboldt, *op. cit.* (ref. 67), 3.
82. *Ibid.*, 17.
83. [W. H. Fitton], "Art. IV. Transactions of the Geological Society, established November, 1807. Vol. III. 4to. pp. 444. W. Phillips, London, 1816", *Edinburgh review,* xxix (1817), 70-94, p. 71.
84. F. W. K. H. A. von Humboldt, *op. cit.* (ref. 67), 461-4.
85. Something of this kind was presented in von Humboldt's paper, "Esquisse d'un tableau géologique de l'Amérique méridionale, par F. A. Humboldt", *Journal de physique,* liii (*an* 9), 30-60, p. 60.
86. F. W. K. H. A. von Humboldt, *op. cit.* (ref. 67), 465.
87. *Ibid.*, 476.
88. See J. Knowlson, *Universal language schemes in England and France: 1600-1800* (Toronto & Buffalo, 1975), ch. 6 ("Pasigraphy in the 1790s").

# XII b

## HISTORICISM AND THE RISE OF HISTORICAL GEOLOGY, PART II

### FRANCE AND SWITZERLAND

Let us turn now to consider some of the geological work that was carried out in France and Switzerland in the eighteenth and early nineteenth centuries. As far back as 1748, in the work of de Maillet,[89] we find a distinction drawn between Primitive and Secondary mountains, justified on the basis of the hypothesis of the gradually receding ocean. But it was this theoretical treatment, linked with an essentially Cartesian schema,[90] which clearly distinguished de Maillet's system from historicist explanations of the state of the globe. The overall cyclic theory of the Earth, on which is imposed a series of six stages, found its origin in the Cartesian theory of vortices rather than the painstaking results of geological field work. The *Telliamed* cannot be represented as an history of the globe.

Somewhat similar remarks can be made about the theories of Buffon, though some commentators such as Rhoda Rappaport[91] have seen the publication of the *Théorie de la terre* in 1749[92] as marking a major watershed in the development of geological thinking in France during the eighteenth century. I shall not give here a general exposition of Buffon's three major treatments of mineralogical and geological matters.[93] Rather, let us address ourselves to the question of whether Buffon may be thought of as representing an historicist outlook in any significant way.

It is true that in some passages in his *Théorie* Buffon explicitly referred to himself as an "historian",[94] and the second discourse of the work as a whole is significantly entitled "Histoire et théorie de la terre". But the "history" is far from historicist in character and "theory" is all important. Buffon postulated a general flow of the oceans from east to west,[95] which would shape mountains and build up new sedimentary layers. This idea was not, however, supported by reference to detailed observations of particular strata, piecing items of evidence together so as to develop a system of historical explanation. Rather, the hypotheses were sustained by subsumption under a general cosmic theory, a description of which formed the section entitled "Preuves de la théorie de la terre".[96] And, excluding the discussions of the actual formation of the planetary system, there is no forward directionalist march. Indeed, Buffon explicitly stated that the strata follow no general rules of arrangement,[97] and Jacques Roger has gone so far as to say that "the *Théorie de la terre* is situated outside time altogether".[98] However, Roger sees the addition of the idea of a slow progressive cooling of the Earth (such as one finds in the *Époques de la nature*) as giving a directionalist aspect to the total scheme, and this he sees as providing a necessary ingredient required for an historically oriented theory.[99]

But this does not mean that the *Époques* can be construed as an historicist work. It is true that Buffon offered a kind of 'absolute' chronology for the cooling Earth, and he significantly broadened the time scale within which the student of the Earth's past might frame his account.[100] Moreover, the very opening sentences of the *Époques* may incline us to interpret Buffon's thinking as being historicist in character:

> Just as in civil history one refers to titles, looks for medals, or deciphers ancient inscriptions, in order to work out the epochs of human revolutions and establish the dates of intellectual events, so also in natural history it is necessary to rummage through the archives of the world, extract ancient monuments from the interior of the earth, collect together their remains, and assemble in a body of proofs all the tokens of physical change which enable us to trace back to nature's various [former] ages.[101]

But at best this passage can only forge a link with the historiographical traditions of Buffon's contemporary France;[102] and these were by no means antagonistic towards natural law doctrines, or favourable towards an historicist mode of explanation. Buffon's arguments about the Earth's past and present circumstances were underpinned by a speculative cosmogonical system and experiments on the law-like cooling of iron spheres, rather than the piecing together of items of evidence concerning the dispositions of individual strata.

Roughly contemporary with the work of Buffon was that of G. F. Rouelle, who included in his chemical lectures extensive discussions of minerals and strata. As is well known, many of Rouelle's theoretical opinions were never formally published, though they may be ascertained by examining copies of his students' lecture notes, or by looking at the account that was given of them by Desmarest in the *Encyclopédie méthodique*.[103] This situation, needless to say, makes it difficult to supply a precise date for the formulation of Rouelle's opinions. There appears, however, to be some indication of the influence of Lehmann on his thinking,[104] so that one can probably say that Desmarest's exposition gives an expression of Rouelle's ideas as they stood in the 1760s.

Rouelle distinguished between rocks of the *"ancienne terre"* and those of the *"nouvelle terre"*.[105] Respecting the latter, he recognized that the fossils they contained were not distributed randomly but that they commonly occurred in particular groupings.[106] He proposed that beds should be named after their commonest fossils.[107] Possibly, the varying associations of fossils might be an indication of climatic changes, attributable to slow changes in the polar axes.[108] Possibly also, there had been "revolutions" of the globe which would account for alterations in the relative distributions of land and sea. In all this we certainly have some of the elements that characterize an historicist approach to studies of the Earth. But Rouelle continued to search for an explanatory theory; he sought a law-like explanation of what was observed.

If we wish to identify the first introduction of the historicist approach to strata in France, in which one finds an attempt to build up the history of an area by a consideration of geometrical interrelationships of individual strata, one may refer to the stratigraphical studies of Lavoisier,[109] or the investigations of Desmarest in the Auvergne district.[110] Guettard does not, I think, qualify for this 'accolade'.[111] In fact, he explicitly stated that it was better to attempt to know all about the *composition* of the Earth, rather than its *formation*.[112]

According to Rhoda Rappaport,[113] "Lavoisier's long-range plan . . . [was] to write not a traditional 'theory of the earth', but rather a geological history of France in which he would examine the state of the crust during successive epochs".[114] But in fact such a project was never accomplished. As is well known, Lavoisier did make important contributions to a geological atlas of France, establishing certain 'type-sections' as a means to this end. But as far as geological publications were concerned, his chief contribution was his discussion of processes and conditions of sedimentation. The promise of a geological or stratigraphical history remained unfulfilled.

So we turn to Desmarest. He, I think, can be said to have traced out an historical sequence of geological events in the Auvergne district, though he was chiefly concerned with the somewhat special case of lava flows, rather than layers of sedimentary strata. Desmarest divided the lavas of the Auvergne into three epochs, according to their relative positions and degrees of weathering and erosion. With the aid of this schema, he was able to draw up an historical sequence of events for the volcanoes and lava flows of the Auvergne, and prepare corresponding geological maps.[115] But in his late memoir of 1806, he maintained that the tripartite division of strata might serve as a general exemplar for work in other parts of the globe,[116] which meshes with the investigations undertaken by von Humboldt at about the same period. A general patterning of the Earth's strata on the basis of a particular area was envisaged.

Desmarest's mature viewpoint may usefully be contrasted with the opinions expressed in his much earlier article "Géographie physique", in Diderot's *Encyclopédie.*[117] (The *Encyclopédie* as a whole contained no article on geology *per se,* which was represented by the essay on physical geography.) On examining this article one does not find any great concern with the history of the globe. "What must be examined?", Desmarest asks rhetorically, and then provides the following reply: "It is necessary to be concerned with external configurations and obviously conspicuous shapes."[118] From an examination of the externals of continents, seas, mountains, beds of fossils, etc., general patterns might begin to emerge, which could be regarded as "the essence drawn from an abundant foundation of observations".[119] Desmarest does consider moving from the present well-known state of affairs to the earlier conditions of things. This, he says,[120] constitutes a study of theories of the Earth, but it is a subject that should be clearly distinguished from physical geography. It appears, then, that in this early publication of Desmarest we do not find a strictly historical system of explanation, such as may be found in the memoirs on the Auvergne. This historical approach must have been assimilated into his methodology during the 1760s and early 1770s. It is interesting to note that Desmarest, writing in an article for the *Encyclopédie méthodique* for 1811, overtly compared natural history with civil history, which he explicitly considered to be something at a higher level of abstraction than mere tables of chronology.[121] This effects a link between his views and the historiographical views of the early nineteenth century, though as we have seen the link or analogy between terrestrial and civil history had been envisaged on numerous occasions since the seventeenth century.

Let us turn now to consider the figures of de Luc and de Saussure in Geneva. Here the connection between geological investigations and the rising interest in nature and mountain scenery becomes particularly significant,[122] and it is natural to look for links with Rousseau,[123] himself intimately associated with the beginnings of the Romantic movement in the French-speaking part of Europe. It is known, for example, that de Luc was personally acquainted with Rousseau, and in his *Lettres physiques et morales* of 1778 de Luc explicitly referred to the influence of Rousseau on his thinking, and the devotion for mountain scenery that it inspired.[124] The *Lettres* are permeated by a certain idyllic sense of euphoria, very likely associated with the presence of a certain "Mlle S" who accompanied de Luc on his journeys through Switzerland in 1774/5.[125]

It is noteworthy that in his *Lettres* (1778) de Luc made a particular point of stating that he was unwilling to employ the word '*géologie*'.[126] But a year later de Saussure displayed no such inhibitions,[127] and de Luc himself used the word without qualms in the 1790s.[128] We have here an indication of the establishment of a new science, and moreover one with an explicitly recognized historical field of enquiry.

De Luc maintained that there was a natural harmony between the moral and the physical aspects of nature. He likened "the history of the Earth" to the "history of Mankind",[129] and claimed that his ideas arose from the pleasure that he had gained from the mountain scenery of Switzerland.[130] For de Luc, one might take delight in and empathize with the Earth, just as one might with one's fellow men or women.

De Luc has been treated with some disdain by certain twentieth century writers,[131] yet in fact he occupies a position of no little importance in the history of geology. It is true, of course, that he was an ardent scriptural geologist, and a considerable amount of his effort was directed towards achieving a reconciliation between the Biblical accounts of the formation of the globe and the observations that may be made by geological investigators. In this respect his intentions were not so very different from those of the seventeenth century physico-theologians. But the means employed were very different. De Luc displays all the characteristic empiricism of the late eighteenth century. Moreover, his approach bears similarities to the situation that existed among the contemporary writers of philosophical histories.

Like Condorcet, for example,[132] de Luc, in his *Geological letters,* divided the history of the globe into a series of epochs,[133] choosing as might be expected, the number six, because of an evident wish to achieve a unification with the Mosaic account. Although de Luc sought to distinguish geology from natural history by the fact that the latter was merely concerned with phenomena, whereas the former considered the causes of phenomena, and although the proposed explanations of causes were couched in somewhat naive (it may seem to us) Neptunist terms, his whole approach was permeated through and through by historical thinking. Numerous quotations may be given in support of this contention. Three must suffice:

> We may observe in some of the abrupt grounds we meet with, sections of great masses of *strata,* where it is as easy to read the history of the Sea, as it is to read the history of Man in the archives of any nation.[134]

> I am here tracing the *History* of the *Earth* itself, from its own Monuments.[135]

> The entire mass of our continents is composed of *strata,* similar in this respect to the regular courses of stone in our buildings. A succession of strata indicates a succession of time for their formation;

> and the change from one species of stratum, to another species placed upon it, indicates a change of cause. Thus is the mass of our continents the product of successive operations, during which the producing causes have undergone successive changes.[136]

It is true that the total *objective* of de Luc's investigations was (or became so during the course of his lifetime) external to geology *per se*. He desired to know the origin and nature of man, and this he sought to achieve through a study of man's history. To this end he thought it necessary to know the history of man's planet; and this he hoped to achieve by studying the "monuments of its revolutions". This meshes well with the historicist mode of *explanation*, as described at the outset of this enquiry, even though we do not have—in de Luc—an exclusive concern with a study of the history of the Earth for *its own sake*, as might be expected if we are to find an ideal analogy with historicist thinking with respect to man's past. And de Luc himself, to some degree, altered the focus of his interests during the course of his lifetime. The young man's concern for the beauties of mountain scenery gradually gave way to a passionate preoccupation with a comprehension of the essence of Man himself, viewed through a pious Christian framework.

It was within such a context that de Luc became involved in his celebrated controversy with the Huttonians. With respect to this controversy, it is important to note that one of his chief objections to Playfair's *Illustrations* was that it wholly excluded historical considerations. In de Luc's view,[138] it was not the obscurity of Hutton's treatise that led to its neglect, as Playfair imagined. Rather, the *Theory of the Earth* was a wholly misconceived intellectual enterprize in that it concerned itself merely with the *laws* regulating the changes of the Earth's surface, rather than the historical changes that had occurred to that surface.[139] And de Luc was probably right in this. Very likely, one of the reasons that caused Hutton's rather chilly reception was the fact that his was essentially an eighteenth century mind, propounding a non-historical geological *theory* of the Earth to historically-minded nineteenth century audiences. As such it did not cohere satisfactorily with the ideas of the majority of his readers, despite the *Theory*'s many original and profound features. The readers of the nineteenth century would have been more receptive to a history of the Earth, rather than a theory, or a statement of the laws of its changes. To be sure Hutton's ideas held considerable appeal to Lyell, but as I shall argue below Lyell's opinions, though highly influential, did not mesh perfectly with those of his contemporaries.

Among the writers who served as a source of inspiration to de Luc, the names of Pallas and de Saussure stand out conspicuously. Werner is also mentioned, but not so frequently. Pallas and Werner have already been treated in our earlier discussions. Let us, therefore, turn our attention briefly to de Saussure's contributions.[140]

Whereas de Luc constantly spoke of de Saussure as his chief mentor and source of inspiration,[141] de Saussure himself maintained that he was deeply indebted to Pallas, whom he eulogized.[142] On the other hand, he claimed that it was not until after the publication of the second volume of his major work, the *Voyages dans les Alpes,* that he became familiar with the works of Werner.[143]

It is clear that de Saussure possessed a profound love for the beauties of his native Switzerland and for wild nature, but unlike de Luc he does not seem to have acquired this from connections with Rousseau, so far as can be established from a reading of the *Voyages.* Rather, it seems that it was through his intercourse with the great Swiss scholar, Albrecht von Haller,[144] that de Saussure became imbued with his deep empathy towards his surroundings.

There are only meagre and scattered theoretical discussions in the *Voyages,* but the theory held by the author seems to have been fairly standard for his times; that is to say it was a Neptunist doctrine similar to that being taught by Werner at Freiberg. Although de Saussure was prepared to read off in a general way events such as "*grandes révolutions*" from his examination of the Earth's strata, he did not draw up a general stratigraphical sequence such as that of Werner; neither did he recommend such an activity in the *Voyages.* However, this work was never intended as a definitive statement of de Saussure's geological principles. Such a work was planned, but owing to ill-health it was never completed, and consequently in some ways the best that can be done towards reaching an understanding of de Saussure's general geological viewpoint is to examine his *Agenda,* which was intended to serve as a series of guide-lines to those who might wish to carry out geological investigations.[145]

The whole programme of enquiry indicated by the *Agenda* appears to be Wernerian in character. Of course, this is not unexpected, for by the 1790s the Wernerian system was making considerable progress in scientific circles, and was in fact becoming paradigmatic in character, even in France.[146] But the section of the *Agenda* (sect. X) dealing with the layering of mountains is chiefly concerned with the determination of the geometry of the beds within a mountain (which was a prime consideration for geognosists), and there was no statement that it was the geologist's task to write an history of the globe.

On the other hand, de Saussure apparently saw a smooth transition between geological history and the history of man on his planet, and consequently we find in the *Agenda*[147] an emphasis on the importance of taking historical records and oral traditions into account when pursuing geological inquiries. Clearly, geology was seen to take over, for histories of the Earth, where written or oral testimonies petered out, just as archaeology could take over and give a greater depth of perspective where historical accounts of human personages or activities became unavailable.

In summary, de Saussure's geological system displays certain elements of historical thinking, with an interconnection perceived between human and Earth histories, but at the same time he is propounding a genetic theory. His chief claim must be that he served as an influential advocate of geological science, encouraging men to move into the remote and inaccessible portions of the globe in order to further their investigations, this being just one aspect of the rising tide of Romantic enthusiasm for the beauties of nature. But de Saussure did not make any special contributions to geological methodology in a technical sense.

Returning to France, consideration should be given to the work of Jean-Louis Giraud Soulavie, often credited with being an independent precursor of William Smith, Brongniart and Cuvier in establishing the leading principles of the science of stratigraphy. Indeed, he was a forerunner of Lyell also, in that he maintained that old and new strata might be distinguished one from another according to the proportion of extant fossils that they contained.[148] Soulavie re-emphasized the significance of the principle of superposition, insisting on its importance for geological enquiries,[149] and arguing that on the basis of this principle one might draw up a chronology of the Earth's crust. Like Buffon, he believed that in the course of the Earth's past there have been "six distinct and separate epochs".[150] The chronological sequence of these epochs could, he claimed, be deduced directly from six sets of "facts".

But what is more interesting for our present enquiry is Soulavie's attempt to give a classification of the mineral kingdom according to a temporalized system. In 1785, the St Petersburg Academy set as a subject for a prize competition the question of how rocks might be classified into genera, species, and varieties, not only by their chemical or their external characteristics, but also according to their modes of origin, and their ages, and according to the "epochs" of nature in which they were formed. The very fact that such a competition was proposed in 1785 is, I suggest, not without significance, for it reveals in a general way how thinking was being directed towards attempting to give the science of mineralogy (or petrology, as we would say today) a temporalized footing.

Soulavie was awarded only honourable mention for his entry, which perhaps was too speculative for the taste of the Russian academicians, though in fact it accorded well with the terms laid down for the competition. Incorporating the results of his geological investigations in southern France into his Essay, he asserted that:

> Each age corresponds to a class, each order to a revolution, each phenomenon to a species, and each occurrence [*accident*] to a variety.[151]

So in effect the taxonomy and nomenclature would depend upon events rather than forms.

As may be imagined, the task that Soulavie set himself was greater than that which could in fact be accomplished. In truth, it is not possible to forge a neat link between lithological or mineralogical character and chronology. Consequently, the scheme was not wholly coherent, and displayed considerable inconsistencies. But is interesting that Soulavie attempted to make clear his theory by reference to *human* history. Thus, for example, an historian might consider successively and separately the Egyptian, Greek, Roman, and European nations and the events occurring to each of these. Similarly, suggested Soulavie, the historian of the Earth would consider the granitic rocks and their changes, the calcareous rocks and the things that had happened to them, and so on. But just as there may have been temporal overlap between the Greek and Roman civilizations, so too there may have been overlap between the events of Soulavie's various "ages". Consequently, lacking any kind of absolute method for dating the ages and the revolutions, his system was by no means coherent or even intelligible. Nevertheless, it occupies an interesting and quite important place in our account. The attempt to impart a temporal and quasi-historical dimension to the study of the mineral kingdom is clearly evident. Yet by thinking in terms of distinct epochs there is a considerable lack of historical continuity in the system, and there are obvious affinities with the somewhat earlier 'genetic' schemes of Buffon.

Somewhat similar to the scheme of Soulavie, was the suggestion of Dolomieu that each of the chemists' simple earths (i.e., lime, magnesia, argillaceous earth, and siliceous earth) might correspond to different epochs of the Earth's history.[152] Thus, it was suggested, all these earths had been present in an initial universal ocean, and each might have been precipitated successively in an order corresponding to the different solubilities, silica coming first and lime last. But this was highly speculative; it afforded a very blunt instrument for the understanding and classification of the rocks of the Earth, and in fact it attracted little attention and negligible following, though Dolomieu's geological opinions were described at length by Delamétherie.[153] In a series of subsequent additions to his paper,[154] Dolomieu attempted to subsume his scheme under a general theory of chemical affinities. He planned to give a systematic account of his geological opinions in a series of articles appearing in the *Journal de physique*, but this intention was never completed, for the enterprize was interrupted by the events of the French Revoloution.[155] Whether Dolomieu realized that he was becoming enmeshed in a fruitless programme of enquiry it is difficult to say, but it is possible that he was not wholly disappointed to find a reason for discontinuing his series of papers.

As previously mentioned, during the 1790s and the first decade of the nineteenth century, Wernerian ideas became increasingly widely known in France, in large measure through the publication of the German's ideas by Delamétherie. But the Wernerian system, though providing the chief basis for stratigraphical enquiries in the early nineteenth century in that it offered a *general* sequence of strata, the confirmation or refutation of which provided a wealth of investigative potential, was far from satisfactory, for ages and lithological characters cannot be joined one with the other by any simple relationship. It is fossils that provide the link between strata and history. And the system of Werner, for all its rigour, ultimately proved as problematic as that of Soulavie or Dolomieu.

As is well known, so far as France was concerned this link was forged more or less successfully by Brongniart and Cuvier in their celebrated investigations of the strata of the Paris Basin.[156] It is perhaps noteworthy that this 'breakthrough' was achieved through the collaboration of an eminent mineralogist and a pre-eminent biologist and palaeontologist.

Brongniart and Cuvier proposed an eleven-fold division of the strata of the Paris Basin.[157] Fossils were used as an aid to the discrimination of beds that were lithologically similar yet chronologically different. Werner's gypsums, for example, were distinguished from those of the Paris area by their relationship with certain shell beds.

On the basis of their carefully revealed and mapped structure of the Paris Basin, the authors were then able to present their readers with graphic descriptions of the sequence of events that had occurred in that part of the world, long before any human beings were present who might have produced written or oral testimonies of the events in question. Thus descriptions were given of the advance and retreat of the seas, the coming and passing of freshwater conditions, the occasions when the waters swarmed with shells and the times when they were barren. The authors showed no hesitation in using the theory-laden Wernerian term 'formation' with respect to the various strata.[158]

A modern historian is never content if his narrative contains abrupt transitions, or is marked by temporal lacunae in which the reader is left unaware of the flux of events. To be sure, it is impossible to eliminate all gaps and omissions, but the competent historian makes every effort to reduce such areas of uncertainty to a minimum. In the case of Brongniart and Cuvier, however, they believed that their formations were deposited "entirely separately, at times that were sharply separated one from another".[159] And they were seemingly not disturbed by this. Indeed, in his subsequent *Discours sur les révolutions du globe,*[160] Cuvier seemed positively to revel in the situation, laying great emphasis upon the importance of these revolutionary episodes in the history of the planet. Such a viewpoint must necessarily have arisen where attention was focussed on a limited area, for the record cannot help but be fragmentary if the interconnections of deposits in different parts of the globe are ignored. In addition, as we have seen, there was a long French tradition which probably predisposed Brongniart and Cuvier to think in terms of revolutionary episodes or discrete epochs. But if we wish to find out how the final stages in the establishment of geology as an historical science occurred, in which a 'smoothing' of the historical description of the Earth was effected, eliminating appeals to sudden revolutions or supposedly catastrophic events, then we must turn our attention to the situation as it was developing in England and Scotland.

### THE EMERGENCE OF HISTORICAL GEOLOGY IN BRITAIN

Perhaps because of the very magnitude of the contributions of figures such as Hutton, Smith and Lyell, the fact that Britain played such an important role in nineteenth century geology, and that much of the history of geology has been written in English, there is perhaps a tendency for English-speaking geologists to imagine that Britain has always played the leading part in the development of the Earth sciences, or at least that this was so until the present century. But this judgement can scarcely be sustained when the situation is examined more closely, at least as far as one can tell from *published* sources. Thus, there was nothing in Britain comparable in scope to the work of Arduino or Lehmann in the middle years of the eighteenth century. It is true that there were several British publications that set out to classify the mineral kingdom on the basis of external characteristics.[161] But there was little in the way of published work that might be thought of as early stratigraphical studies, and which might be relevant to our present enquiry. By the 1760s and 1770s, however, some significant changes were occurring. Studies such as those of Catcott[162] and Whitehurst[163] were undertaken with an eye to developing theories of the Earth that would cohere with received religious opinions, and the authors did not offer much in the way of significant stratigraphical *principles.* However, Neve and Porter[164] draw attention to the fact that such studies were based on a good deal of carefully-planned fieldwork, the notes of which were never published but which indicate that attention was firmly fixed on *strata,* which could be construed as relics of the Earth's past. Nevertheless, there is not very much that needs to be said respecting historical thinking in British geology before the work of Hutton and Smith. Schemes were directionalist, and in that limited sense could be said to be historical, but essentially they were tied to a Judaeo/Christian or Augustinian temporal and historical framework, and did not stand independently as geological contributions.

The case of James Hutton, however, is very different. As is well known, Hutton's theory was essentially *cyclic* and *dynamical* in character,[165] displaying no directionalist tendencies. There was no spiral twist to the Huttonian cycle that might make it resemble in some manner the theory of history of a writer such as Vico. It may be said, therefore, that Hutton's theory revealed an eighteenth century preoccupation with *order,* though it was through the agency of a continuous cycle of events that Hutton's theory displayed the order of nature. And we do not find in the *Theory of the Earth*[166] that he sought to explain phenomena by individual and discrete historical antecedents and circumstances. It would therefore be quite inappropriate to refer to Hutton as an historicist.

On the other hand, while modern commentators have properly concentrated their attentions on the cyclic character of Hutton's work, and it has perhaps been customary to think that he showed little interest in fossils or in the stratigraphical sequences to be found in the rocks, all attention being focussed on a general theory of Earth dynamics, it should be remembered that Hutton did in fact have an interest in fossils.[167] In addition, in recent years some of Hutton's field sections (as prepared by John Clerk) have come to light,[168] revealing that Hutton was, in fact, capable of high-level stratigraphical work.

More importantly, with the idea of a continuous geological cycle, Hutton was developing a system which later was to play a considerable role in closing the stratigraphical gaps in the system of Brongniart and Cuvier in France—and of William Buckland in Britain. Nevertheless, our judgement must be that Hutton was essentially an eighteenth century figure who did not seek to provide an *history* of the globe but aimed rather to identify the *processes* according to which the non-directional changes occurred on and through the Earth.[169]

Hutton's 'illustrator', John Playfair, writing at the beginning of the nineteenth century, perhaps adopted a marginally more historicist attitude towards the study of the Earth when he wrote

> To trace the series of these revolutions, to explain their causes, and thus to connect together all the indications of change that are found in the mineral kingdom, is the proper object of a THEORY OF THE EARTH,[170]

and when he saw the Earth as "the theatre of many great revolutions".[171]

The word 'revolutions' here may lead one to think of Playfair as having catastrophist tendencies, but in this context he seems to have been referring to cycles of Earth history rather than sudden interruptions in the course of geological events.

The case of William Smith is of considerable interest, for opinions about his geological work have begun to shift quite radically in recent years, and we do not, perhaps, even now, have a firm understanding of his theories and their role in the history of geology.

In his *Stratigraphical system of organized fossils* of 1817, Smith claimed priority for his method, and stated that "the identification of Strata by the help of organized Fossils, becomes one of the most important modern discoveries in Geology".[172] In addition, he went on to say:

> Organized Fossils are to the naturalist as coins to the antiquary; they are the antiquities of the earth; and very distinctly show its gradual regular formation, with the various changes of inhabitants in the watery element.[173]

Evidence of this kind might lead us to the conclusion that by means of his celebrated stratigraphical principle Smith was enabled to think immediately in historicist terms, in the manner of a typical nineteenth century geologist.

But on closer examination the situation becomes less clear. There is, in fact, a good deal of evidence to suggest that Smith was much more concerned with the determination of the geometry of the Earth's strata, than with historical or temporal relationships. For example, as an old man he attempted to set down his general opinions concerning theoretical geology, and some copies of the proofs of this work are still in existence, though it was never published.[174] Smith's opening thoughts in this so nearly published work were:

> The principles of Geology like those of geometry must begin at a point, through two or more of which the Geometrician draws a line, and by thus proceeding from point to point, and from line to line, he constructs a map, and so proceeding from local to general maps, and finally to a map of the world. Geometricians founded the science of Geography, on which is based that of Geology.[175]

Much of the rest of the text reveals that Smith had changed little of his basic thinking since the end of the eighteenth century when he first enunciated his well-known theoretical principles, and that by the end of his life he was way behind his contemporaries in theoretical matters.[176] But the emphasis given to *geometry* in the passage quoted gives us cause to question whether Smith was really much concerned with history at any level deeper than that of the general principle of the superposition of strata.

It may be noted also that Smith's view of the Earth was in many ways quite static. For example, in a note contained among his unpublished papers, dated 4 September 1824, we read:

> Though we may not yet be enabled to account for . . . the position of the strata, we have, I think, sufficient ground to believe that the Strata were formed much as they are and that since their formation the crust of the Earth has not been much distorted or disturbed except at the Deluge and then only superficially.[177]

And in the manuscript of a projected work, a prospectus of which was issued in 1801, we find the following very interesting passage:

> [My] numberless observations . . . made on the Strata . . . [have] made me confident of their uniformity throughout this Country & [have] led me to conclude that the same regularity . . . will be found to extend to every part of the Globe for Nature has done nothing by piecemeal there is no inconsistency in her productions the Horse never becomes the Ass nor the Crab an Apple by any mixture or artificial combination whatever nor will the Oak ever degenerate into an Ash or Ash into an Elm however varied by Soil or Climate the species will still be distinct on this ground then I argue that what is found here may be found elsewhere[.] When proper allowances are made for such irregularities as often occur and the proper situation and natural agreement is well understood I am satisfied there will be no more difficulty in ascertaining the true quality of the Strata and the place of its possition [*sic*] than there is now in finding the true Class and Character of Plants by the Linean [*sic*] System . . . .[178]

It would seem from this that Smith's opinions had much in common with the static eighteenth century world-view; and the remark about Linnaeus is particularly striking in this respect.

Recently, Dr Rachel Laudan has claimed that Smith, in his actual geological investigations and map work, made use of the principle of the correlation of strata by means of their fossil contents to only a limited degree and that his normal procedure was to map basically in terms of lithologies and according to the prominent exposures of the strata.[179] This is not the place to expand upon this interpretation, for which Dr Laudan has provided a considerable amount of evidence. It may be said, however, that it is not really a surprising conclusion. If strata are to be identified and correlated from their fossil contents, it is necessary—in order to arrive at Smith's principle in the first place—that *some* strata be recognized as independent entities by means of their lithologies. Without such recognition one could not gain a foothold on which to erect the principle.

There is much more to be said on this question, but it would not be proper to devote too great a portion of this study to Smith alone. Briefly, however, it would appear that he saw as his leading contribution the idea of a regular and universal succession of strata with a constant order of superposition. For *practical* purposes, the strata were generally determined by their lithologies and their features of outcrop, though in certain cases (most notably "the upper and lower rocks of freestone in the Hills near Bath"),[180] Smith undoubtedly did use palaeontological criteria and the principle that is associated with his name. Thus it was the *geometrical* problem of the strata that interested Smith most particularly, and also the possible utilitarian aspects of his discoveries. In later life he also showed strong leanings towards natural theology, but this seems to have been something that stood outside his actual geological practice. To be sure, Smith believed that strata were laid down successively in a temporal sequence, but this notion does not seem to have exercised any particular influence on his actual procedures. Above all, he sought to reveal the geometrical order of the strata, and this primarily for utilitarian reasons.

Let us leave Smith, then, and turn to consider the more general picture of British geology in the first quarter of the nineteenth century. There was during this period widespread debate on questions such as the validity of Werner's doctrine of universal formations, the use of palaeontological as opposed to lithological criteria in stratigraphical investigations, and, of course, the relative adequacies of the Huttonian and Wernerian theories as a whole. Slowly, during this period a general methodology was evolved on directionalist, and historicist (but not necessarily gradualist) principles, the cyclic views of Hutton making little headway.

Jameson, zealously urging the geognostic approach in Edinburgh, believed that:

> To execute . . . [a geognostic] investigation one must trace correctly the bounds of all the formations; then delineate the most striking features in the disposition of their stratification; next must describe their different strata and beds, and lastly the individual rocks. But in doing this, we must adopt a determinate order, we must not begin indifferently with any formation, but should adhere to the order which nature appears to have followed in their disposition, that is . . . proceeding from the oldest to the newest.[181]

He followed German authors such as von Humboldt, von Buch,

Freiesleben and Grüner in thinking that the succession of the Floetz strata at least was quite general.[182] And in his lectures he argued against the Huttonian position on the grounds that a cyclic theory was incompatible with the observed "great regularity in the succession of its [the Earth's] strata".[183]

But other writers were less certain on this score. W. H. Fitton considered that "the several parts of the globe are, in their geological structure, analogous but not the same".[184] Bakewell gave an outright denial of the general validity of the doctrine of universal formations;[185] and Macculloch, writing in 1821, referred to the doctrine (or "dogma" as he called it) as a "phantom" that was slowly fading.[186] Most writers, therefore, whether or not they were followers of the Wernerian scheme, agreed that for the present it would be wise "to persevere with patience in the accumulation of useful facts".[187] And, as is well known, it was primarily to this task that the early members of the Geological Society addressed themselves.

Greenough's geological map of Britain was very much less obviously theory-laden than that of Smith, and in his *Critical examination of the first principles of geology* Greenough questioned the general validity of Smith's stratigraphical principle, rightly pointing out that by actualist arguments it would be unreasonable to expect fossils to be distributed universally in rocks of a particular age.[188] Yet it was through the work of the early members of the Geological Society that we see some of the first examples of the characteristically nineteenth century stratigraphical methods in operation, with a piecing together of stratigraphical sequences observed in separate regions, so as to construct a composite historical picture. The renunciation of theory, and the emphasis on the collection and synthesis of discrete items of stratigraphical information accords well with what I regard as an historicist approach to the Earth's past. And it manifested itself at about the same time that the writing of human history was acquiring a distinctive historicist aspect in Britain.

In his important paper on the geology of the Isle of Wight and the London Basin published in 1814,[189] Thomas Webster, curator of the museum, draughtsman, librarian and subsequently secretary of the Society, made a bold attempt to correlate the strata of the island with those observed by Brongniart and Cuvier in the Paris Basin, and he used primarily palaeontological criteria for this purpose. In addition, he developed an account of the geological history of the area, including presentation of the periods of marine transgression and regression, the kinds of fauna that were found at different periods, and the conditions of deposition (e.g., marine or freshwater) that obtained when the strata were laid down. He reconstructed the occurrence of massive Earth movements and the subsequent widespread erosion and transport of materials to give the present topographical configurations. The paper was illustrated by suitable coloured maps and sections. It was, I suggest, through the wide extension of work of this kind in the nineteenth century that the stratigraphical history of the globe was gradually unravelled. The doctrine of universal formations, useful though it had been in the late eighteenth century, was fairly quickly rejected in the nineteenth, and we find, then, the first steps in the stratigraphers' seemingly endless task of piecing together a multiplicity of discretely observed stratigraphical sections, with the constantly associated problem of weeding out stratigraphical synonyms. Thus the contemporary Irish writer, William Richardson, succinctly described the perceived task of the geologist in the following terms:

> Enumeration of the strata of which any country is composed, with an account of the changes they have undergone, and the operations that have been performed upon them, forms the true geological history of that country, and an accumulation of many of these detached histories affords the best materials for a general history of the physical world.[190]

### LYELL'S HISTORICISM

There remains one major question for discussion: the relationship of all this to the ideas of Lyell. In the work of the 1820s, the surviving eighteenth century notion of discontinuities in the thread of Earth history was still to be found. A general historical, directionalist approach towards a study of the Earth had become established by the second decade of the nineteenth century, but this history was still envisaged as proceeding in a saltatory manner. The theory was non-gradualistic; it contemplated with equanimity a doctrine of 'catastrophes'. The influential text of Conybeare and Phillips provides us with good examples of this kind of thinking, as when Conybeare wrote, with respect to the mountains of the European Alps:

> It is impossible to conceive a more striking picture of derangement, apparently resulting from the operation of violent convulsions, than is here presented; and however we may doubt whether the details of the section [from the Jura to Mont Blanc] are perfectly ascertained, it is impossible that such a representation could have been formed, did not the greatest disturbance actually exist.[191]

How, then, we may ask, was this picture modified by Lyell? And was the historicism of the 'catastrophists' preserved in the gradualistic synthesis of the *Principles of geology?*

In his biography of Lyell, Wilson has drawn attention to a section in Lyell's notebooks in which he specifically wrote on the "Analogy of Geology and History".[192] Here, Lyell emphasized that the past history of the globe must inevitably leave its traces to be perceived by present observers. In other words, if we wish to understand the nature of the world around us, we must take into account its previous conditions and circumstances.

Similar views were expressed on the very first page of the *Principles*. An historian studies mankind's past in order to achieve a greater understanding of the present circumstances. Something similar might be said with respect to the Earth's historians, for

> As the present condition of nations is the result of many antecedent changes, some extremely remote and others recent, some gradual, others sudden and violent, so the state of the natural world is the result of a long succession of events, and if we would enlarge our experience of the present economy of nature, we must investigate the effects of her operations in former epochs.[193]

This might well be called paradigmatic historicism. Lyell is saying that a knowledge of Earth history is a necessary condition for understanding her nature. Explanation and understanding are to be achieved by recourse to history. Geikie's adage is stood on its head, so to speak, becoming: "The past is the key to the present."

Examples of historical metaphors are to be found frequently in Lyell's writings, particularly in the *Manual of elementary geology*. For example, it is in this work that we come across the following significant statement:

> Series of formations are like volumes of history, in which each writer has recorded the annals of his own time, and then laid down the book, with the last page uppermost, upon the volume in which the events of the era immediately preceding it were commemorated. In this manner a lofty pile of chronicles is at length accumulated; and they are so arranged as to indicate, by their position alone, the order in which the events recorded in them have occurred.[194]

Then again, Lyell states that fossils, for geologists, have "that authority which belongs to contemporary medals in history".[195] Or, in his discussion of the establishment and use of 'type sections' (to employ our present-day term), Lyell utilizes the metaphors of "contemporary annals" and "monuments".[196] And in general he reckons that the best basis for the classification of the Earth's rocks is in terms of relative ages.[197] All this indicates that Lyell considered his aims and methods as analogous to those of the historians of his time.[198]

Yet there are a number of distinctive features in Lyell's outlook that require further discussion. Despite what has been said above, Lyell's basic methodological procedure was to take his bearings from the present, extrapolating from the present to the past. The well-known method of correlation of the Tertiary strata by consideration of the percentages of the still-extant organic forms provides a prime example of this particular approach.

It should be noted that this way of proceeding was a special feature of Lyell's work, rather than of uniformitarians in general, for, as has been shown by Rudwick in his study of the work of Scrope, uniformitarianism of a substantive kind was advocated by some writers after, and on the basis of, particular examinations of the Earth's strata.[199] This was Scrope's procedure. Lyell's approach was somewhat different. He used the *present* as a *map* or *sign-post* to guide him into the past, and his assumption of uniformity was of a methodological character. But once the general character of the past was understood with the help of the principle of uniformity, then one might investigate its historically individual aspects and thereby one could understand the features of the Earth's present situation, and also her general structure.

The upshot of all this was that in the Lyellian scheme the historicism of the earlier nineteenth century stratigraphers was retained, but the saltatory aspects of the previous schemes were replaced by a smooth gradualism. "Nature moves with dignity" might have been Lyell's dictum. However, while achieving a gradualistic system, and employing an historical approach to geological investigations, Lyell abandoned the directionalist characteristics of the earlier schemes such as those of the Wernerians, Brongniart and Cuvier, or Conybeare and Phillips. Lyell believed that conditions on the globe might fluctuate randomly about a mean, so that if the same conditions happened by chance to be restored, so too might the flora and fauna of an earlier age recur. In well-known passages he spoke of the possible return of iguanodon,[200] should conditions again become suitable for the existence of this creature. From the present-day standpoint, this may appear incongruous. And, as Bartholomew has shown,[201] Lyell's contemporaries were themselves unwilling to accept his non-directionalist thesis, though in time they came to approve his gradualism and his actualistic methodology. Eventually, in Darwin's synthesis, directionalism was restored, but—following Lyell—it returned in a gradualistic mode, without recourse to saltations in either the organic or the inorganic realms. So it is in Darwin that we eventually find all the elements of nineteenth century historicism deployed in relation to studies of the Earth.

Before bringing this enquiry to a close, I should like to consider Lyell once again in relation to the views of his day respecting philosophies of history. If we go back to Hume in the eighteenth century, it may be recalled that he considered that human nature was largely unchanging, and for this reason one might hope to be able to make valid judgements about men's actions in the past.[202] With this belief in the general uniformity of human nature, Hume can be seen to evince the ahistorical attitude of the scholar of the Enlightenment. Hume did not write history as we understand it, though he did, of course, write books dealing with man's past.[203] By contrast, nineteenth century historians such as Macaulay certainly thought very differently from Hume on the question of man's nature.[204]

Now it may be argued that there was something rather 'Humean' about Lyell's methodology. The assumption of uniformity of geological process (or non-directionalism) paralleled Hume's assumption of the uniformity and law-like character of human nature, and both Lyell and Hume were taking the present as their keys to the past. This reveals that though Lyell undoubtedly considered it the task of a geologist to write an *history* of the globe he was not a nineteenth century historicist in a total sense. After all, that man of the Enlightenment, James Hutton, was his chief mentor.

So I suggest that Lyell's somewhat 'Humean' stance can go some way towards explaining the opposition that the *Principles* initially encountered. A rigid uniformitarianism, or non-directionalism, ran counter to the view prevailing in his day, and Lyell himself had to forego some elements of his uniformitarianism in his later work as, for example, when he eventually accepted the theory of the ice ages and the Darwinian theory itself; and ultimately, as has been pointed out by Bartholomew, he came to accept a directionalist system.[205] But in his early work, it might be said that Lyell's attitude towards history was rather conservative, aligned somewhat towards the stance of writers of the previous century. However, this did not prevent his seeing the geologist's task as being that of writing the history of the Earth, and in this respect his ideas agreed with the general outlook of his day. Lyell was concerned to build up an historical account of the Earth's past by synthesizing a multiplicity of items of stratigraphical information, ignoring speculative theories as to origins and dynamical behaviour.[206]

## GEOLOGY AND THE ROMANTIC MOVEMENT

One last point. It is often said that the Romantic movement, as one of its manifold characteristics, displayed a yearning for and attempted empathy with the past. So one might ask whether evidence may be found for such attributes amongst nineteenth century geologists, for if this were possible it would give support to our general hypothesis that parallels are to be found between thinking with respect to man's past and thinking with respect to the Earth's.

In fact, the major geological memoirs of the early nineteenth century do not provide us with much evidence of this kind, though writers such as William Buckland believed that geological investigations allowed the "unfolding [of the] records of the operations of the Almighty Author of the Universe, written by the finger of God himself, upon the foundations of the everlasting hills".[207] So for Buckland, and other writers in the tradition of natural theology, geology provided an appropriate means whereby man might seek some understanding of the nature of his putative Maker. Lyell also wrote within the tradition of natural theology, though he was less sanguine than Buckland in his hopes for using geological investigations to divine the Creator's plan. But a clear statement that geological investigations could attune man's mind to the Earth's past is not to be found in the more serious and weighty publications. Nevertheless, if we look in the more popular geological literature of the nineteenth century, we find plenty of evidence of this kind. Thus Granville Penn, in his widely-read *Conversations on geology,* specifically wrote of geology as a "Romantic Science":

> I . . . call Geology romantic, because it not only leads us to travel among the wildest scenery of nature, but *carries the imagination back* to the birth and infancy of our little planet, and follows its history of deluges and hurricanes and earthquakes, which have left us numerous traces of their devastations. Would you not think it romantic to travel, as must be done by the geological inquirer, among mountains and valleys, where the tempests have bared and shattered the hardest rocks, and where alternate rains and frosts are crumbling the solid materials of mountains, while the springs and rivers wash away the fragments to deposit them again in the various stages of their course? And would you not think it romantic *to dream about the young world emerging from darkness,* and rejoicing in the first dawn of created light? *To think of the building of mountains,* the hollowing out of the valleys, and the gathering together of the great waters of the ocean? *And will it not be romantic to discover the traces of the ancient world* before the time of Noah, in every hill and valley which you examine?[208]

And this can give us a good indication of the attitude of mind of the early nineteenth century geological investigators, even if they did not choose to reveal such sentiments in their formal publications.

Thus we may assert that there was a strong resemblance of form between investigations of man's history and investigations of the Earth's history throughout the period that we have examined in this enquiry.

*Acknowledgements*

I am indebted to Paul McCartney and Joan McCartney for the translation of certain Italian passages. Professor Bert Hansen has very kindly supplied me with a copy of his unpublished translation of Füchsel's *Historia* (ref. 50). Roy Porter has been most helpful in his comments on an earlier version of this paper, and an anonymous referee has also offered similar assistance. I am indebted to Professor Vincent of the Department of Geology at Oxford University for permission to quote from the Smith archives in his care; and to Mrs Joan Eyles for kindly showing me the manner of arrangement of these documents and for guiding me to the items of particular relevance to my enquiries. The Librarian of the Verona Public Library has furnished me with photocopies of certain manuscript maps and sections from the Arduino archives, for which assistance I am most grateful.

REFERENCES

89. For a convenient English translation, see A. V. Carozzi (trans. and ed.), [B. de Maillet], *Telliamed, or conversations between an Indian philosopher and a French missionary on the diminution of the sea* (Urbana, 1968). And for an analysis of the work, see A. V. Carozzi, "De Maillet's *Telliamed* (1748): An ultra-Neptunian theory of the Earth", in C. J. Schneer, *op. cit.* (ref. 52), 80-99.
90. For an account of the Cartesian geological theories, see my paper, "Mechanical mineralogy", *Ambix,* xxi (1974), 157-78.
91. R. Rappaport, "The geological atlas of Guettard, Lavoisier and Monnet: Conflicting views on the nature of geology", in C. J. Scheer, *op. cit.* (ref. 52), 272-87, p. 274.
92. G. L. L. de Buffon, *Histoire naturelle, générale et particulière, avec la description du cabinet du roi* (44 vols & atlas, Paris, 1749-1804), i, *Histoire et théorie de la terre.*
93. (a) *Ibid.* (b) *Ibid., Supplément, Tome cinquième* (*Des époques de la nature* (Paris, 1778), 1-254; (c) *Ibid., Histoire naturelle des minéraux par M. le Comte de Buffon* . . . (5 vols, Paris, 1783-88).
94. G. L. L. de Buffon, *op. cit.* (ref. 93, 1749), 98.
95. *Ibid.*
96. *Ibid.,* 127 ff.
97. *Ibid.,* 257.
98. J. Roger (ed.), *Buffon, Les époques de la nature* (Mémoires du Muséum nationale d'histoire naturelle, Série C, Science de la terre, xx, Paris, 1962), xxi.
99. *Ibid.,* xxvii-xxviii. Roger suggests that Buffon may have arrived at this new opinion as a result of reading the *Dissertations sur la glace* of J. J. Dortous de Mairan (ref. 93, 1749), 98.
100. This work, depending on the celebrated experiments on the cooling of iron spheres, is well described in F. C. Haber, *op. cit.* (ref. 7), 116-18.
101. G. L. L. de Buffon, *op. cit.* (ref. 93, 1778), 1.

102. The work of A. R. J. Turgot makes a useful analogue: "Second discours sur les progrès successifs de l'esprit humain (11 decembre 1750)", in *Oeuvres de M. Turgot, . . . précédées et accompagnées de mémoires et de notes sur sa vie, son administration et ses ouvrages* (9 vols, Paris, 1908-11), i.

103. N. Desmarest *et al., Encyclopédie méthodique: Géographie-physique* (5 vols & atlas, Paris, *an* 3—1828), i, 409-31.

104. *Ibid.*, 424.

105. *Ibid.*, 410.

106. *Ibid.*, 416.

107. *Ibid.*, 417.

108. *Ibid.*, 419.

109. A. L. Lavoisier, "Observations générales, sur les couches modernes horizontales, qui ont été déposées par la mer, et sur les conséquences qu'on peut tirer de leurs dispositions, relativement à l'ancienneté du globe terrestre", *Histoire de l'Académie Royale des Sciences. Année M.DCC.LXXXIX. Avec les mémoires de mathématique et de physique pour la même année* (Paris, 1793), 351-71. For an analysis of Lavoisier's stratigraphical ideas, see A. V. Carozzi, "Lavoisier's fundamental contribution to stratigraphy", *Ohio journal of science,* lxv (1965), 71-85.

110. N. Desmarest, "Mémoire sur l'origine & la nature du basalte à grandes colonnes polygones, déterminées par l'histoire naturelle de cette pierre, observée en Auvergne", *Histoire de l'Académie Royale des Sciences. Année M.DCC.LXXI* . . . (Paris, 1774), 705-75; "Extrait d'un mémoire sur la détermination de quelques époques de la nature par les produits des volcans, & sur l'usage de ces époques dans l'étude des volcans", *Observations sur la physique, sur l'histoire naturelle et sur les arts,* xv (1779), 115-26; "Mémoire sur la détermination de trois époques de la nature par les produits des volcans, et sur l'usage qu'on peut faire de ces époques dans l'études des volcans", *Mémoires de l'Institut National des Sciences et Arts. Mémoires de la classe des sciences mathématiques et physiques* (1806), 219-78.

111. R. Rappaport, *op. cit.* (ref. 91), 279, points out that although Guettard apparently had more than sufficient empirical information which he might well have used for the purpose of developing a history of the Earth, in fact he made no definite move in this direction. In Dr Rappaport's view, "he tried hard to avoid thinking of the Earth as having much of a history". See also R. Rappaport, "Lavoisier's theory of the Earth", *The British journal for the history of science,* vi (1973), 247-60.

112. J. E. Guettard, *Mémoires sur la minéralogie du Dauphiné* (2 vols, Paris, 1779), i, vj.

113. R. Rappaport, *op. cit.* (ref. 91); "Lavoisier's geological activities, 1763-1792", *Isis,* lviii (1967), 375-84; and "Lavoisier's theory of the Earth", *op. cit.* (ref. lll).

114. R. Rappaport, *op. cit.* (ref. 91), 283.

115. N. Desmarest, *op. cit.* (ref. 110, 1774), 774.

116. N. Desmarest, *op. cit.* (ref. 110, 1806), 220.

(1934), 408-16. For the relevance of travel literature to the present theme, see G. de Beer, *Early travellers in the Alps* (London, 1930), and G. B. Parks, "The turn to the romantic in the travel literature of the eighteenth century", *Modern language quarterly*, xxv (1964), 22-33.

117. [N. Desmarest], "Géographie physique", in D. Diderot *et al.* (eds), *Encyclopédie, ou dictionnaire raisonné des sciences, des arts et des métiers, par une société de gens de lettres* (17 vols, supplements, plates & tables, Paris, 1753-80), vii (1757), 613-26.

118. *Ibid.*, 615.

119. *Ibid.*, 618.

120. *Ibid.*, 626.

121. N. Desmarest, *op. cit.* (ref. 103), iv (1811), 352.

122. On this topic, see Marjorie Hope Nicolson, *Mountain gloom and mountain glory: The development of the aesthetics of the infinite* (Ithaca, 1959); W. Kirchner, "Mind, mountain, and history", *Journal of the history of ideas*, xi (1950), 412-47; K. Chorley, "The romantic approach to scenery", *The month* (May, 1952), 263-73. Also of relevance here is R. A. Aubin, "Grottoes, geology, and the Gothic revival", *Studies in philology*, xxxi (1934), 408-16. For the relevance of travel literature to the present theme, see G. de Beer, *Early travellers in the Alps* (London, 1930), and G. B. Parks, "The turn to the romantic in the travel literature of the eighteenth century", *Modern language quarterly*, xxv (1964), 22-33.

123. This is mentioned by Rousseau in his *Confessions* (pt ii, bk 8) and in a letter from Byron to John Murray, dated 9 April, 1817: *The confessions of Jean-Jacques Rousseau translated from the French with a preface by Edmund Wilson* (2 vols, London, 1924), ii, 63; *The letters of Lord Byron selected and edited by R. G. Howarth* (London & New York, 1936), 165-7.

124. J. A. de Luc, *Lettres physiques et morales, sur les montagnes et sur l'histoire de la terre et de l'homme: adressées à la reine de la Grande Bretagne, par J. A. de Luc, citoyen de Genève* (La Haye, 1778), 193.

125. According to de Beer (*op. cit.* (ref. 122), 127), the young lady in question was Miss Schwellenberg, a lady-in-waiting to Queen Charlotte, whom de Luc served as 'reader'.

126. J. A. de Luc, *op. cit.* (ref. 124), vii. He wrote: "Je n'entends ici par *Cosmologie* que la connoissance de la *Terre*, & non celle de l'Univers. Dans ce sens, Géologie eût été le mot propre; mais je n'ose m'en servir parce qu'il n'est pas usité."

127. H. B. de Saussure, *Voyages dans les Alpes, précédes d'un essai sur l'histoire naturelle des environs de Genève* (4 vols, Neuchâtel, 1779-96), i, v.

128. J. A. de Luc, "Geological letters, addressed to Professor Blumenbach. Letter I. On the phaenomena characterizing the causes that have formerly operated upon this terrestrial globe; and particularly on those which fix the date of the origin of our present continents", *The British critic*, ii (1793), 231-8, 351-8, p. 231. It should be noted that the term 'geology' was in fact in use well before the end of the eighteenth century, but not in its modern sense. For example, Benjamin Martin, in his *Philosophical grammar* (London, 1735, p. 189), used the word to mean "the general doctrine of the globe, of its various divisions and subdivisions, of the vicissitudes of seasons, and other

general affections". His five subdivisions of the section of his work on 'geology' were: (1) a philosophical view of the globe; (2) the philosophy of earths, stones, etc; (3) the philosophy of water (seas, rivers, springs, etc.); (4) the philosophy of plants and vegetation; (5) the philosophy of animal bodies (p. 187). For a recent discussion of the etymology of 'geology', see D. R. Dean, "The word 'geology' ", *Annals of science*, xxxvi (1979), 35-43.

129. J. A. de Luc, *op. cit.* (ref. 124), xx.
130. *Ibid.*, xiv.
131. C. C. Gillispie, *Genesis and geology: A study in the relations of scientific thought, natural theology, and social opinion in Great Britain, 1790-1850* (paperback edn, New York, 1959), has been particularly scathing in his criticisms of de Luc.
132. See A. N. de Condorcet, *Sketch for a historical picture of the progress of the human mind* (London, 1955). The first French edition was published posthumously in 1795: *Esquisse d'un tableau historique des progrès de l'esprit humain.*
133. J. A. de Luc, "Geological letters, addressed to Professor Blumenbach. Letter III. Containing the history of the Earth, from the origin of all that is now observed upon it, to the production of the strata of sand-stone", *The British critic*, iii (1794), 467-78, 589-98; "Letter IV. The history of the Earth, from the formation of the sand-stone strata, to the last period in which the sea remained in its first bed; the space of time in which the volcanic eruptions began, and the strata of coals and rock salt were formed", *The British critic*, iv (1794), 110-20, 328-36.
134. J. A. de Luc, "Geological letters, addressed to Professor Blumenbach. Letter II. An analysis of the geological phenomena, leading to their origin", *The British critic*, iii (1794), 110-18, 226-37, p. 226.
135. J. A. de Luc, *op. cit.* (ref. 133), Letter III, 598.
136. J. A. de Luc, *op. cit.* (ref. 128), 232.
137. J. A. de Luc, "Geological letters, addressed to Professor Blumenbach. Letter VI. Being a physical commentary on the eleven first chapters of Genesis", *The British critic*, v (1795), 197-207, 316-26, p. 325.
138. J. A. de Luc, *An elementary treatise on geology: determining fundamental points in that science, and containing an examination of some modern geological systems, and particularly of the Huttonian theory of the Earth .... Translated from the French manuscript, by the Rev. Henry de la Fite, M.A. of Trinity College, Oxford* (London, 1809), 14.
139. *Ibid.*
140. On de Saussure's life and work, with particular reference to his career as a mountaineer, see D. W. Freshfield & H. F. Mortimer, *The life of Horace Benedict de Saussure* (London, 1920). See also J. Senebier, *Mémoire historique sur la vie et les écrits de Horace Bénédict Desaussure, pour servir d'introduction à la lecture de ses ouvrages* (Geneva, *an* 9).
141. J. A. de Luc, *op. cit.* (ref. 138), 41.
142. H. B. de Saussure, *op. cit.* (ref. 127), i, xix.
143. *Ibid.*, iv, 583.
144. *Ibid.*, ii. 460-1.

145. H. B. de Saussure, "Agenda, ou tableau général des observations et des recherches dont les résultats doivent servir de base à la théorie de la terre; par M. de Saussure, de Genève", *Journal des mines,* iv, no. 20 (*an* 4), 1-70. The Agenda were also published in the fourth volume of the *Voyages.* See H. B. de Saussure, *op. cit.* (ref. 123), iv, 467 ff.

146. See, for example, the references listed in refs 63, 64 and 70. Delamétherie kept up a steady flow of information on the Wernerian system in the pages of his *Journal de physique,* and several Wernerian manuals by French authors such as d'Aubuisson, Brochant de Villiers and Lucas were published in this period. Werner visited Delamétherie in Paris in *an* 10 and gave him an up-to-date account of his opinions, which were duly published in the *Journal de physique.* (See ref. 70 above.)

147. H. B. de Saussure, *op. cit.* (ref. 127), iv, 474.

148. J. L. Giraud Soulavie, *Histoire naturelle de la France méridionale, ou recherches sur la minéralogie du Vivarais, du Viennois, du Valentinois, du Forez, de l'Auvergne, du Velai, de l'Uségois, du Comtat Venaissin, de la Provence, des Diocèses de Nismes, Montpellier, Agde, &c. Sur la physique de la Mer Mediterranée, sur les météores, les arbres, les animaux, l'homme & la femme de ces contrées . . .* (7 vols, Nîmes & Paris, 1780-84), i, 161-3, 317-32.

149. J. L. Giraud Soulavie, "La géographie de la nature, ou distribution naturelle des trois regnes sur la terre. Description d'une carte du Vivarais dressée en relief où cette distribution est enluminée selon la nature de sol & les variétés des êtres organisés. Méthode pour rendre par des reliefs la forme du sol d'une province dont on écrit l'histoire physique, & pour l'enluminer selon la nature du terrain . . .", *Observations sur la physique,* xvi (1780), 63-73, p. 65. The author laboured the point somewhat by saying that in the instance of a church built on the top of a hill, the hill came first, then the foundations of the church, followed by the upper levels of the building, and finally the tower.

150. *Ibid.,* 65.

151. J. L. Giraud Soulavie, *Les classes naturelles des minéraux et les époques de la nature correspondante à chaque classe. Ouvrage qui a remporté le seconde accessit sur la question proposée par l'Académie Impériale des Sciences de St. Petersbourg, pour le prix de 1785* (St Petersburg, 1786), 5.

152. D. G. S. T. G. de Dolomieu, "Mémoire sur les pierres composées et sur les roches; par le commandeur Déodat de Dolomieu", *Observations sur la physique,* xxxix (1791), 374-407.

153. J. C. Delamétherie, *Leçons de géologie, données au Collège de France* (3 vols, Paris, 1816), iii, 222-3.

154. D. G. S. T. G. de Dolomieu, *op. cit.* (ref. 152); and "Suite du mémoire sur les pierres composées et sur les roches . . .", *Journal de physique,* xl (1792), 41-62, 203-18, 372-403, 481.

155. *Ibid.* The last entry ended with the following poignant sentences: "I am, sir, too sensible of the misfortunes of my country, I am too shocked by the outrages of supporters of the various factions, I am too preoccupied with the dangers that surround the hereditary representatives of the nation, to have the mental disposition needed to undertake the scientific work. My duty and my will dedicate my time and my energies to the

safety of the king, consequently I interrupt my memoir on the composite stones until such time as my country is delivered from the enemies that conspire for its destruction."

156. G. Cuvier & A. Brongniart, "Essai sur la géographie minéralogique des environs de Paris. Par MM. Cuvier et Alex. Brongniart", *Journal des mines,* xxiii (1808), 421-58; and *Essai sur la géographie minéralogique des environs de Paris, avec une carte géognostique, et des coupes de terrain* . . . (Paris, 1811).

157. *Ibid.* (1811), 8. The divisions established were: chalk, plastic clay, coarse limestone with marine sandstone, siliceous limestone, gypsum and marl with osseous remains and the first fresh-water formation (*terrain*), marine marls, sand and sandstone with shells, upper marine sandstone, millstone formation without shells and argillaceous sand, second fresh-water formation, and lastly mud, rolled pebbles, pudding-stones, black argillaceous marls and peat.

158. *Ibid.*, 8-9 (note).

159. *Ibid.*, 254.

160. G. Cuvier, *A discourse on the revolutions of the surface of the globe, and the changes thereby produced in the animal kingdom* . . . (London, 1829), 17. (Here we read the well-known passage: "The thread of the operations is broken; the march of nature is changed; and not one of her agents now at work would have sufficed to have effected her ancient works.")

161. E.g., J. Hill, *A general natural history, or new and accurate descriptions of the animals, vegetables, and minerals of the different parts of the world; with their virtues and uses, . . . Vol. I: A history of fossils* (London, 1748); E. Mendes da Costa, *Natural history of fossils . . .*, i, pt 1 (London, 1757).

162. A. Catcott, *A treatise on the deluge* . . . (2nd edn, London, 1768).

163. J. Whitehurst, *An inquiry into the original state and formation of the Earth; deduced from facts and laws of nature. To which is added an appendix, containing some general observations on the strata in Derbyshire. With sections of them, representing their arrangement, affinities, and the mutations they have suffered at different periods of time. Intended to illustrate the preceding inquiries, and as a specimen of subterraneous geography* (London, 1778).

164. M. Neve & R. Porter, "Alexander Catcott: Glory and geology", *The British journal for the history of science,* x (1977), 37-60; R. Porter, "The role of fieldwork in the history of geology in Britain" (unpublished paper, delivered at a meeting of the British Society for the History of Science, Chelsea, 3 January, 1976).

165. For emphasis on these aspects, see particularly S. I. Tomkeieff, "James Hutton and the philosophy of geology", *Proceedings of the Royal Society of Edinburgh* (Section B), lxiii (1948-49), 387-400; R. H. Dott, "James Hutton and the concept of a dynamic Earth", in C. J. Schneer, *op. cit.* (ref. 52), 122-41; P. A. Gerstner, "James Hutton's theory of the Earth and his theory of matter", *Isis,* lxiii (1968), 26-31; F. Ellenberger, "La thèse de doctorat de James Hutton et la renovation perpetuelle du monde", *Annales Guébhard-Séverine,* xlix (1973), 1-37. See also J. E. O'Rourke, "A comparison of James Hutton's *Principles of knowledge* and *Theory of the Earth*", *Isis,* lix (1978), 5-20.

166. J. Hutton, *Theory of the Earth with proofs and illustrations* (2 vols, Edinburgh & London, 1795).
167. See V. A. & J. M. Eyles, "Some geological correspondence of James Hutton", *Annals of science,* vii (1951), 316-39.
168. These are now held by the Library of the Royal Society of Edinburgh and have recently been published: G. Y. Craig (ed.), *James Hutton's Theory of the Earth: The lost drawings* (Edinburgh, 1978).
169. It may be added here that Hutton's cyclic Earth theory does not have much in common with the general theories of history that were being promulgated in his time, which most commonly were directionalist and sometimes saltatory, or at least conceived in terms of fairly distinct epochs. Possibly this lack of congruence with the leading contemporary view of history may do something to explain the poor reception accorded to Hutton's *Theory.* But the emphasis of David Hume on the uniformity of human nature (see below) should not be overlooked.
170. J. Playfair, *Illustrations of the Huttonian theory of the Earth* (Edinburgh & London, 1802), 2.
171. *Ibid.*
172. W. Smith, *Stratigraphical system of organized fossils, with reference to the specimens of the original geological collection in the British Museum: Explaining their state of preservation and their use in identifying the British strata* (London, 1817), vii.
173. *Ibid.,* ix-x.
174. W. Smith, *Abstract views of geology.* Page proofs of this unpublished work are held in the library of the Department of Geology at the University of Oxford. The work has been discussed by L. R. Cox, "New light on William Smith and his work", *Proceedings of the Yorkshire Geological Society,* xxv (1942), 1-99.
175. W. Smith, *op. cit.* (ref. 174), i.
176. It is not, I suggest, 'fair' to form any kind of judgement upon Smith's capacities as a geologist merely by recourse to the *Abstract,* and this is not the intention here. The document is, however, useful for the light that it throws upon some of Smith's more deep-seated intellectual commitments.
177. Smith MSS, Geology Department, Oxford University. This document is apparently an addendum to the notes that Smith prepared for a lecture on geology given at Scarborough in August, 1824. See J. M. Edmonds, "The geological lecture-courses given in Yorkshire by William Smith and John Phillips, 1824-1825", *Proceedings of the Yorkshire Geological Society,* xl (1975), 373-412.
178. W. Smith, *Natural order of the strata in England and Wales accurately delineated and described with practical demonstrations thereon containing many geological remarks on the extent of such regular arrangement and a particular description of the fossil and mineral productions of each stratum by Wm. Smith Landsurveyor and Drainer 1801,* f.7.v. (Smith MSS, Oxford).
179. R. Laudan, "William Smith. Stratigraphy without palaeontology", *Centaurus,* xx (1976), 210-26.
180. Letter: W. Smith to J. Phillips, Hackness, 24 January, 1831, giving an account of the gradual development of Smith's ideas (Oxford MSS). Evidently, the information furnished by Smith was used by Phillips in

the preparation of the biography of his uncle: *Memoirs of William Smith, LL.D., author of the "Map of the strata of England and Wales", by his nephew and pupil, John Phillips, F.R.S., F.G.S....* (London, 1844).

181. R. Jameson, *A mineralogical description of the County of Dumfries* (Edinburgh & London, 1805), 42.

182. R. Jameson, "Mineralogical queries, proposed by Professor Jameson (Read 9th April and 14th May 1808)", *Memoirs of the Wernerian Natural History Society*, i (1808-10), 107-25.

183 C. G. B. Daubeny, "Notes taken at Jameson's lectures on natural history 1816-17 with additional matter extracted from other sources" (Daubeny MSS 386, Magdalen College Library, Oxford (reproduced here by courtesy of the President and Fellows of the College)).

184. [W. H. Fitton], "Art. III—1. *A delineation of the strata of England and Wales* . . . . By W. Smith . . . London . . . 1815. 2. *Geological section from London to Snowdon* . . . . By W. Smith, 1817. 3. *A memoir to the map and delineation, &c.* By W. Smith . . . London, 1815. 4. *A series of county maps, . . . from documents in Mr. Smith's possession* . . . London, 1817. 5. *Strata identified by organized fossils* . . . London, 1816. 6. *Stratigraphical system of organized fossils* . . . by W. Smith . . . London . . . 1817", *Edinburgh review*, xxix (1818), 310-37, p. 325.

185. R. Bakewell, *An introduction to geology illustrative of the general structure of the Earth; comprising the elements of the science, and an outline of the geology and mineral geography of England: by Robert Bakewell* (London, 1813), 43.

186. J. Macculloch, *A system of geology, with a theory of the Earth, and an explanation of its connection, with the sacred records* (2 vols, London, 1831), i, 268. (Macculloch states in his preface that the text of this work was written in 1821, though it remained unpublished for ten years.)

187. J. Kidd, *A geological essay on the imperfect evidence in support of a theory of the Earth, deducible either from its general structure or from the changes produced on its surface by the operation of existing causes* (Oxford, 1815), 269.

188. G. B. Greenough, *A critical examination of the first principles of geology; in a series of essays* (London, 1819), 287-8.

189. T. Webster, "On the freshwater formations of the Isle of Wight, with some observations on the strata over the chalk, in the south-east part of England", *Transactions of the Geological Society of London*, ii (1814), 161-254. Webster had been engaged by Sir Henry Englefield to conduct a geological survey of the island. Webster's series of letters to Englefield (dated between 21 May, 1811 and 11 February, 1813) were published in H. C. Englefield, *A description of the principal picturesque beauties, antiquities, and geological phaenomena, of the Isle of Wight. . . . With additional observations on the strata of the island, and their continuation in the adjacent parts of Dorsetshire. By Thomas Webster, esq.* (London, 1816).

190. W. Richardson, "On the strata of mountains", *Philosophical magazine*, xxxvii (1811), 367-9, p. 368.

191. W. D. Conybeare & W. Phillips, *Outlines of the geology of England and Wales, with an introductory compendium of the general principles of that science, and comparative views of the structure of foreign countries* . . . (pt 1, London, 1822), xxi.

192. L. G. Wilson, *Charles Lyell: The years of controversy; the revolution in geology* (New Haven & London, 1972), 215-16.

193. C. Lyell, *Principles of geology, being an attempt to explain the former changes of the Earth's surface, by reference to causes now in operation* (3 vols, London, 1830-33), i, 1.

194. C. Lyell, *A manual of elementary geology: or, the ancient changes on the Earth and its inhabitants as illustrated by geological monuments* (3rd edn, London, 1851), 96-97.

195. *Ibid.*, 98.

196. *Ibid.*, 106. (Note also the sub-title of the *Manual.*)

197. *Ibid.*, 2.

198. On this point, see the recent discussion of M. J. S. Rudwick, "Historical analogies in the geological work of Charles Lyell", *Janus*, lxiv (1977), 89-107. Rudwick sugests that Lyell's thinking may well have been influenced by the work of Niebuhr. He also cites the following remark of Lyell from L. G. Wilson, *op. cit.* (ref. 192): "We err as much when we judge of a political constitution without considering the pre-existent state of the laws from which it has grown". This places Lyell within the ambit of the historicist school of law, advocated by von Savigny, and referred to above.

199. M. J. S. Rudwick, "Poulett Scrope on the volcanoes of the Auvergne: Lyellian time and political economy", *The British journal for the history of science*, vii (1974), 205-42. For S. J. Gould's use of the term 'substantive uniformitarianism', see his paper: "Is uniformitarianism necessary?", *American journal of science*, cclxiii (1965), 223-8.

200. C. Lyell to G. A. Mantell, 15 February, 1830, in K. M. Lyell (ed.), *The life, letters and journals of Sir Charles Lyell, Bart* (2 vols, London, 1881), i, 262; C. Lyell, *op. cit.* (ref, 193), i, 123.

201. M. Bartholomew, "The non-progress of non-progression: Two responses to Lyell's doctrine", *The British journal for the history of science*, ix (1976), 166-74.

202. D. Hume, *An enquiry concerning human understanding*, sect. VIII, *pt* I, §65, in L. A. Selby-Bigge (ed.), *Enquiries concerning the human understanding and concerning the principles of morals by David Hume* (2nd edn, Oxford, 1902), 83-84.

203. See, for example, D. Hume, *The history of England . . . in six volumes* (London, 1841). Hume's *History* might today be regarded as 'mere chronicle' rather than 'history proper'. Indeed, by giving a year-by-year account of events in the history of England, his work had something of the character of the earlier 'annalists'.

204. Macaulay, in his famous critical attack of 1829 upon James Mill's *Essay on government*, emphasized the extraordinary diversity of human beings, and maintained that it was quite impossible to deduce a "science

of government" from a knowledge of the principles of human nature (*i.e.*, the principle of self-interest). See G. M. Young (ed.), *Macaulay: Prose and poetry* (London, 1952), 579-609.

205. M. Bartholomew, *op. cit.* (ref. 201), 167.

206. It should be noted here that my interpretation of Lyell as one definitely concerned with Earth history—his uniformitarianism notwithstanding—runs counter to the interpretation offered by Professor R. Hooykaas, who sees the "steady-state system [of the uniformitarians as] . . . essentially ahistorical" (*Science and belief: From Copernicus to Darwin . . . . The new outlook for science* (Milton Keynes, 1974), 13). I contend, however, that it is possible to give an historical account of the Earth's past even within a uniformitarian framework; for the particular course of geological events may be traced out, in accordance with the directives of a uniformitarian methodology.

207. W. Buckland, *Geology and mineralogy considered with reference to natural theology* (2 vols, London, 1836), i, 8.

208. G. Penn, *Conversations on geology; comprising a familiar explanation of the Huttonian and Wernerian systems; the Mosaic geology, as explained by Mr. Granville Penn; and the late discoveries of Professor Buckland, Humboldt, Dr. Maculloch, and others* (London, 1840), 6-7 (emphasis added).

# XIII

## By Grid and Group Divided: Buckland and the English Geological Community in the Early Nineteenth Century

N. A. RUPKE, *The great chain of history: William Buckland and the English school of geology* (1814–1849). Oxford: Clarendon Press, 1983. xii + 322 pp. 14 plts. £22·50.

M. RUDWICK, 'Cognitive Styles in Geology', in, *Essays in the sociology of perception*, edited by M. DOUGLAS, London, Boston and Henley: Routledge & Kegan Paul, 1982. pp. 219–41.

Nicolaas Rupke has written an excellent book, offering a novel way to write the history of geology for a particular intellectual community in a given period—that is, by examining a major figure in detail, and using this figure as a key to the understanding of the period in question, both in its intellectual and its social aspects. The period involved is approximately the first half of the nineteenth century. The intellectual community is that of the *English* (as opposed to the Scottish) school of geology, encompassing Oxbridge academics, members of the Geological Society, 'practical' men, and a host of amateur observers and clerics with an interest in geology. And the chosen central figure is the Oxford scholar, William Buckland (1784–1856).

Since Rupke has sought to display the social formations within which, and according to which, Buckland worked, as well as his specific scientific contributions, I have chosen to consider his book in the light of an important recent essay by Martin Rudwick. This also examines the geological community in Britain in the early nineteenth century, but does so in a way that is still fairly new to historians of science, namely in terms of Mary Douglas's method of 'grid/group' analysis.[1] Thus, by juxtaposing the work of Rupke and Rudwick, we have a means whereby we may appraise the merits of Rupke's historiography, and also the adequacy of Rudwick's application of grid/group theory to the history of geology.

First, let me mention some of the more salient features of Rupke's book. Buckland, of course, has long been represented as an exponent of Cuvier's catastrophism, and his identification of the last of Cuvier's catastrophes with the Deluge is well known.[2] By this identification, Buckland sought to align geological and scriptural history. He is also known for his wish to have man as a late entrant to the geological scene,

[1] M. Douglas, *Natural Symbols: Explorations in Cosmology* (London, 1970); *Cultural Bias* (London, 1978); and 'Introduction' to *Essays in the Sociology of Perception*, edited by M. Douglas (London, 1982). For a slim bibliography of works in the history of science that have deployed grid/group analysis, see S.S. Shapin, *History of Science*, 20 (1982), 157–211 (p. 211).

[2] For example, C.C. Gillispie, *Genesis and Geology: A Study in the Relations of Scientific Thought, Natural Theology, and Social Opinion in Great Britain, 1790–1850* (New York, 1959), p. 110.

supposedly being, as Porter has put it, 'fenced off from geological history'.[3] Thus, we have the legend of the 'Red lady of Paviland', with the Oxford don apparently unwilling to accept good empirical evidence for the antiquity of man. As a result, Buckland has sometimes been regarded as an obscurantist by reason of his theological prejudices, and he was formerly the object of some degree of derision among historians of geology.[4] But this has presented historians of geology with an 'extraordinary paradox',[5] for one might think that Buckland, as scriptural geologist, would be pleased to find evidence of humans who had died as a result of the Deluge. Scheuchzer, after all, had been glad to find '*homo diluvii testis*';[6] so why the fuss about the 'Red lady of Paviland'?

The 'paradox'[7] is quickly dispelled by Rupke's analysis. Cuvier's theory assumed an exchange of positions of land and sea at the time of the Deluge.[8] So the remains of antediluvian man lay beneath the modern oceans; and previous supposed 'discoveries' of antediluvians, like that of Scheuchzer, were illusory. But in any case, Rupke explains, Buckland was *not* simply a spokesman for Cuvierian theory. In fact, he *did* have hopes of finding samples of fossil antediluvian man, perhaps in Asia, if not in Britain (to which country man might not have penetrated in antediluvian times). And on Buckland's version of diluvial theory, as stated in his *Reliquiae Diluvianae*, while Cuvier's last great catastrophe did correspond with the Deluge, no general land/sea reversal was presumed.[9] So the finding of fossil man was not excluded; and the discovery of fossil cattle bones in diluvial gravel gave Buckland cause to hope for the eventual discovery of fossil human remains—man and cattle being supposedly 'linked' by divine intent.[10] So on Rupke's view, Buckland's interpretation of the remains of the 'Red lady' was an indication of 'professional prudence', not obscurantism.[11] For he appreciated how loose material could be reworked by floods, and how buried human remains could easily be given mistaken stratigraphical locations. There was, therefore, nothing peculiarly stupid about the 'Red lady' episode. However, Buckland was never persuaded that bona fide human fossil remains had been found in true association with *extinct* animal remains, and consequently he could maintain that Genesis and geology remained essentially in accord.

Even so, some of Buckland's contemporaries found his work intellectually disturbing. While his investigations of the 'hyena dens' of Kirkland Cave and elsewhere contributed to his high reputation among geologists, the cave work was not to the liking of all orthodox diluvialist theorists, for it gave a picture of *antediluvian* conditions

[3] R. Porter, 'Creation and Credence: The Career of Theories of the Earth in Britain, 1660–1820', in *Natural Order: Historical Studies of Scientific Culture*, edited by B. Barnes and S. Shapin (Beverley Hills and London, 1979), pp. 97–123 (p. 116). (Porter is here quoting from an unpublished paper by W. F. Bynum.)

[4] Gillispie (footnote 2), 109–10. A more balanced account of Buckland's excavations at Paviland was given earlier in the 1940s: F. J. North, 'Paviland Cave, the "Red Lady", the Deluge, and William Buckland', *Annals of Science*, 5 (1942), 91–128. Regrettably, my own book, *Darwinian Impacts* ...(1980), p. 229, leaned on Gillispie's discussion of Buckland more than I would now wish.

[5] Porter (footnote 3), 116.

[6] J. J. Scheuchzer, *Homo Diluvii Testis et Θεοσκοπός, publicae συζητήσει expositus a Joh. Jacobo Scheuchzero*,...(Zurich, 1726).

[7] There is no 'paradox' implicit in the even-handed discussion of Buckland by W. F. Cannon in the *Dictionary of Scientific Biography*.

[8] G. Cuvier, *Essay on the Theory of the Earth. Translated by Robert Kerr, with Mineralogical Notes by Professor Jameson* (Edinburgh, 1813), p. 131.

[9] W. Buckland, *Reliquiae Diluvianae; or, Observations on the Organic Remains Contained in Caves, Fissures, and Diluvial Gravel, and other Geological Phenomena, Attesting the Action of an Universal Deluge* (London, 1823), pp. 162–3.

[10] Rupke cites evidence fron the Buckland papers at the University Museum, Oxford, on this point.

[11] Rupke, p. 92.

and tended to detract from the importance of the Flood as *the* agent of geomorphological change. There was some displeasure, therefore, when Buckland virtually ignored the Flood in his Bridgewater Treatise and explicitly denied that it could be held responsible for all the layers of stratified rocks, with their succession of fossil contents.[12]

However, this did not provide Buckland with a solution to the problem of the origins of the 'diluvial' superficial deposits, with which he was particularly concerned from his earliest investigations. So it comes as no surprise that after some initial scepticism, he adopted Agassiz's glacial theory to explain the phenomena that he had previously accounted for in diluvialist terms.[13] The gradual elaboration of Buckland's views on these points is now considerably clarified by Rupke's researches, with the benefit of his examination of Buckland's Oxford lecture notes.

This brings us to consider the way in which Rupke links his story to the institutional framework provided by early nineteenth-century Oxford science. As geology was established in the academic curriculum it had to accommodate itself to the entrenched traditions of classical learning.[14] For such an accommodation, diluvial geology served most satisfactorily. It provided a link between the 'textual' evidence offered by Biblical history and the history of the Earth that was beginning to be worked out by nineteenth-century stratigraphers. Thus, consideration of Buckland's social environment helps to explain why diluvial theory was so important to him. But this should not be taken as an indication of some kind of intellectual charlatanry on his part, as Gillispie's Whiggish *Genesis and Geology* might seem to suggest. It is more profitable to see Buckland as attempting to 'legitimate' scientific studies in the Oxford curriculum, seeking to 'ingraft' (his term) geology and mineralogy with traditional classical studies.[15] We also see a resolute effort to 'vindicate' ('legitimate'!) geology in terms of the design argument, with natural theology running as a guiding thread through Buckland's life work. All this does not betoken intellectual insincerity or incompetence. It was a natural way for geological researches to be accommodated to their social surroundings.

As to the specifics of Buckland's work in geochronology, it should be recognized as perfectly characteristic of his time. He was an active participant in the early nineteenth-century efforts to build a composite international history of the Earth's strata, based upon correlations using paleontological and lithological criteria.[16] His tabular 'Order of Superposition of Strata in the British Islands' (prepared for teaching purposes at Oxford) was widely distributed, and was, Rupke reports, highly regarded by Continental geologists.[17] Thus, while it is true that Buckland saw an important

[12] W. Buckland, *Geology and Mineralogy Considered with Reference to Natural Theology*, 2 vols (London, 1836), I, 16–7.

[13] W. Buckland, 'Memoir on the Evidences of Glaciers in Scotland and the North of England', *Proceedings of the Geological Society of London*, 3 (1842), 332–7, 345–8.

[14] Chemistry had to do likewise. See D. R. Oldroyd and D. W. Hutchings, 'The Chemical Lectures at Oxford (1822–1854) of Charles Daubeny, M. D., F. R. S.', *Notes and Records of the Royal Society of London*, 33 (1979), 217–59.

[15] W. Buckland, *Vindiciae Geologicae; Or the Connexion of Geology with Religion Explained* (Oxford, 1820), p. 2. Adam Sedgwick undertook a somewhat similar enterprise in his well-known 'Discourse', given in 1832 at Trinity College, Cambridge. See A. Sedgwick, *A Discourse on the Studies of the University... With an Introduction by Eric Ashby and Mary Anderson* (New York and Leicester, 1969), pp. 22–5 and passim.

[16] See D. R. Oldroyd, 'Historicism and the Rise of Historical Geology', *History of Science*, 17 (1979), 191–213, 227–57.

[17] Rupke, p. 123. Buckland had several such synoptic tables printed. One that I have examined, thought to have been published in 1817, shows Buckland's attempts to integrate Werner's tables, based chiefly on German observations, with those of British stratigraphers, notably William Smith.

stratigraphical distinction between 'antediluvial' and 'postdiluvial' materials and circumstances,[18] he extended his investigations on chronology deep into 'antediluvian' times, and did work that meshed well with the current stratigraphical programme in Britain and Europe. His work also accorded with the prevailing 'directionalist' theory of Earth history, with an assumed progressive fossil succession that permitted stratigraphical correlation.[19] Moreover, Buckland envisaged an ample geological time-scale. Rupke presents evidence that from early in his career, Buckland was thinking of the Earth as being of very great age, and Rupke denies that this view was forced on Buckland by the arguments of Huttonians such as Lyell.[20] However, it was only in the fairly late Bridgewater Treatise (1836) that Buckland began to publish ideas about the Earth being millions of years old, before the first 'day' of the Mosaic narrative.[21] So possibly Rupke underestimates the influence on Buckland's thinking of arguments such as those advanced by Lyell.[22] Buckland, we may note, seems, like Lyell, to have held that there were successive creations and extinctions (e.g. the dodo) of living forms. But for Buckland the changes were 'progressive', being related to a gradually cooling Earth; for Lyell they were not.

Buckland's work exemplified a number of the ideals of the Romantic era, and his empathetic re-creation of past worlds—as inhabited by giant extinct lizards—greatly appealed to the public imagination. Rupke shows how Buckland's work was intelligently guided by Paleyesque design arguments, so that, for example, he was able to show—contrary to what Cuvier had supposed—that the skeleton of the *megatherium* was not that of some ill-designed, mis-shaped monster. The creature was splendidly suited to its life on the Pampas, where it could obtain its root diet with the help of feet, admirably fitted for digging.

Such matters are of concern not only to historians of science, but also to students of English literature. An attractive feature of Rupke's book is his analysis of the influence of geological thought on painting and poetry. The huge canvasses of John Martin are shown to have been influenced by diluvialist geological theory. And, Rupke argues, the extensive critical literature that has grown up which sees evidence of Lyellian geology and anticipations of Darwinian notions in Tennyson's *In Memoriam* and *The Princess* is so much nonsense! For if one looks at these poems forgetting all about Lyell or Darwin, one sees, sure enough, the ideas of Buckland and Mantell: creationism, progressionism and diluvialist theory; monsters and aeons of time.

But Rupke's book will probably be of greatest interest to those concerned to analyse the social and intellectual structure of geology in the early nineteenth-century. For Rupke, the main division in British geology at that time ran between the 'Scottish school' and its 'English' counterpart. The Scottish school—in its Huttonian and Wernerian (Jamesonian) versions—is generally well known.[23] Lyell, of Scottish blood, belonged essentially to this grouping, which, on Rupke's view, was not strictly

[18] Buckland (footnote 9), 2.

[19] See M. J. S. Rudwick, 'Uniformity and Progression: Reflections on the Structure of Geological Theory in the Age of Lyell', in *Perspectives in the History of Science and Technology*, edited by D. H. D. Roller (Norman, 1971), pp. 209–27.

[20] Rupke, pp. 165–6.

[21] Buckland (footnote 12), I, 21–2.

[22] See M. J. S. Rudwick, 'Lyell on Etna, and the Antiquity of the Earth', in *Toward a History of Geology*, edited by C. J. Schneer (Cambridge, Mass., 1969), pp. 288–304.

[23] A. Geikie, 'The Scottish School of Geology', in his *Geological Sketches at Home and Abroad* (New York, 1892), pp. 250–71. (This essay formed Geikie's Inaugural Lecture for the Class of Geology and Mineralogy in the University of Edinburgh, 6 November, 1871.)

'historical' in outlook. The English school, now properly characterized for the first time, had the following seven features: (1) its own self-characterization as an identifiable entity; (2) its interest in the Secondary rocks of England and Wales; (3) its emphasis on fossil-based stratigraphical studies, with particular interest in fossil monster reptiles; (4) diluvialism and progressive changes in species types, meshing with a changing environment produced by a cooling Earth; (5) an attempted integration of geological and human history with the help of the Mosaic record; (6) a commitment to natural theology; and (7) a disdain for economic geology.[24]

Such a scheme accords with previous accounts of the early nineteenth-century English geological community—though the existence of an 'English school' *per se* has largely passed unnoticed. But some readers may be startled at Rupke's suggestion that 'no such thing as a catastrophist synthesis [with Buckland as its chief architect] ever existed in early nineteenth-century England'.[25] To be sure, the English geologists believed there had been catastrophes in the Earth's past. But this did not, in Rupke's view, have special methodological and theoretical significance, as uniformitarianism did for Lyell. So the old polarity of catastrophism versus uniformitarianism is repudiated by Rupke. Rather, he argues, the important thing to recognize is that the 'English school was one of historical geology'.[26] This agrees with my own characterization of the geological thought of the age[27] and also with that of Gabriel Gohau.[28] However, our concern has been to draw distinctions between the 'historicist' approach of nineteenth-century geologists, using '*archives pétrographiques*' and '*archives paléontologiques*' (Gohau's terms), and the earlier 'genetic' theories of the Earth (such as that of Leibniz) or 'steady-state' systems (such as that of Hutton). Clearly, then, some other kinds of discrimination are required in order to comprehend the different modes of geological thinking of early nineteenth-century geological thought. For this, the recent work of Martin Rudwick commands our attention.[29]

Rudwick has characterized four '*styles*' of geological work: 'abstract', 'agnostic', 'concrete' and 'binary'. The *abstract* style is exemplified by the work of grand theorizers such as Hutton and Lyell: 'big men' and strong theoreticians interested in grand intellectual syntheses. Sedgwick and Buckland are offered as instances of the *concrete* style—focusing their attention on 'the concrete order of strata and other rocks, rather than the temporal earth-history that they represent'.[30] Their chief concern was to identify formations, and their order and structural relationships. Contrary to the thinking of (say) Hutton, time was regarded as a function of strata, rather than the other way round. And geologists displaying the 'concrete' style did not evince much interest in causal agents—to account for the phenomena attributed to catastrophes, for example.

In the *agnostic* style, such as that of George Greenough, there was a kind of positivistic renunciation of theoretical generalizations. Such geologists marked different geological strata on maps, but were reluctant to synthesize their observations

[24] Rupke, pp. 16–18.
[25] Ibid., p. 193.
[26] Ibid.
[27] Oldroyd (footnote 16).
[28] G. Gohau, 'Idées anciennes sur la formation des montagnes', *Cahiers d'Histoire et de Philosophie des Sciences*, No. 7 (Paris, 1983), 45 and passim.
[29] Rudwick, 'Cognitive Styles in Geology'.
[30] Ibid., 224.

into any orderly sequence, in the light of some guiding theory. Finally, in the *binary* style, authors seemingly envisaged a 'unique boundary-event in the past'[31]—sharply separating the present and past order of things. The earlier period was regarded as undifferentiated and chaotic, while the latter was ordered and corresponded with present conditions. Scriptural geologists and modern 'creation scientists' can, it is suggested, be pigeon-holed here. Perhaps there is some reason to include Buckland in this category, on account of his reliance on scriptural evidence for geological work, his distinction between 'antediluvial' and 'postdiluvial' epochs, and his distinction between the pre-human and the human periods of Earth history. Agassiz's ice-age might offer another example of singularity in the Earth's geological past, dividing the past and present order of things, thereby instancing the 'binary' style of geological thinking. Each style, Rudwick suggests, corresponded to the characteristic social formations of its adherents. Hence we have a tool for analysing the relationships between 'internal' and 'external' factors in the history of geology.

As stated above, Rudwick's scheme is based upon the method of 'grid/group analysis'. According to this procedure, the social anthropologist seeks to classify persons first, according to their degree of commitment to the social *group* to which they belong and by which their actions are shaped and determined; and second, according to the intensity of social control to which they are subjected by reason of the social categories and concomitant roles that obtain within the society of which they are members (i.e. *grid* control). Plotting 'grid' and 'group' intensities on two perpendicular axes, we have four 'pigeon holes' for describing a person's social environment and associated social control. The four categories may be named (if one can tolerate the sociological jargon) as those of 'individualism' (*A*), 'atomized subordination' (*B*), 'ascribed hierarchy' (*C*), and 'factionalism' (*D*).[32] Thus we have:

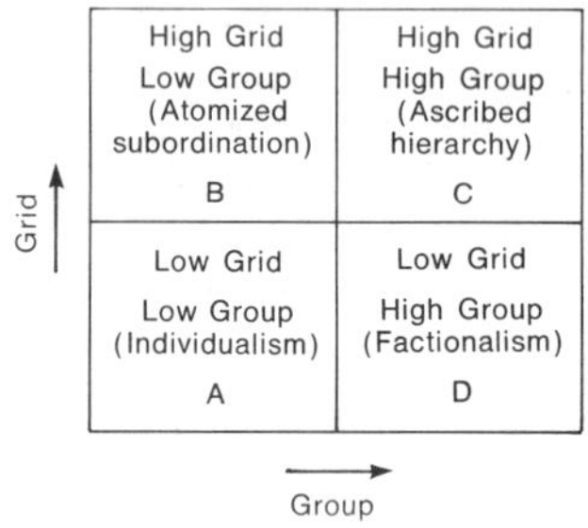


Figure 1.
Rudwick's grid/group scheme of social environment and control.

To illustrate the general idea, we may offer some homely examples. Box *A* might be inhabited by 'rugged individualists'—frontiers-people! *B* would be occupied, for example by the unwilling conscripts of some army, subject to hierarchical controls imposed by the system in which they find themselves, but with minimal allegiance to that system. (Mary Douglas's telling example is the Edwardian female domestic.) In *C*

[31] Ibid., 229.
[32] This terminology is due to David Ostrander. See Douglas, editor, *Essays* (footnote 1) 4.

we might find members of a Japanese corporation, with a strong hierarchical social structure, which is, however, willingly accepted by its members, who have, amongst themselves, a high degree of group solidarity. The members of some kinds of religious sects might occupy Box *D*. They have negligible social hierarchy within their sect, but divide the world as a whole into regions of 'them' and 'us'—between those who are within or beyond the pale.

Now, *if* it be accepted that there are four such broad social formations in which one may find oneself, and *if* one accepts the sociology-of-knowledge hypothesis that one's thinking, or one's cognitive structures, are strongly influenced (nay determined) by the social formations within which one functions and with which one interacts—*then* one may expect to find four basic 'cosmologies'[33] associated with these four broad types of social formation. Such a mode of analysis has been applied to a wide variety of social circumstances, both historical and contemporary, and it is Rudwick's contribution to seek to apply the method to the history of geology. He claims to identify the four different cognitive styles, briefly described above, in the early nineteenth-century British geological community. And then he suggests that these are just what might be anticipated as a result of the application of the method of grid/group analysis. To this end, he has prepared several diagrams, indicating four different attitudes to time, four different ways of classifying strata in relation to time, and four possible responses to anomalies that theories may encounter. To take all this in at a glance, we may combine Rudwick's separate diagrams, yielding a composite picture for a four-fold classification of geological 'styles', corresponding with the four broad divisions arising from the method of grid/group analysis (see Figure 2).

Given this analytic scheme, and Rupke's work, we may now set the two together to see whether they are mutually compatible with one another. It may be noted, first, that Rudwick actually mentions Buckland as occupying *two* regions within his map of the intellectual and social terrain of early nineteenth-century geology, *viz.* squares *C* and *D*.[34] However, Rudwick also states that diluvialists represented a 'less pure' form of the binary style; in fact, they may be regarded as 'intermediate between the "pure" binary style and the concrete style'.[35] So perhaps we should locate Buckland close to the *C–D* borderland of the grid/group map of the geological community.

However, we need not require Buckland to occupy a fixed location throughout his lifetime, and Rupke offers us some opportunity to trace Buckland's 'world-line', or intellectual/social trajectory through time. But to do this is difficult, since *The great chain of history* does not provide a year-by-year account of Buckland's personal and intellectual development. However, I take it that if Rudwick's schema is to be accepted a person must, at any given time, occupy a position *somewhere* on the map, even if he shifted around somewhat from time to time, owing to changes in his social context.[36] Can we say that this requirement is satisfied for Buckland, given what is now revealed by Rupke's work?

[33] In Douglas's terminology, people's 'cosmologies' correspond to the ultimate justifying ideas that they invoke as a warrant for the 'natural order' of things, as they comprehend it according to their social environment (*Essays* footnote 1, 5).

[34] Rudwick, pp. 224, 229.

[35] Ibid., p. 229.

[36] Buckland may, for example, have shifted his position within the grid/group map according to whether he was under the immediate influence, of say, Greenough or Agassiz. There is, it seems, some tendency for the grid/group method of analysis to regard the social environment as static. But one needs to consider possible changes in social context, which may yield different cognitive styles in a scientist's work at different periods of his career.

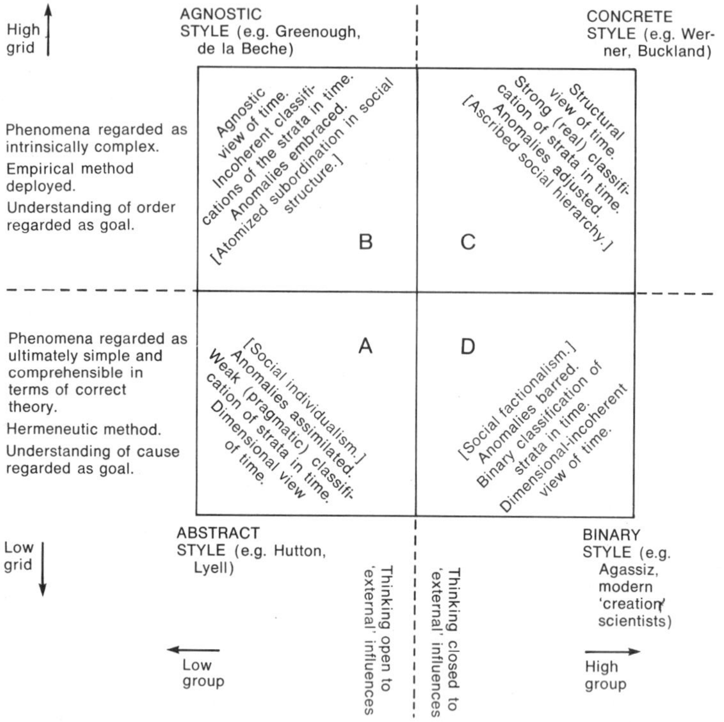


Figure 2.
Combination of Rudwick's schema and classification of geological styles.

Considering Buckland's preparation of synoptic stratigraphical tables, it would seem appropriate to locate him in Box *C*. But given the tendency towards binary thinking, supposedly manifested by Buckland's distinction between antediluvial and postdiluvial geology and his subsequent acceptance of Agassiz's glacial theory, we also have reason to locate him in Box *D*. Rudwick's location of Buckland in the *C–D* border region, with a style that was 'transitional binary/concrete',[37] therefore seems to serve quite well. But then, we may ask, are Rudwick's other characteristics for 'concrete' and 'binary' styles satisfied by Buckland, as he is revealed by Rupke's analysis? And can we find any arguments for thinking of Buckland as belonging to *A* or *B*?

Well, as indicated above, Buckland did have a binary view of Earth history in *some* measure. But his construal of antediluvial time was not 'incoherent'. It was clearly 'structured' in terms of the observable layers of the stratigraphical column, as revealed by his stratigraphical tables. And while Buckland may have thought of Earth history in terms of millions of years, we have no indication that he had anything more than a *relative* scale to deal with chronology in antediluvial times. For Buckland, geological

[37] Rudwick, p. 232.

time was defined by strata in the antediluvial period. Thus, so far as his attitude to geological time was concerned, he seems to have been primarily a Box *C* man.

Rudwick sees Box *C* as being inhabited by people in mining schools, universities or geological surveys, content to become experts in some specialized field, and belonging to some invisible college with a tacitly acknowledged pecking-order therein. Their work is 'group-oriented' and 'routinized'.[38] To my mind, this category has some analogy with the 'normal scientists' of Kuhn. By contrast, we find the 'system builders' (Hutton, Lyell, Darwin *etc.*) in Box *A* doing (it may be suggested) Kuhnian 'revolutionary science'. So, for Rudwick, the 'stable' axis of geological theory and practice is that of *A–C*. *C–D* is the 'unstable' axis,[39] with an arid positivism at one end[40] and a blind adherence to a faith, arrived at independently of geological considerations, at the other. On this interpretation, the history of geology can be seen as a dialectical interplay between the poles of the stable *A–C* axis, with *C–D* making a relatively insignificant contribution. Buckland, then, as we have said, hovers on the border region between *C* and *D*, but, though showing some tendency towards the unstable 'binary' style, may be regarded as belonging to the 'useful' *A–C* axis.

To judge further the adequacy of this placement, we may ask how Buckland sought to deal with 'anomalies'. Did he 'bar' them or 'adjust' them? According to traditional historiography (e.g. Gillispie), Buckland 'barred'[41] the anomaly of the 'Red Lady of Paviland' by means of '*ad hoc* barring'. However, in my view, Buckland's actions exemplify his doing 'normal' science in his area of expertise, namely the excavation of cave remains. Again, we have more reason to regard Buckland as a Box *C* man, away from the 'unstable' Box *D*, where, I believe, traditional historiography of science would be inclined to place him (had it had access to some kind of conceptual analytic scheme such as that of Rudwick).

We should note that Rudwick regards the inhabitants of Boxes *C* and *D* as being relatively closed to external influences. This doubtless applies well enough to modern 'creation scientists', and perhaps also to the 'normal' inhabitants of Box *C*. But was it true of Buckland? How willing was he to change his mind in the face of arguments contrary to his view? Well, given his gradual abandonment of the Deluge as the major event of geological history, despite the high reputation of the *Reliquiae Diluvianae*, we seem to have evidence of Buckland's leaning more towards the 'concrete' style than the 'binary', or out of the right-hand side of the grid/group diagram altogether. For his part, however, Rudwick sees Buckland's willingness to entertain Agassiz's glacial theory as an indication of 'binary' tendencies within Buckland's geological style.[42] But

[38] Ibid., p. 235.

[39] Here Rudwick follows David Ostrander, who sees environments *A* and *C* as potentially stable and offering the possibility of enduring social structures. *C* and *D* are held to form the 'unstable diagonal', 'not capable of generating an enduring social structure'. See D. Ostrander, 'One- and Two-Dimensional Models of the Distribution of Beliefs', in Douglas, editor, *Essays* (footnote 1), 14–30 (p. 26).

[40] Rachel Laudan has argued that the methodology adopted by Greenough and his associates in the Geological Society at its first formation was, with its agnosticism towards geological theory, not conducive to successful research. See her paper 'Ideas and Organization in British Geology: A Case Study in Institutional History', *Isis*, 68 (1977), 527–38.

[41] The question of attitude to anomalies—whether they are 'barred' or embraced, adjusted or assimilated—is of critical significance to some exponents of grid/group analysis. See D. Bloor, 'Polyhedra and the Abominations of Leviticus: Cognitive Styles in Mathematics', in Douglas, editor, *Essays* (footnote 1), 191–218. Supposedly the 'barring' of anomalies is related to the social formations that one finds in Box *D*—with a clear demarcation between those who are, or are not, of the 'faith'. Rudwick would also see a parallel in attitudes towards time. Box *D* people create a rigid distinction between the past condition of things (e.g. antediluvial times) and the present, where the known laws of nature prevail (postdiluvial times).

[42] Rudwick, p. 232.

equally one might see Buckland's theoretical shift here as a willingness to be open to ideas lying outside his established line of thinking. On the other hand, it might be said that Buckland's acceptance of the glacial theory marked a return to the 'binary' fold. Perhaps of greater significance is a point to be found in his 'Life and Letters', in which he is described as acquiescing to an objection made against one of his doctrines by Lyell at a meeting of the Geological Society. 'This', wrote Mrs Gordon, 'is only one of the many instances of Buckland's large-mindedness and of the heartiness with which he always welcomed any opinion which seemed a nearer approach to the truth than he had himself... been able to form'.[43] This hardly betokens membership of Box *D*.

So let us suppose that we think it right, for the most part, to situate Buckland in Box *C*. Is it then correct to regard his work as 'group oriented' and 'routinized'? Well, it was certainly group oriented. As Rupke emphasizes, Buckland was a leading member of the English *School*, and he was engaged in a co-operative effort to unravel the British stratigraphy, in concert with his fellow academics and members of the Geological Society, and in consultation with his European colleagues. But whether his work could be said to be 'routinized' is another matter. As an Oxford don, Buckland had a good deal of social autonomy, and his well-known eccentricities of character do not betoken a 'routinized' social formation. On the other hand, he was never a lone system builder. His systematic investigation of cave deposits could, in a sense, be said to be 'routinized'; but *he* determined the routine. Thus it appears that Buckland's position on the grid/group map cannot be specified with precision, even with the help of all the information that Rupke has gathered together for our consideration. But this, we may suggest, is a deficiency of grid/group theory, not Rupke's historiography.

The foregoing discussion may bring to view some of the problems that face those who wish to deploy grid/group analysis for studying the history of geology. In addition, it may draw to our attention certain differences of emphasis amongst grid/group theorists. For example, while making valuable use of grid/group theory, K. L. Caneva warns against facile 'base–superstructure determinism', reductions to social structures, or 'empty "correlationism" between independent realms'.[44] Mary Douglas herself seeks to display general 'cosmologies' manifesting themselves in a wide range of human activities as a result of grid/group characteristics.[45] But she does not try to explain the intellectual work of *individuals* on this basis. On the other hand, Celia and David Bloor have sought to display causal connections between young scientists' social formations and their attitudes towards scientific theory and practice.[46] Presumably this approach would not commend itself to Caneva.

[43] [E. O.] Gordon, *The Life and Correspondence of William Buckland, D.D., F.R.S., Sometime Dean of Westminster, Twice President of the Geological Society, and First President of the British Association* (London and New York, 1894), p. 85.

[44] K. L. Caneva, 'What should we do with the Monster? Electromagnetism and the Psychosociology of Knowledge', in *Sciences and Cultures*, edited by E. Mendelsohn and Y. Elkana (Dordrecht, 1981), pp. 103–31 (pp. 124–5).

[45] For example, in her *Cultural Bias* (footnote 1), she attempts to give an account, in terms of grid/group theory, of different attitudes towards travel, gardening, cookery, medicine and health, youth, old-age, space, time, personal relations, justice, punishment. and much else besides! The theory is very accommodating, to say the least.

[46] C. B. Bloor and D. Bloor, 'Twenty Industrial Scientists: A Preliminary Exercise', in Douglas, editor, *Essays* (footnote 1), 83–102. Also relevant is: J. Hampton, 'Giving the Grid/Group Dimensions an Operational Definition', in ibid, 64–82.

As for Rudwick, he is evidently willing to apply the grid/group method to individuals, in a manner that appears compatible with the so-called 'strong programme' in the sociology of knowledge, though he does so with the caveat that his analysis stands in need of 'more detailed illustration and documentation'.[47] For myself, I am persuaded that the application of grid/group analysis can help us to understand, in some measure, the general character of the work of individual scientists. However, there remains the question of whether the social structure 'produces' the scientist, or whether scientists of particular temperaments tend to gravitate towards particular social environments. Or is there, more plausibly, a reciprocal interaction between the scientist and his social milieu? We seem to have here yet another version of the nature/nurture problem.

But leaving aside these more general theoretical issues, it seems that, in broad terms, Rudwick's analysis performs its task, for the time and place for which it was designed, quite satisfactorily. And Rupke's new data do not clash radically with Rudwick's proposals. But it must be recognized that Buckland was one of the figures specifically considered by Rudwick in drawing up his scheme. So it is not so very surprising that the Buckland data fit the theory quite nicely. We need to look at other figures such as McCulloch, Murchison, Mantell, and so on, to judge the case more securely. Also, we need to ask how universal are the putative relationships between social formations, in terms of grid and group, and geological theory. For example, may not Box *B* 'normal' geologists one day come to hold a Lyellian view of time? Yet their social formation might seem to require them forever to hold a 'structural' view of time. It may be that this point can be accommodated by the theory, though it seems to raise the question of whether it can deal satisfactorily with the dynamics of theoretical change in geology.

But even without going beyond the first half of the nineteenth-century, further lines of enquiry suggest themselves. For example, the relationships, within the 'English school' of geology, between the academics (e.g. Buckland and Sedgwick), the practical men (e.g. Smith and Farey), the men of the Geological Society (e.g. Greenough and Webster), and the men of the Survey (e.g. de la Beche and Murchison), require closer analysis. Also, if the 'English school' was located somewhere on the *C–D* borderland of Rudwick's grid/group map, where did the 'Scottish school' belong? And if there were differences, why was this so? For such studies, detailed prosopographical investigations would appear to be necessary. Again, analysis of the geological community might usefully be undertaken in terms of various statements about scientific method made by its members, and thereby interesting groupings might be discerned.[48] There are various possibilities. Meanwhile, we are indebted to the two illuminating studies of Rupke and Rudwick, which generally reinforce one another, rather than being mutually incompatible. But there are some loose ends that still need tidying up.

[47] Rudwick, p. 238. He further cautions that it is 'important not to apply the grid/group theory in a Procrustean manner to the historical or sociological data'.

[48] My colleague, D. P. Miller, is presently engaged in a study of this kind. I should express here my thanks to Dr Miller for his most useful comments on the first draft of this review.

# XIV

# Sir Archibald Geikie (1835–1924), Geologist, Romantic Aesthete, and Historian of Geology: The Problem of Whig Historiography of Science


## Summary

The characteristics of inductivist historiography of science, as practised by earlier scientist/historians, and Whig historiography, as practised by earlier political historians, are described, according to the accounts of Agassi and Butterfield. It is suggested that the writings of Geikie on the history of geology allow us to characterize him as a Whig/inductivist historian of science who formulated anachronistic judgements. It is further suggested that his writings have had a considerable long-term effect on interpretations of the history of geology. The character of Geikie's historiography is related to his social, political and religious views, his historicism, and his romantic enthusiasm for Nature. His methodological pronouncements are examined: he believed that the present is the key to the past, and also that the past is the key to the present. He was an empiricist and inductivist. These epistemological and methodological views impinged on his historiography of science. If one attempts to criticize Geikie's historiography, though one may try to judge his work according to the norms of his own day and age, historical anachronism cannot be avoided entirely and one may oneself be charged with acting 'Whiggishly'. Historians of science, in the process of professionalization, have accepted the historiographical norms of general historians (perhaps for socio-economic reasons), but in so doing have inherited a problem that arises whenever they contemplate earlier historical writings. It is suggested that there may be more room for Whiggish historiography of science than is presently deemed acceptable. Alternatively, one may wish to draw a careful distinction between science and its meta-discipline, the historiography of science. Whig historiography conflates the two, and the work of Geikie (a scientist/historian *par excellence*) provides a good illustration of this.


## Contents

## 1. Introduction

This paper is concerned with certain theoretical issues of interest and importance in the historiography of science, namely the nature of Whig historiography and also the associated problem of historical anachronism. But the problems will not be addressed in a purely abstract manner. Rather, they will be treated through a consideration of the work of one of the best-known and most influential writers on the history of geology: Sir Archibald Geikie (1835–1924). It is hoped that this approach will offer a two-fold advantage. Firstly, the discussion of a particular person may facilitate the presentation of the theoretical problems in a congenial manner. Secondly, it allows a discussion of some of the overall features of Geikie's work as a geologist and his general methodological principles. This, in itself, is a worthwhile exercise, for Geikie, despite his important role in the history of geology, has not hitherto attracted much attention from historians of science.

## 2. Sir Archibald Geikie

Geikie was undoubtedly one of Britain's most eminent scientists in the nineteenth century—a man of remarkable industry with great gifts as teacher, administrator, field-worker, popular writer and historian. His interests ranged over numerous fields and his abilities were fully recognized in his own day, so that from an initial position in the Geological Survey (1855) he became successively F.R.S.E. (1861), F.R.S. (1865), Director of the Scottish branch of the Survey (1867), Professor of Geology at Edinburgh University (1871), Director-General of the Geological Survey (1881), President of the Geological Society of London (1890), President of the British Association (1892), President of the Royal Society (1908), and President of the Classical Association (1919). He was knighted in 1891 and received the Order of Merit in 1913. There were many other awards and memberships of learned societies. Thus during the course of a long and fruitful career Geikie moved smoothly to the very top of the British scientific establishment.

Besides his contribution to geology through his pioneering survey work in Scotland, his direction of the Geological Survey, and his various lecture courses and text-books, Geikie's scientific reputation today chiefly rests on his contributions to geomorphology, particularly in his synthetic work *The scenery of Scotland*,[1] and to vulcanology, notably in his investigations of the igneous rocks of Great Britain, described in detail in *The ancient volcanoes of Great Britain*.[2] These researches will not, however, be the subject of the present paper, despite the fact that Geikie has not yet found his modern biographer, and his scientific work has not yet been the subject of detailed critical analysis.[3] Rather, we shall examine his work as an historian of geology, for it raises a number of historiographical problems of general interest. And although Geikie's well-known *Text-book of geology*[4] has quite recently been described as 'still indispensable',[5] it is his historical and literary writings that one finds still in

[1] A. Geikie, *The scenery of Scotland viewed in connection with its physical geology* (1865, London).

[2] A. Geikie, *The ancient volcanoes of Great Britain* (2 vols., 1897, London and New York).

[3] For a convenient summary of Geikie's work, see: J. Challinor, 'Geikie, Archibald', in C. C. Gillispie (ed.), *Dictionary of scientific biography*, vol. 5 (1972, New York), 333–338, which gives a list of the more important secondary sources. Geikie himself wrote a most interesting autobiography: *A long life's work: an autobiography* (1924, London). A very useful bibliography of Geikie has recently been published: E. Cutter, 'Sir Archibald Geikie: a bibliography', *J. Soc. Bibl. Nat. Hist.*, **7** (1974), 1–18. This lists 246 items.

[4] A. Geikie, *Text-book of geology* (1882, London).

[5] J. Challinor (footnote 3), 335. The remark is made in reference to the two-volume fourth edition of 1903.

print today;[6] and it is through the lens of the widely read and admired *Founders of geology*[7] that most English-speaking geologists maintain a vision of the early history of their discipline.

But before attending to Geikie's historiographical work, and its relation to his overall views on society on the one hand and the practice of geology on the other, it may be helpful to remind the reader of certain general problems associated with the historiography of science.

### 3. Inductivist historiography of science

The historiography of science is not yet over-burdened with theoretical discussions, but of those we have, Joseph Agassi's *Towards an historiography of science*[8] has perhaps been the most influential.[9] Agassi, it will be recalled, drew attention to the way in which historians of science customarily view the history of science from particular philosophical perspectives, which, he maintained, tend to colour the historical accounts that they develop. He considered two particular examples—inductivism and instrumentalism/conventionalism—and sought to demonstrate the allegedly deleterious historiographical consequences that flowed from adherence to these two brands of philosophy of science.[10] Inductivists, we are informed, envisage progress in science arising from the gradual increase in empirical knowledge, for in their view truer theories can be educed from wider factual bases. Thus the inductivist historian of science has a simple ready-made formula for the presentation of the results of his investigations. He envisages the past history of science as a progressive ascent, leading to the present state of knowledge. Generally speaking, recent opinions will be more 'correct' than those resulting from earlier investigations; and consequently one may award 'black' and 'white' marks, so to speak, to former scientists, according to the extent to which they represented the 'modern' view in their thinking, or contributed to development of the present state of affairs in science.

Agassi was doubtless correct in attributing such historiographical practices to inductivist philosophical proclivities; but he might have added that they were natural enough at a time when most history of science was written by scientists, for such persons imported the value system of the practising scientist into their historiographical work. And the scientist is naturally dissatisfied with the earlier conditions of his subject, and believes, rightly or wrongly, that the investigations which he and his colleagues make *do* result in greater knowledge and theoretical understanding; that we *do* know more now than in the past; and that his appointed task is the prosecution of empirical and theoretical enquiry and the promotion of the *progress* of his chosen science.

[6] A. Geikie, *The founders of geology* (1962, New York); *Charles Darwin as geologist* (1977, New York); *Life of Sir Roderick I. Murchison* ... (1972, Farnborough); and *Types of scenery and their influence on literature* (1970, New York).

[7] A. Geikie, *The founders of geology* (1897, London and New York).

[8] J. Agassi, *Towards an historiography of science* (1967, Middletown)—first published as 'Beiheft 2' of *History and theory* (1963).

[9] The only other monograph specifically devoted to problems in historiography of science with which I am familiar is M. A. Finocchiaro, *History of science as explanation* (1973, Detroit). But this volume, written from the stance of a dialectician rather than an historian, has had much less impact than Agassi's study.

[10] Agassi himself, when writing his essay, was in a strongly Popperian, falsificationist frame of mind, and advocated the application of a falsificationist methodology in the historiography of science.

This being the case, so long as history of science was largely written by scientists, and so long as scientists adhered to the inductivist philosophy of science, it is not surprising that their work possessed the character that Agassi described. However, since history of science has to some degree come within the purview of the general historian,[11] and has attained some measure of professionalization, its historiographical presuppositions have changed markedly; and so too has the character of the history of science that is published. Perhaps even in 1963 Agassi was directing his criticism in part towards men of straw.[12]

## 4. Whig historiography

Critics were quick to seize on the ammunition lent them by Agassi's forthright remarks. And it became clear that inductivist historiography of science had much in common with the 'Whig' history that had earlier been the subject of complaint by the general historian, Herbert Butterfield.[13] Butterfield, it will be recalled, had taken the nineteenth-century Whig historians to task for supposing that there was an ineluctable condition of social, political and economic progress—and even a progress in wisdom—such that they were fortunate to occupy a specially privileged vantage-point from which they might appraise men and deeds of the past. Inevitably, this entailed acute historiographical anachronism, for *past* events were appraised in terms of present circumstances and values, rather than in terms of their own time. Such anachronism was inadmissible, Butterfield maintained. And his criticisms were widely accepted by his colleagues.[14]

Thus both inductivist historiography of science and Whig political and social history are beset by the same problem: *anachronism*. Indeed, historiography of science is even more likely to be subject to the charge of anachronism than is traditional Whig political history, for 'progress' in science is perhaps more evident than in society at large. Be this as it may, historians of science were quick to take up the plaint against 'Whiggery' and regard it as a cardinal error: science should be judged in the context of its *own* time, not the terms of the present. Perhaps, indeed, Agassi was following such trends, rather than being their prime mover.

## 5. Geikie as a Whig/inductivist historian of science

As indicated above, the writings of Sir Archibald Geikie are to be used as a basis for a wider discussion relating to the problems of Whig/inductivist historiography of science and historiographical anachronism. It is not difficult to establish that Geikie's historical writings provide a good illustration of work of this character, and just a few examples should suffice for this purpose.

Geikie, a Scotsman, was profoundly influenced by the work of his countrymen, James Hutton, John Playfair, and Sir James Hall, whereas he found the 'Saxon' theories of Abraham Werner, espoused by Robert Jameson, an utter anathema. So

[11] This process had previously been initiated by Herbert Butterfield in his *Origins of modern science: 1300–1800* (1949, London).

[12] It may be noted that in reviewing *Towards an historiography of science* C. C. Gillispie objected that historians of science were in many instances not guilty of the historiographical solecisms with which they were charged by Agassi (see *Isis*, **55** (1964), 97–99).

[13] H. Butterfield, *The Whig interpretation of history* (1931, London; republished 1973, Harmondsworth).

[14] For a recent account of the history of Whig historiography, see P. B. M. Blas, *Continuity and anachronism: parliamentary and constitutional development in Whig historiography and in the anti-Whig reaction between 1890 and 1930* (1978, The Hague).

we find Geikie constantly bestowing bouquets ('white marks') on the Huttonians and deprecating the Wernerians (awarding them 'black marks'). He wrote many times on this topic;[15] his language was perhaps the strongest, most Whiggish, and anachronistic in an address read to the Royal Physical Society of Edinburgh in 1880,[16] in which he explicitly compared the investigational methods of Jameson in the early nineteenth century with the methods, utilizing petrographic analysis, employed by the officers of the Geological Survey in the latter part of the century.

Hutton, we are informed, treated geological dynamics with the 'bold, broad, far-reaching hand of a consummate master'.[17] His gifts as geologist did not receive the immediate recognition that was their due because of his 'somewhat cumbrous style' and 'chiefly because he was so far in advance of his age'.[18] But he was fortunate to have two friends who were 'earnestly and loyally sympathising with his doctrine';[19] the 'genial' Playfair, with 'facile and graceful' pen, and the 'quick, original, [and] inventive' Hall, with 'an enthusiastic love of science, and a determination to take nothing on trust, but to subject every statement where possible to the test of experiment'.[20] By contrast, Werner 'framed a theory not less crude and unfounded than the wildest of those which he opposed'.[21] His 'crude notions' took their rise 'within the narrow confines of Saxony', and bore witness to the 'limited experience of their founder'. His dicta were 'expressive of authoritative certainty'. And by introducing the Wernerian doctrines into the Scottish domain, Jameson produced an effect 'on the progress of geological science ... [that] was disastrous'.[22] Wernerianism, under Jameson, almost extinguished 'true geology' in Scotland. The Whiggish character of Geikie's historiography is perfectly well known, and need not be laboured here.

## 6. Geikie's influence as an historian of geology

It would be a task beyond the boundaries of the present paper to attempt to establish the full extent of the influence of Geikie's historiography amongst geologists, teachers and the general public. But I believe there is a strong *prima facie* case for the claim that such influence has been very considerable in the English-speaking world, amongst those who, for one reason or another, are concerned with the history of geology. His book *Founders of geology* is still in print, and continues to be widely read by professional geologists. Where examination questions on the history of geology are set within the secondary-school curriculum, they continue to

[15] A. Geikie, *The Scottish school of geology. The inaugural lecture at the opening of the class of geology and mineralogy in the University of Edinburgh* (1871, Edinburgh) (republished in *Geological sketches at home and abroad* (1882, London), 286–311); *The ancient volcanoes of Great Britain* (2 vols., 1897, London and New York), vol. 1, viii–ix; *The founders of geology* (2nd ed. 1905, London and New York), chapter 9 and *passim*; 'The centenary of Hutton's "Theory of the Earth"', in *Landscape in history and other essays* (1905, London and New York), 158–197 (originally given as the Presidential Address to the Edinburgh meeting of the British Association in 1892); and 'The state of geology at the time of the foundation of the Geological Society', in W. W. Watts (ed.), *The centenary of the Geological Society of London, celebrated September 26th to October 3rd, 1907* (1909, London), 107–131.

[16] A. Geikie, 'The "Pitchstone" (vitreous basalt) of Eskdale: a retrospect and comparison of geological methods', *Proc. Roy. Phys. Soc. Edinburgh*, **5** (1878–80), 219–255.

[17] *Ibid.*, 220.

[18] *Ibid.*, 221.

[19] *Ibid.*, 222.

[20] *Ibid.*

[21] *Ibid.*, 223.

[22] *Ibid.*, 231.

reinforce Geikie's view of Hutton's geological work.[23] The efforts of writers such as Ospovat[24] and Seddon[25] to reinstate the reputation of Werner and the Freiberg School have by now more or less achieved their intentions amongst historians of science, but not, I suspect, amongst practical geologists. I might add also that until relatively recently the training of professional historians of science still bore the mark of some of Geikie's interpretations.[26]

Thus the time-lag between changes in historical interpretation among professional historians of science and the wider public is very considerable. The present repudiation of Whig historiography by professional historians of science is now almost universal. But this change of outlook has not made much headway in, say, secondary education. Whereas history teachers, directly 'exposed' to professional historians during their tertiary education, are likely to be aware of possible historiographical solecisms, and will be wary of Whig historiography and historical anachronism, science teachers are still in large measure inductivists, and have little contact with recent historiography of science. If they read any texts in the history of geology, Geikie is still most likely to be their first choice. Thus the influence of his historiographical work is by no means entirely dissipated, even today. For this reason, if no other, it is worth examining some of the basic suppositions that

[23] For example, the following question was set in the '2 Unit A' examination for the Higher School Certificate in New South Wales in 1976:

> What Newton achieved in the field of mechanics and mathematics, Hutton achieved in the field of geology. It was he who provided geology with a theory, and here we must take theory, not as something opposed to fact, but in its original meaning in Greek, namely *comprehension* [*sic*].
>
> 'The [*sic*] Theory of the Earth' as stated by Hutton is the core of the modern science of geology.
>
> Write an account of the life and contribution to science of James Hutton, justifying the statements that have been made about him.

[24] Ospovat has written widely on this topic. See, for example, his *Abraham Gottlob Werner: short classification and description of the various rocks translated with an introduction and notes by Alexander M. Ospovat* (1971, New York).

[25] G. Seddon, 'Abraham Gottlob Werner: history and folk-history', *J. Geol. Soc. Australia*, **20** (1973), 381–395.

[26] Consider, for example, the following questions from the M.Sc. examinations in History and Philosophy of Science at University College, London, kindly supplied to me by Dr. W. A. Smeaton:

> 'One may be an excellent mineralogist . . . but if he is to reach a true conception of the structure and origin of the globe he must "Take the world for his parish"'. Discuss this statement with particular reference to the work of Werner and Hutton. [1958]
>
> Discuss the statement that the rival theories put forward in the Neptunist-Vulcanist controversy were profoundly influenced by the geology of the localities in which their originators worked. [1961]
>
> 'The geological record furnishes a mass of evidence which no arguments drawn from other departments of Nature can explain away, and which, it seems to me, cannot be satisfactorily interpreted save with an allowance of time much beyond the narrow limits which recent physical speculation would concede' (GEIKIE, 1892). Set this statement in its historical context in a discussion on the age of the Earth. [1968]

The quotation in the first of these questions is actually from F. D. Adams's *Birth and development of the geological sciences* (2nd ed. 1954, New York), 246, but it expresses a view that Geikie himself espoused (see, for example, *Founders* (footnote 6), 206, 213, and the account of von Buch's gradual emancipation from Wernerism on pp. 245–252). Geikie's view of the matter was itself orthodox in British historiography of geology, being evident, for example, in Lyell's *Principles* (8th ed. 1850, London), 49. The second question meshes well with Geikie's general view that human history and character are related to geographical and geological circumstances, and also with various remarks in *Founders* (for example, pp. 213–222, 228). In the third question, of course, the quotation of Geikie is perhaps best regarded as a primary, rather than a secondary, source. As such, it is characteristically 'Geikiean' in that the 'departments of Nature' suggest an economic or bureaucratic metaphor, with the various aspects of Nature likened to the branches of a firm or the divisions of a civil service.

underlay his work. But more generally we are concerned here with the wider historiographical problems associated with writings such as those that flowed so smoothly from Geikie's active pen.

## 7. The character of Geikie's historiography in relation to his social and intellectual milieu

From what has been said in Sections 5 and 6 above, it might appear that the professional historian of science of today would naturally repudiate Geikie's work, regardless of whether or not it is appropriate for 'popular' history of geology to wear a Whiggish, 'Geikiean' garb. Geikie stands condemned for his Whiggery. Away with him! But to adopt this position would seemingly place *us* within the Whig camp ourselves, for we would be regarding ourselves as knowing more history and philosophy of science than did Geikie, and we should be judging him in our terms, not his own. Let us, therefore, for the time being at least, seek to eschew Whiggery in any form. Let us shun anachronism and attempt to view Geikie within the context of his own time. Let us try to determine some of the reasons that may have led him to regard the history of his subject as he did and portray it in his writings in his own distinctive manner. Such a task is considerably facilitated by Geikie's detailed and felicitously written autobiography, which gives us a great deal of information concerning his extraordinarily industrious career and ostensibly amiable character.[27]

Geikie wrote very widely on the history of his discipline. Besides the well-known *Founders of geology*, there are major biographies of Murchison,[28] Ramsay[29] and Forbes,[30] and briefer accounts of the work of Darwin,[31] Michell[32] and Prestwich.[33] Geikie also edited, or was associated with the publication of, works by Hutton,[34] Faujas de Saint-Fond[35] and Seneca.[36] All these were linked in one way or another with theories of vulcanism, which was perhaps the aspect of geology that most

[27] If we judge by Geikie's autobiography (footnote 3), he was a man of benign temperament, always able to see the best in other men, with a wide circle of friends, and able to smooth over any difficulties in the running of the Geological Survey with a few tactful words. His evident love of Nature would lead us to suppose that he was a man with a 'beautiful soul'. But Sir Aubrey Strahan, Director-General of the Survey from 1914 to 1920, wrote in his obituary of Geikie that 'A clear if somewhat cold judgement controlled his actions, but in his biographical work the coldness was masked by a studied kindliness of expression. Though he made many friends at home and abroad, his sympathies with his fellow men were somewhat overshadowed by his love of nature and passion for work. He did not seek collaboration, but preferred to work single handed, nor could he brook criticism' (*Annual report of the Smithsonian Institution 1925* (1926, Washington), 591–598 (p. 598)).

[28] A. Geikie, *Life of Sir Roderick I. Murchison ... based on his journals and letters with notices of his scientific contemporaries and a sketch of the rise and growth of palaeozoic geology in Britain ... in two volumes* (1875, London).

[29] A. Geikie, *Memoir of Sir Andrew Crombie Ramsay* (1895, London and New York).

[30] G. Wilson and A. Geikie, *Memoir of Edward Forbes, F.R.S., late Regius Professor of Natural History in the University of Edinburgh* (1861, Cambridge, London and Edinburgh).

[31] A. Geikie, *Charles Darwin as geologist: the Rede Lecture given at the Darwin Centennial Commemoration on 24 June 1909* (1909, Cambridge).

[32] A. Geikie, *Memoir of John Michell* (1918, Cambridge).

[33] A. Geikie, 'Summary of the scientific work of Sir Joseph Prestwich', in G. A. Prestwich, *Life and letters of Sir Joseph Prestwich* (1899, Edinburgh and London), 402–421.

[34] J. Hutton (ed. A. Geikie), *Theory of the earth with proofs and illustrations*, vol. 3 (1899, London).

[35] *A journey through England to the Hebrides in 1784 by B. Faujas de Saint Fond: a revised edition of the English translation edited with notes and a memoir of the author by Sir Archibald Geikie* (2 vols., 1907, Glasgow).

[36] *Physical science in the time of Nero being a translation of the Quaestiones naturales of Seneca by John Clarke ... with notes on the treatise by Sir Archibald Geikie* (1910, London).

interested Geikie. Besides these major historical writings, there were several shorter articles dealing with particular aspects of the history of geology, and especially with the history of the Geological Survey. In addition, he liked to preface his geological monographs with illustrative historical remarks.[37]

Why should Geikie have laid such emphasis on the historical aspects of his discipline? Was it simply a form of relaxation from the sterner tasks of field-work and administrative duties? Perhaps. But I suggest that the reasons ran very much deeper than this. Geikie sought to be, and was, a man of very wide culture, so that his interests ran the whole gamut from classics, through general literature (particularly English poetry), history, natural history, cartography, geography, physical geology, stratigraphy and vulcanology, to petrology, with an eye directed towards the educational needs of his countrymen in all aspects of this range. This great breadth of learning, coupled with an impeccable prose style, undoubtedly appealed to Geikie's peers in an age when the 'gentlemanly' approach to science was still widely esteemed; and I suggest that his unimpeded ascent to the very peak of the scientific establishment was greatly facilitated by his impressive literary powers, complementing his unquestionable prowess as field geologist and administrator.

Let us note the pivotal place that history occupied within the totality of Geikie's interests and achievements. It provided a crucial link between his literary and scientific concerns, and in this respect, I suggest, it formed a natural preoccupation for the nineteenth-century scholar.[38] Like his predecessors such as de Saussure,[39] Lyell[40] and Hugh Miller[41] (whom, we are told, he revered in his early years 'as a kind of demigod'[42]), Geikie envisaged a smooth continuum between history, geography and geology, so that geological investigations took over where historical and archaeological studies petered out through lack of data. Thus, 'Geography and history are . . . inseparably interwoven, and neither is fully comprehensible without the other.'[43] Likewise, 'Between the neolithic and the present period no line can be drawn. They shade insensibly into each other, and the materials from which the history of their geographical and climatic vicissitudes, their changes

[37] A full list of Geikie's minor historical writings, including the many obituaries that he wrote, may be found in E. Cutter (footnote 3).

[38] Elsewhere, I have examined the rise of historical geology in the late eighteenth and early nineteenth centuries in relation to the contemporary historicist attitude that was manifest, for example, in the historicist school of law, considering this relationship up to the middle of the nineteenth century. Geikie, I suggest, meshes with an extrapolation of this pattern almost perfectly. See D. R. Oldroyd, 'Historicism and the rise of historical geology', *History of science*, **17** (1979), 191–213, 227–257, where I have sought to define the meaning of the term 'historicism' in the required context. The same meaning is intended here.

[39] H. B. de Saussure, *Voyages dans les Alpes, précédés d'un essai sur l'histoire naturelle des environs de Genève* (4 vols., Neuchâtel, 1779–96), vol. 4, 474.

[40] On Lyell's views on the relationship between history and geology see M. J. S. Rudwick, 'Historical analogies in the geological work of Charles Lyell', *Janus*, **64** (1977), 89–107.

[41] Compare H. Miller, *Sketch-book of popular geology* (1886, Edinburgh), 1–2 (posth.). Miller wrote:

> Geological history may be regarded as ending where human history begins. The most ancient portions of one piece on to the most modern portions of the other. But their line of junction is . . . not an abrupt, but a shaded line; so that, on the one hand, the human period passes so entirely into the geological, that we find our conclusions respecting the first human inhabitants rather on what may be deemed *geologic* than on ordinary *historic* data; and, on the other hand, some of the later . . . geologic changes have taken place in periods comparatively so recent, that . . . we are able to catch a glimpse of them in the first dawn of history proper—that written history, in which man records the deeds of his fellows.

[42] A Geikie (footnote 3), 24.

[43] A. Geikie, *The teaching of geography: suggestions regarding principles and methods for the use of teachers* (2nd ed. 1892, London), 185 (1st ed. 1887, London).

of fauna and flora, and their human migrations and development, form a common ground for the labour of the archaeologist, the historian, and the geologist'.[44] Such views cohered well with the character of nineteenth-century thought with its historicist perspective.

In pursuance of this general link between geology, geography and history, Geikie liked to emphasize the role that physical geography might play in the actual historical development of man's affairs. The geographer could show how geographical features have 'influenced the history of political events, ... directed the migration of peoples, guided or arrested the tide of conquest, moulded national character, or given ... colouring to national mythology and literature.'[45]

Geikie believed that the contrasted character of Celtic peoples in Ireland and Scotland might be attributable to geological factors;[46] thus the 'crystalline rocks' and the Old Red Sandstone formerly marked a rough boundary between the regions of Celts and Teutons in Britain, and produced marked differences in architecture and language which exist to this day.[47] Such 'environmental determinism' has not been without subsequent advocates. For example, in Australia the admirable writings of the geographer T. Griffith Taylor were shaped in this mould.[48]

Besides seeking to demonstrate the influence of geological and geographical features on the historical development of human culture, Geikie was also a strong advocate of thinking in historical terms when seeking to comprehend geological phenomena. Again, he was probably taking his cue from Hugh Miller, who supposed Nature to be like 'a vast tablet, inscribed with signs, each of which has its own significancy, and becomes poetry in the mind when read; and geology is simply the key by which myriads of these signs, hitherto undecipherable, can be unlocked and perused, and thus a new province added to the poetical domain'.[49] Very similarly, Geikie, in his early popular work, *The story of a boulder, or gleanings from the note-book of a field geologist*, supposed that 'in reading off the geological structure of a country, we ascertain its history during many thousand ages long prior to that of man. We become, as it were, interpreters of hieroglyphics, and historians of long-perished dynasties'.[50]

The novice had to learn to read the history of the boulder (a glacial erratic) just as an historian might read an ancient text. The boulder might be regarded as 'a curious volume, regularly paged, with a few extracts from older works'.[51] It was, 'as it were, a quaint, old, black-letter volume of the middle ages, giving an account of events that were taking place at the time it was written, and containing on its earlier pages numerous quotations from authors of antiquity. The scratched surface, to complete the simile ... [might] be compared to this old work done up in a modern binding.'[52]

[44] A. Geikie, *A class-book of geology* (1886, London), 476.

[45] A. Geikie (footnote 43), 2.

[46] A. Geikie, 'The geological influences which have affected the course of British history', in *Geological sketches at home and abroad* (1882, London), 353–382 (first published in *Macmillan's magazine*, **45** (1882), 364–375).

[47] *Ibid.*, 371.

[48] T. Griffith Taylor, *Sydneyside scenery and how it came about* ... (1958, Sydney).

[49] H. Miller (footnote 41), 87.

[50] A. Geikie, *The story of a boulder or gleanings from a note-book of a field geologist* (1858, Edinburgh and London), xi-xii. Geikie offered a somewhat similar didactic account in a lecture delivered in Manchester in 1872 ('The ice age in Britain', in *Science lectures for the people, fourth series* (1873, Manchester), 1–19).

[51] *Ibid.*, 4. The 'few extracts from older works' were analogous to xenoliths, one may suppose.

[52] *Ibid.*, 5.

In all this we may readily perceive the emphasis which Geikie placed on geology as a palaetiological science (to use William Whewell's term[53]). Geological explanation entailed a knowledge of the *history* of the Earth. By the same token, then, students of geology needed to be familiar with the historical development of their subject in order to understand its theories and concepts.

It is true that Geikie was, perhaps, a little over-sanguine as to the ease with which the geologist might read the 'book of nature'. He supposed that 'provided only we proceed cautiously, reasoning from positive facts, and striving as far as possible to exhaust what Bacon calls the "negative instances", our deductions possess all the certainty of truth.'[54] And 'reasoning from positive facts' he concluded that the 'true solution of the problem'[55] of his erratic boulder's position was to be found in the action of a floating iceberg. Yet five years later he found himself required to admit that 'The iceberg hypothesis . . . must . . . be abandoned [for] it will not account for the phenomena. We cannot conceive of a set of ice-rafts moving for ages in one persistent direction within a given area of the sea'.[56]

Such a turnabout may make one call in question the adequacy of Geikie's epistemology and his views on the methodology of science. But if we make this criticism we may find ourselves guilty of the sin of historical anachronism—of being wise after the event. So for the moment let us eschew criticism and consider further the general characteristics of Geikie's thought in relation to that of his contemporaries, with the intention of understanding how and why he wrote his history of science as he did. The question of his philosophy of science will be taken up in sections 8 and 9 below.

For Geikie, the apprehension of the past condition of things was an essentially aesthetic experience. It was, he tells us, 'the exuberance of . . . [his] joy in the active and continuous prosecution of geological investigation',[57] and the deep joy which the study of Nature afforded that led him to write *The story of a boulder*. Moreover,

> Assuredly there is something thrilling to even the calmest imagination in contemplating the results of vast and sudden upheavals, in picturing the solid crust of the earth heaving like a ground-swell upon the ocean, in tracing amid "Crags, knolls, and mounds confusedly hurl'd, the fragments of an earlier world;" and in conjuring up visions of earthquakes, and frightful abysses from which there ever rose a lurid glare as hill after hill of molten rock came belching up from the fires below.[58]

Geikie's romanticism is further evinced in numerous passages in his autobiography, from which we may extract the following passage by way of illustration. Here he describes the attributes of the artist:

> He sits face to face with nature, and his soul, if he has one, communes with her. He looks at every detail of her features and sees on every side manifold variety and beauty of form and of colour that escape most eyes. A landscape means more to him than most people, especially if he be also a geologist. As he

[53] W. Whewell, *The philosophy of the inductive sciences, founded upon their history* (2nd ed. 2 vols., 1847, London), vol. 1, 637.
[54] A. Geikie (footnote 50), 181.
[55] *Ibid.*, 8.
[56] A. Geikie, *On the phenomena of the glacial drift of Scotland* (1863, Glasgow), 74.
[57] A. Geikie (footnote 3), 66.
[58] A. Geikie (footnote 50), 164–165.

> tries, with more or less success, to paint it, its lineaments are printed on his memory and remain there as a possession for ever.[59]

The geologist, through his field-work, and his historical empathizing with the past, buoyed up by a romantic enthusiasm for Nature, can achieve something of the aesthetic experience of the artist. Indeed, geological expertise can furnish a useful reinforcement to the artist's total aesthetic range.

One also finds Geikie's world-view meshing with that of the nineteenth-century British gentleman when we look at his religious views, which displayed the characteristic natural theology of his day:

> Geology lifts off for us the veil that shrouds the past, and lays bare the monuments of successive creations that had come and gone long ere the human race began. She traces out the plan of the Divine working during a vast cycle of ages, and points out how the past dovetails with the present, and how the existing condition of things comes in as but the last and archetypal economy in a long progressive series. By thus revealing what has gone before, she enables us more fully to understand what we see around us now. Much that is incomplete she restores; much that is enigmatic she explains. She teaches us more fully man's true position in the created universe, by showing that in him all the geologic ages meet—that he is the point towards which creation has ever been tending.[60]

Geikie concluded his *Class-book of geology* with a somewhat similar passage, likewise an indication of his ready acceptance of the nineteenth-century version of the 'idea of progress':

> It is ... the greatest triumph of geological science to have demonstrated that the present plants and animals of the globe were not the first inhabitants of the earth, but that they have appeared only as the descendants of a vast ancestry, the latest comers in a majestic procession which has been marching through an unknown series of ages. At the head of this procession we ourselves stand, heirs of all the progress of the past and moving into the future wherein progress towards something higher and nobler must still be for us, as it has been for all creation, the guiding law.[61]

Essentially the same position emerges in Geikie's most explicit statement of his views on the relation between science and religion, given in his Presidential Address to the Edinburgh Geological Society in 1869.[62] The geological record, he argued, gives a clear indication of change and progress. And the history of man provides one of the most 'signal illustrations' of the general law of progress, with the development of man's religious nature forming a particularly important feature of the overall process. 'The outer crust of prejudice' encasing 'the grand essential truths' is, Geikie informed his listeners, being gradually stripped away, thus providing 'the happiest augury for the future of our race'.[63] Consequently, science and religion mutually enhanced one another, and strictly ought to be of 'one flesh'.[64]

[59] A. Geikie (footnote 3), 173.
[60] A. Geikie (footnote 50), 261–262.
[61] A. Geikie (footnote 44), 478. See also *Elementary lessons in physical geology* (1880, London), 365–366.
[62] A. Geikie, 'Opening address [to the thirty-sixth anniversary meeting of the Edinburgh Geological Society]', *Trans. Edinburgh Geol. Soc.*, **2** (1869–70), 1–13.
[63] *Ibid.*, 12.
[64] *Ibid.*, 9.

In a somewhat similar vein, we find that towards the end of his career Geikie expressed the view that 'students of Nature' strive 'towards the erection of that great Temple of Truth wherein all genuine knowledge shall be enshrined'.[65] Thus science itself is almost deified; certainly it is the means whereby 'those secrets the knowledge of which has proved so beneficent to the mental, moral, and material well-being of man' are revealed.[66]

Thus Geikie's historiography can be seen to have had a moral, as well as an intellectual, function. It provided a revelation of the progress of mankind within a particular branch of science. In this way, it illustrated the overall process of divinely inspired cosmic progress. Geikie elegantly concluded his address of 1869[67] with some nicely chosen lines from Tennyson, to which, presumably, he was happy to subscribe:

> Let us continue our work with the most perfect freedom and fearlessness, yet with modesty and in the spirit of love, guided by a clear-sighted faith in
> 'One God, one law, one element,
> And one far-off divine event
> To which the whole creation moves'.[68]

Next, we may examine the economic background to Geikie's investigations, and to his historiography. Although the British government's intention in funding the Geological Survey was in the first instance motivated by economic considerations, rather than by reasons of altruism,[69] and Geikie spent the greater part of his working life in association with the Survey, we do not find that he was overly interested in economic geology, and the greatest efforts of the Survey during his directorship were aimed at the elucidation of theoretical problems associated with geological structures in north-west Scotland, rather than, for example, the stratigraphy of the coal-bearing regions. Thus, although he commented that in the early days of the Survey 'the immediate economic or industrial value [of geology] was not ... sufficiently insisted upon',[70] his interest in geology was chiefly that of the pure scientist, heightened by a romantic enthusiasm for Nature. It happened that he was able to satisfy his interests through the medium of an organization that was underpinned by economic considerations. But this, we may conjecture, Geikie considered to be his good fortune, rather than an indication that he should devote his attention chiefly to the economic aspects of his science.

Nevertheless, despite the fact that Geikie was not conspicuous for his interest in economic geology, when we examine his historiographical work we find that he turned quite naturally to the use of 'economic' or 'commercial' metaphors. And in this, no doubt, he represented the thinking of his age in a characteristic manner. For example, in his biography of Murchison, Geikie wrote:

> His [Murchison's] 'Siluria' became, in his eyes, a kind of personal property, over which he watched with solicitude. Or, it might rather perhaps be compared to a vast business which he had established, of every original detail of which he was complete master, and which he laboured to extend into other

[65] A. Geikie, *English science and its literary caricaturists in the 17th and 18th centuries* ... (1914, Haslemere), 45.

[66] *Ibid.*

[67] A. Geikie (footnote 62), 13.

[68] A. Tennyson, 'In memoriam A. H. H. obiit MDCCCXXXIII', in *The works of Alfred Lord Tennyson Poet Laureate* (1887, London and New York), 247–286 (p. 286).

[69] See E. B. Bailey, *Geological Survey of Great Britain* (1952, London), 26.

[70] A. Geikie (footnote 3), 45.

countries, while he kept up through life a close correspondence with those by whom the foreign extensions were so abundantly and successfully carried out.[71]

But more frequently it was the military metaphor that was preferred. Thus, according to Geikie's account, Murchison in his early investigations of 1831 'struck northwards into the hills beyond the coal-field, and there began to invade the Transition rocks of South Wales . . . It was near the town of Llandeilo that Murchison first broke through into these older rocks with the purpose of making them disclose their true place and order in the geological series'.[72]

Overall, Geikie envisaged the progress of geology as analogous to that of empire-building, and it was the 'imperial' character of the science that furnished one of the most cogent reasons for writing its history:

> A retrospect of this kind [such as is presented in *Founders of geology*] leads to a clearer realization of the precise position at which . . . [geology has] arrived, and a wider conception of the extent and limits of the domain of knowledge which has been acquired. On the other hand, by enabling . . . [the readers] to comprehend how, foot by foot, the realms of science have been painfully conquered, it furnishes suggestive lessons as to tracks that should be avoided, and fields that may be hopefully entered.[73]

Perhaps in accordance with such a model, Geikie represented himself in his historiographical work as an exponent of the 'great man' theory of history.[74] Thus, in introducing *Founders of geology*, he wrote:

> If all history is only an amplification of biography, the history of science may be most instructively read in the life and work of the men by whom the realms of Nature have been successively won. I shall therefore dwell on the individual achievements of a few great leaders in the onward march of geology, and indicate how each of them has influenced the development of science. . . . [I shall] dwell rather fully on the labours of a few of the early masters, who . . . laid the broad foundations of the science, . . . [but] touch only lightly on the work of some of their less illustrious contemporaries.[75]

From the foregoing illustrations, we may gain an idea of the way in which Geikie, and his science, fitted into the general social context of his times. Indeed, it seems reasonable to suppose that this 'goodness of fit'—together with his very considerable scientific skills and his indefatigable labours—may readily account for his smooth and uninterrupted ascent to the pinnacle of the scientific establishment. In bringing forward the various exemplifications of Geikie's thought, I have consciously sought to avoid historical anachronism, to write in non-Whiggish terms, and to present him quite neutrally as an exemplar of his day and age—just as the rules of modern historiography require. Whether I have been successful in this the reader may judge; whether it is the most appropriate approach we shall consider below. But first let us make a few remarks concerning his views on methodology, within the domain of geological enquiry.

[71] A. Geikie (footnote 28), vol. 1, 243.
[72] *Ibid.*, 182.
[73] A. Geikie (footnote 6), 2.
[74] It is worth noting that Geikie recorded in his autobiography two occasions when he met Thomas Carlyle ((footnote 3), 132 and 140).
[75] A. Geikie (footnote 73), 4–5.

## 8. Geikie as a methodologist

Some points relating to Geikie's views on methodology have already been raised in connection with his historicism, but we may take the matter somewhat further at this juncture.

Geikie is well known for his use of the maxim 'the present is the key to the past'[76] in his discussion of the 'uniformitarian'[77] methodology of James Hutton. Yet this may appear to mesh somewhat awkwardly with his doctrine of progress and his nineteenth-century historicism, for which an appropriate motto might have been: 'the past is the key to the present'. How did Geikie seek to minimize this tension?

In fact, Geikie did not give a great deal of attention to questions of methodology, but there are a number of passages in his writings which allow us to build up a simple 'identikit' picture. He was undoubtedly an empiricist, though not perhaps a very reflective one. He supposed that 'we are surrounded with phenomena that ceaselessly press themselves upon our notice'.[78] A 'study of the present must come first', but it 'must necessarily be superficial and imperfect, until it has been connected with an inquiry into the past'.[79]

So the geologist must first acquire familiarity with a wide range of present-day geological processes. Somehow, *these* are to be used as a 'key that unlocks the secrets of the past'.[80] Thus, one should proceed 'from the known and familiar to the less known and foreign'.[81] The process was seemingly unproblematical, such that any sensible mason (Hugh Miller perchance?) might undertake. Thus, by way of illustration of the processes of geological reasoning, Geikie imagined that the Firth of Forth had been 'raised above the level of the sea and covered with verdure, and that ...a mason...[had] excavate[d] on the lately-born land the foundation for a dwelling-house'.[82] This mason might have dug down through the soil and rocks, observed the changes in substance at different levels, and from this evidence he might have developed 'a very interesting history'. His reasoning could be reconstructed as follows:

> 'There', says he, 'is a layer of peat with the rush-stalks and moss-fibres matted together in the soft brown mouldering substance, exactly as I have seen them a hundred times in the peat-cuttings on the moors, and I cannot but believe that they must both have had the same origin, that is, that they grew in swampy hollows of the land. There, too, lies a stratum of fresh-water shells identical with those that occur in our ponds and marshes. Although mouldering now, they are evidently not fragmentary, but entire and unbroken; ... such shells could only live in fresh water, therefore the spot where I stand must have been at one time a fresh-water lake. 'There again,' he continues, 'is a bed of

[76] *Ibid.*, 299.

[77] The term 'uniformitarian' was, of course, coined by Whewell, not Hutton. See Whewell's review of the second volume of Lyell's *Principles* in *Quarterly review*, **47** (1832), 103–132 (p. 126).

[78] A. Geikie (footnote 43), 2.

[79] *Ibid.*, 132.

[80] A. Geikie (footnote 44), 7.

[81] A. Geikie (footnote 43), 182.

[82] A. Geikie (footnote 50), 179. Geikie, we may note, was rather preoccupied with changes in the level of the Firth of Forth at about this period of his life. See his 'A rise of the coast of the Firth of Forth within the historic period', *Edinburgh new philosophical journal*, **14** (1861), 102–112; and 'The raised beach of Britain, and how Scotland has risen in the world', *Macmillan's magazine*, **5** (1862), 311–319. The title of this latter meshes rather well with Geikie's nationalistic zeal.

oysters which cannot have been transported hither, for their valves are together, lying just as they do in our present oyster-beds. This green field, therefore, must have been at one period a muddy sea-bottom'.[83]

The mason, Geikie assures us, is an 'intelligent, honest fellow, and uses his eyes and his head',[84] so that there is really 'no hypothesis or speculation' in his reasoning. Thus, 'After this manner and upon this kind of evidence must all inquiries into the past changes of the earth's surface be conducted'.[85]

An important aid to reasoning in the process of geological enquiry was, of course, the principle of superposition, and this, together with the information provided by palaeontology (which 'can be applied with infallible success'[86]), should enable one to build up a composite picture of the Earth's history—much as did the honest mason in Geikie's parable. The process might also be compared with that by which a student of architecture, familiar with the peculiarities of particular architectural styles, might examine the structure of an ancient cathedral, and thereby decipher the story of the many stages of construction and demolition in the history of the extant building.

Geikie was not entirely oblivious to the problems linked with the use of 'the present is the key to the past' maxim, as illustrated in the foregoing fable. In his *Text-book of geology*, he wrote:

> While . . . the present condition of things is . . . employed [as the basis of investigation], we must obviously be on our guard against the danger of unconsciously assuming that the phase of nature's operations which we now witness has been the same in all past time. . . . Of course we may assume this uniformity of action, and use the assumption as a working hypothesis. But it ought not be allowed any firmer footing, nor on any account be suffered to blind us to the obvious truth that the few centuries wherein man has been observing nature form much too brief an interval, by which to measure the intensity of geological action in all past time.[87]

Yet having recognized this difficulty, Geikie simply passed it by and went on to consider the problems inherent in the incompleteness of the geological record.

The task of the 'Geikiean' geologist, then, was to use the observations of present *processes* as a means of comprehending past processes. Then, having a general idea of geological processes (with the aid of the uniformitarian principle), one should build up an *history* of the globe, as a means of explaining the present, observed, contingent conditions of things—that is, the actual nature, distribution and disposition of the presently visible rocks. Certain experimental procedures (such as those of Sir James Hall) might be used as an additional guide to the understanding of present and past geological processes.

Thus, the final product of the geologist's investigations should be a *geological history of the globe*. And in Geikie's various writings this was customarily what he

[83] A. Geikie (footnote 50), 180–181.
[84] *Ibid.*, 180.
[85] *Ibid.*, 181.
[86] *Ibid.*, 183.
[87] A. Geikie (footnote 4), 3.

attempted to present to his readers.[88] He placed rather little emphasis on the role of hypotheses in the prosecution of scientific enquiries, and as we have noticed earlier in this section he sometimes failed to make a clear distinction between hypothetical statements and 'deductions which possess all the certainty of truth'. However, the following passage from the end of *Founders of geology* indicates that he did attach some significance to the role of hypotheses in science: 'While availing ourselves freely of the use of hypothesis as an aid in ascertaining the connection and significance of facts, we must be ever on our guard against premature speculation and theory, clearly distinguishing between what is fact and what may be our own gloss or interpretation of it'.[89] Nevertheless, I am inclined to think that he underestimated the importance of hypothesis formulation in science and the extent to which theoretical presuppositions may mould geologists' observations.

For the purposes of our present enquiry, it is important to note the general relationship between Geikie's methodological views and his work as an historiographer. Just as the geologist should be familiar with present geological processes, and use them in order to build up an historical account of the earth, using for this purpose the principle that 'the present is the key to the past' and thereby gaining an *understanding* of the earth's nature, so too the historian of geology is to take his bearings from the present and use this as a guide to the study of the earlier stages of the development of the science of geology. Reciprocally, by knowing the history of geology the student gains an understanding of the present state of development of the subject. It was for this reason that Geikie opened his *Founders of geology* with the pronouncement that 'In science, as in all other departments of enquiry, no thorough grasp of a subject can be gained, unless the history of its development is clearly appreciated'.[90]

## 9. The function of Geikie's epistemology, methodology and social values in his historiography

It is interesting to identify the ways in which Geikie's epistemology and methodological principles shaped his historiographical work. One can offer a number of examples of this, but the account given of Hutton's work is particularly striking. Thus, according to Geikie's view of the matter,

> Instead of invoking conjecture and hypothesis, he [Hutton] proceeded from the very outset to collect the actual facts, and to marshall these in such a way as to make them tell their own story. Unlike Werner, he had no preconceived theory about the origin of rocks, with which all the phenomena of nature had to agree. His theory grew so naturally out of his observations that it involved no speculation in regard to a large part of its subject.[91]

[88] *The ancient volcanoes of Great Britain*, the *Text-book of geology*, and the *Class-book of geology* all adopt the technique of presenting general principles first, followed by a geological history. Such an approach was, of course, fairly standard in nineteenth-century British texts, though Lyell separated general theory and history into two separate books: *Principles* and *Manual* (or *Elements*). Some writers, however, such as G. G. Richardson, inverted the 'history', and worked their way down from the Tertiary Period to the Plutonic rocks (*An introduction to geology* ... ('New edition, revised and considerably enlarged', 1851, London)).

[89] A. Geikie (footnote 6), 473.

[90] *Ibid.*, 1.

[91] *Ibid.*, 299.

Thus, Geikie would have us believe, Hutton did not speculate or formulate hypotheses. He developed 'fundamental generalizations',[92] and started with a 'grand conception'.[93] It was in such terms that the nineteenth-century inductivist/empiricist historian viewed that man of 'Classical' enlightenment, James Hutton, with his deep metaphysical commitment to design, and his tendency towards a 'Kantian' idealism[94]—aspects that Geikie quite failed to apprehend or appreciate, or about which he remained quite silent.

In general, the reader of *The founders of geology* will find that its author conspicuously singled out for approbation all instances where *observation* was thought to play a particularly important role in the history of geology. William Smith is described as 'eagerly scrutinizing every field, ridge, and hill'.[95] Sedgwick had 'eagle eyes [that] seemed as if they must instantly pierce into the very heart of the stiffest geological problem'.[96] Cuvier 'began [his work] by gathering together as large a series as he could obtain of skeletons of living species of vertebrate animals'.[97] (For Cuvier, also, the present was seemingly the key to the past.[98]) Many other examples may be cited. Unquestionably, Geikie's empiricism provided a general framework within which he composed his history of geology.

Just as Geikie's epistemology emerges in his historiography, so too we can discern a link between his historiography and geological methodology. For Geikie, the whole process of historical enquiry was essentially unproblematical. Just as the 'honest mason' might dig down through the layers of soil round the Firth of Forth and without difficulty build up an account of the past events in that region, so too could the historian of geology dig down through the layers of books in libraries and manuscripts in archives and build up a ready picture of past events. Past and present snugly fitted. The present was the key to the past; and the past was the key to the present.

Meshing with Geikie's *social* philosophy, we find him presenting the history of science as a praiseworthy 'humanizing' activity, with certain worthwhile moral adjuncts, such as the recognition of 'the absolute necessity of avoiding dogmatism'.[99] It enabled one to 'look back with gratitude and pride to what has been

[92] *Ibid.*, 291.

[93] *Ibid.*, 299.

[94] The literature on Hutton in relation to his metaphysics is now growing rapidly. See, for example, G. L. Davies. *The earth in decay: a history of British geomorphology 1578–1878* (1969, London), chapter 6; F. Ellenberger, 'La métaphysique de James Hutton (1726–1797) et la drame écologique du xx$^e$ siècle', *Revue synth.*, **93** (1972), 267–283; D. R. Dean, 'James Hutton on religion and geology: the unpublished preface to his Theory of the Earth', *Ann. sci.* **32** (1975), 187–193; R. S. Porter, *The making of geology: earth science in Britain 1660–1815* (1977, Cambridge), 184–196; and J. E. O'Rourke, 'A comparison of James Hutton's *Principles of knowledge* and *Theory of the earth*', *Isis*, **69** (1978), 4–20. By contrast, E. B. Bailey's *James Hutton—the founder of modern geology* (1967, Amsterdam, London and New York) represents the somewhat earlier tradition of the scientist/historian, in which emphasis is placed on Hutton as an acute observer. But note that Davies, Ellenberger and O'Rourke are themselves—or have been—practising scientists; so the influence of the general historian is now penetrating beyond the ranks of professional historians of science.

[95] A. Geikie (footnote 6), 385.

[96] *Ibid.*, 433.

[97] *Ibid.*, 364.

[98] This was in large measure true, despite Cuvier's well-known statement that 'the thread of operations is . . . broken; the march of Nature is changed; and none of the agents which she now employs, would have been sufficient for the production of her ancient works' (G. Cuvier, *Essay on the theory of the earth* (5th eds. 1827, Edinburgh and London), 24). See also pp. 23 and 4–6, where certain features of a 'uniformitarian' methodology are manifest.

[99] A. Geikie (footnote 6), 472.

achieved by our forerunners in the race, ... [so that] while we labour to emulate their devotion, let us hold high the torch of science, and pass it on bright and burning to those who shall receive it from our hands.'[100] It was 'an eminently human and instructive experience',[101] for 'few more interesting records of human endeavour and achievement can be found than that presented by the advance of ... science'.[102] Verily, the study of the history of science allowed one to pass through the portals of the 'Temple of Truth'.[103]

### 10. Two evaluations of Geikie as an historiographer of geology

Having thus given a general conspectus of Geikie's philosophical suppositions, his historiography, and the interaction of the two, so that we may feel that we have a reasonable understanding of his historiographical enterprise, we may now venture further and attempt some appraisal of his contributions to geological historiography. His historiography may, of course, be appraised from two points of view: in terms of the later part of the twentieth century or in the terms of his own day and age.

From the present-day vantage-point, we quickly note that Geikie judged his geological precursors in large measure by the criteria of his own day and age. He was 'guilty', therefore, of anachronism. He believed in the 'idea of progress', and saw the history of his science as a developmental process leading to the state of knowledge that existed in his own day, the process being chiefly facilitated by the acquisition of increments of factual information. I feel justified, therefore, in describing Geikie as a Whig historian of science; and perhaps, therefore, we may wish to judge him 'guilty' on this count.

Geikie was also, it seems, a naïve inductivist, who—though outstandingly successful as a practising scientist—had little understanding of the 'true' nature of scientific method and theory from a philosophical point of view. On this ground, likewise, we may feel moved to 'condemn' him, dismissing him as philosophically naïve. His major methodological principle—that 'the present is the key to the past'—involved him in circularity, and was at odds with the overall historicist approach to geological interpretation that his writings displayed.

If such criticisms are taken seriously, then one may well feel inclined to repudiate Geikie's historiography of science as being of little worth. Certainly it has to be rejected if the various arguments propounded by Agassi are taken to heart.

On the other hand, if Geikie's work as an historian of geology is considered in the light of his own proclaimed intentions, then I think *Founders of geology* and his other historical works (including, for example, his various summaries of the 'state of the art' prepared for presentation to the British Association, and his *Annual reports* for the Geological Survey) may be regarded as largely successful. For they subserved the interests of students of geology, or practising scientists, rather than professional historians of science; and they meshed with nineteenth-century social views, not those of the present day. But this being the case, we cannot expect the modern historiographer to rest content with Geikie's performance.

[100] *Ibid.*, 473.
[101] *Ibid.*, 1.
[102] *Ibid.*, 3.
[103] Compare footnote 65.

## 11. The paradox of Whiggery

If we judge Geikie's historiographical work in late twentieth-century terms, and reject it because of its manifest inductivist/Whiggish/anachronistic character, then we also may find ourselves in difficulties, for like Geikie we are formulating anachronistic judgements.[104] Indeed, the same might be said of Agassi himself, for he was being 'unfair' (anachronistic) in his judgement of earlier twentieth-century inductivist historians of science by his employment of the historiographical presuppositions (underpinned by the falsificationist methodology of science) of a later age. Agassi, for example, launched a strong attack against the historiography of Douglas McKie.[105] Yet in doing so he was himself seemingly being 'unhistorical' for failing to note the particular historical circumstances in which McKie worked, and which moulded his historiographical practices. But then I suppose *I* ought to remember the particular historiographical circumstances under which *Agassi* worked, and treat his remarks accordingly.

Is there not some paradox here? If we are determined to eschew all historiographical anachronism, how may we exercise any critical faculty with respect to earlier historiography?

It would seem, then, that we may have to forego criticism of Geikie's historiography. We note its success, judged in its own terms, and as measured by the sales of Geikie's books, his lasting influence on our understanding of the history of geology, and his own remarkable social success. But having decided to forswear Whiggery and historiographical anachronism we are precluded from making any judgements of our own as to the quality of Geikie's historiographical work. We can only examine how he was received by his contemporaries.

Yet this may leave one unsatisfied. If (according to the modern view) Geikie was really 'guilty' of so many historiographical solecisms, and of anachronism in particular, how could he successfully teach geological concepts through what we now regard as a distorted account of the history of geology? For example, might not the inability of geologists for a long time to deal satisfactorily with the 'granite problem'[106] have been due to the fact that they had been 'indoctrinated' by Geikie into believing that the Huttonian theory was essentially 'the truth'?

More significantly, we may wish to criticize Geikie for inadequate historiographical performance, judged strictly according to his own programme. For example, it might be contended that *Founders of geology* gives grossly inadequate representation of the contributions of Charles Lyell to nineteenth-century geology,[107] so that important aspects of geological theory, as it stood in Geikie's day, would be ill understood by the reader because of this deficiency. To say this would not entail a criticism of Geikie's general programme, but only of its particular performance,

[104] In this section, I am drawing on ideas to be found in C. Condren, 'An historiographical paradox', in F. McGregor and N. Wright (eds.), *European history and its historians* (1977 Adelaide), 85–95.

[105] J. Agassi (footnote 8), 5–6. McKie held the chair of History and Philosophy of Science at University College, London, from 1957 to 1964. Certainly his historiography of science was inductivist in character, according to Agassi's characterization. This led to his being singled out for a rather vicious attack by the philosopher.

[106] On this, see H. H. Read, *The granite controversy: geological addresses illustrating the evolution of a disputant* (1957, London).

[107] Really only three pages were devoted to Lyell in *The founders of geology* (A. Geikie (footnote 6), 403–405), He also wrote two short obituaries of Lyell (*Nature*, **12** (1875), 325–327; and *Proceedings of the Royal Society, London*, **25** (1877), xi–xiv); and an essay on his work in Scotland ('Charles Lyell and Forfarshire geology', in *Handbook and guide to Dundee district* (1912, Dundee), 417–422).

judged in nineteenth-century terms, and according to Geikie's own avowed intentions.

However, it *may* be the case that the full significance of Lyell's work has only become apparent to us with our greater degree of hindsight. (Should we *blame* nineteenth-century biologists and historians for failing to recognize the importance of Mendel?) So perhaps no censure should be passed against Geikie for his seemingly insufficient treatment of Lyell in *Founders of geology*. Or maybe Geikie knew best; perhaps we over-rate Lyell.

Yet surely we feel that *we* know best! *We* have the advantage of hindsight. Surely we can criticize Geikie's sketchy treatment of Lyell, for it was inadequate within the terms of reference of Geikie's own historiographical programme. But this does not really resolve our difficulty. For, I suggest, the criteria of historiographical adequacy that we apply are really our own. We may well find good reasons *why* it was that Geikie gave Lyell such perfunctory treatment. Nevertheless, we do not suppose that he was succeeding as an historian when he did so.

The paradox, therefore, is this. As up-to-date historians of science we have determined to eschew Whiggery. Yet we find ourselves, as it were, hoist with our own petards as soon as we contemplate the earlier work in the history of science. In fact, the very activity of being a professional historian of science inevitably makes one in some degree dissatisfied with earlier writings in the history of science. If this were not so, there would be little point in writing history of science. So, just as the scientist is customarily Whiggish when he contemplates former science, so also is the historian of science Whiggish when he contemplates earlier historiography of science. Or so it would seem.

In this respect, however, the historian of science may be in good company, for the general historians, while ostensibly opposed to Whig history, become Whiggish in their turn when they review the history of historical writing. For example, no less than Herbert Butterfield, one of the first to recognize the 'problem' of Whiggish interpretations of history, was in fact Whiggish himself in high degree when he criticized the Whig historians; and his volume on the history of historiography, *Man on his past*, is, I suggest, remarkable for its Whiggery.[108]

So historians of science who, in the post-World War II period, have established their own special academic niche with reasonable success, and who in so doing have turned their backs on the value systems of the earlier 'scientist/historians', preferring the historiographical canons of their colleagues the general historians, have thereby taken upon themselves what appears to be an uncomfortable historiographical paradox. In the process of their professionalization, historians of science have chosen to eschew Whiggery,[109] for to do so is one of the rules of professional conduct of the general historian, whose attitudes and practices they seek to emulate (the general

[108] H. Butterfield, *Man on his past: the study of the history of historical scholarship* (1955, Cambridge). Incidentally, it is interesting and really quite important to note that in 1944 (in wartime circumstances) Butterfield repudiated his earlier critique of Whig historiography. He wrote: 'We shall treat . . . [the Whig interpretation of history] not as a thing invented by some particularly wilful historian, but as part of the land-scape of English life, like our country lanes or our November mists or our historic inns' (*The Englishman and his history* (1944, Cambridge), 2; and 'We will celebrate this whig inheritance of ours with a robust but regulated pride; observing the part which an interpretation of history has played in building up the centuries and creating the England that we know' (*ibid.*, 4).

[109] For an example of an historian of geology formally repudiating the historiographical practices of the scientist/historians, see the preface to M. J. S. Rudwick, *The meaning of fossils; episodes in the history of palaeontology* (1972, London and New York).

historians being well established in the educational system, with all the concomitant professional and academic appurtenances, such as historians of science might wish to secure for themselves). Yet by engaging in the practice of writing the history of science as a professional activity the historians of science must inevitably suppose that they can 'advance' their discipline, and consequently judge its previous efforts as in some measure unsatisfactory. Otherwise their efforts must seem rather futile.

## 12. Suggestions for resolution of the dilemma

The dilemma may, I suggest, be met in two ways. If the general historians are cheerfully anachronistic when they examine the history of historiography, and if historiographers of science act in a similar manner with respect to earlier writings on the history of science, then so be it. Perhaps Whiggery and historiographical anachronism are but venial sins, not mortal. If this be the case, and if science does progress in some degree, then perhaps there *is* a place in the historiography of science for the formulation of anachronistic comparisons, provided they are not accompanied by censorious moralizations. After all, it seems that the repudiation of Whiggery is as much a social phenomenon as anything else.

An alternative approach is to take much more careful note of the 'levels of discourse' that are involved at every stage of the discussion.[110] A scientist carries out his professional activities, and in so doing he may wish to examine some of the earlier work in his field to establish what has and what has not been done. In so doing he may notice gaps, deficiencies, or inadequacies in the earlier work, and this may act as a stimulus to his own efforts. The doctoral dissertation, prefaced by a 'review of the literature', illustrates this kind of activity, though in truth some moral posturing often enters into such prefatory exercises. But strictly such literature surveys are not *historical*; rather, they are part of the activity of science itself.

Historiography of science, on the other hand, should be regarded as a 'meta-discipline' with respect to science. It is *not* part of science itself. But just as the scientist may review earlier science, so the historiographer of science may review earlier historiography of science, note its strengths and weaknesses, and use it to develop new work in the field. Such appraisal should not be regarded as anachronistic, but an essential part of the intellectual activity of the historiography of science. However, when writing of *science*, and its historical development, the historiographer *should* avoid anachronism.

Then, if we wish, we may erect a meta-meta-level (with respect to science itself), in which we write *about* historiography of science. Here, we should avoid anachronistic judgements with respect to historiography of science. (For example, we would not appraise Geikie in terms of recent historiographers of geology such as Davies or Porter.) But within the meta-meta-level, we may wish to notice deficiencies in earlier 'histories of histories' and attempt to plug up the gaps, or generally improve upon previous work in this domain. Needless to say, for those who do not suffer from vertigo, there is no reason why one should not ascend to yet more exalted meta-levels, as one pleases.[111]

[110] The following suggestions concerning a hierarchy of meta-levels arise from suggestions made by my colleague Dr W. R. Albury.

[111] Dr Grattan-Guinness has pointed out to me that from a logical standpoint there is no objection to the employment of different methodologies at different levels of discourse. Thus one is not involved in a logical contradiction if, for example, one serves as an empiricist scientist, a Popperian historiographer, and an inductivist meta-historian.

But all this may appear extremely artificial. And the problem is that one tends to drift from one level to another, often without being aware of it. Indeed, I suggest that this was essentially Geikie's problem (as it now appears in retrospect!). He was, in part, 'doing science' when he wrote his historical essays, for one of his avowed intentions was to use the history of geology as an aid to the teaching of geology. And much of his work in the history of science was specifically addressed to professional scientists, as in his various presidential addresses, prefaces to his monographs, and so on. Yet at the same time (and particularly in *Founders of geology* of course) Geikie was functioning at the meta-level of historiography of science. But because of his double commitment to both science and its historiography he was very often anachronistic *qua* historiographer; for his two levels of discourse were frequently conflated.

That this should have occurred is in no way surprising. Geikie was, *par excellence*, a scientist/historian. Moreover, the notion of a hierarchy of meta-levels is very much a twentieth-century device for evading the various logical paradoxes of self-referrent statements. We would not expect Geikie himself to have given any thought to the problem.

I conclude, therefore, by making the following points. If we set our faces firmly against *any* kind of anachronism we are stuck in an eternal present and are precluded from criticizing earlier work in the historiography of science. But this is an absurdity. So we have to accept that appraisal of earlier work in the history of science is part of the legitimate activity of historiography of science, provided the level of discourse is maintained at a 'fixed level'. But in so doing we have admitted an element of anachronism. Whiggery, *sensu strictu*, emerges when anachronistic judgements are made by the *historiographer* of science about *scientists*.[112] And this one may decide (for reasons that in the last analysis are social, rather than logical) to eschew. On the other hand, it is quite open to one to espouse a Whiggish historiography. One may use the 'bird's-eye-view' of the meta-level of discourse to run back and forth in time, with all manner of historiographical anachronisms; and if the object of the exercise is regarded as the understanding of the process of the development of science this may perhaps be no bad thing, though in the present climate of opinion it may be a bold man that dares to say so.

The study of the historiographical work of Sir Archibald Geikie provides us with an excellent illustration of some of the issues that are involved when one seeks to decide upon the kind of historiography of science that one is engaged in, or in which one wishes to engage.

### Acknowledgements

I am indebted to the University of London for permission to reproduce three questions from former examination papers for the M.Sc. degree in History and Philosophy of Science (see footnote 26). Suggestions have been received from Dr W. R. Albury, Professor G. L. Davies, Dr I. Grattan-Guinness and Dr R. S. Porter during the preparation of this paper, and their assistance is gratefully acknowledged.

112 Earlier in this paper, I have spoken of the formulation of 'Whiggish' anachronistic statements with respect to earlier historiography of science. If the foregoing analysis concerning meta-levels of discourse be accepted, then one may not wish to use the term Whiggish in this context, but reserve it for statements which conflate different levels of the hierarchy of discourse.

# INDEX